Nutrition and Eye Health

Nutrition and Eye Health

Special Issue Editors

John Lawrenson
Laura Downie

MDPI • Basel • Beijing • Wuhan • Barcelona • Belgrade

MDPI

Special Issue Editors
John Lawrenson
University of London
UK

Laura Downie
The University of Melbourne
Australia

Editorial Office
MDPI
St. Alban-Anlage 66
4052 Basel, Switzerland

This is a reprint of articles from the Special Issue published online in the open access journal *Nutrients* (ISSN 2072-6643) from 2018 to 2019 (available at: https://www.mdpi.com/journal/nutrients/special_issues/Nutrition_Eye_Health).

For citation purposes, cite each article independently as indicated on the article page online and as indicated below:

LastName, A.A.; LastName, B.B.; LastName, C.C. Article Title. *Journal Name* **Year**, *Article Number*, Page Range.

ISBN 978-3-03921-990-2 (Pbk)
ISBN 978-3-03921-991-9 (PDF)

Contents

About the Special Issue Editors

John Lawrenson is Professor of Clinical Visual Science and Research Lead for the Applied Vision Research Centre at City, University of London. He has an active research interest in ophthalmic public health, with particular emphasis on sight-threatening eye diseases e.g. diabetic retinopathy, glaucoma and age-related macular degeneration (AMD). His work has received funding from a variety of sources including: the National Institute for Health Research (NIHR), the International Glaucoma Association and the College of Optometrists (UK). He trained as an optometrist at Aston University and Moorfields Eye Hospital, London. He completed a PhD in Visual Science at City, University of London, which was followed by a Postdoctoral Fellowship in Neuroscience at University College London. In 2014 he was awarded a Master's degree in Evidence-based Health Care from the University of Oxford. He has authored, or co-authored, more than 100 peer-reviewed publications, including 3 highly cited Cochrane Systematic Reviews on nutrition and AMD. Professor Lawrenson's contributions have been acknowledged through the award of a Life Fellowship by the College of Optometrists and a number of research prizes including the BCLA Dallos Award and the Arthur Bennett Prize. He edits the eye and orbit sections of Grays Anatomy, is an Editor for the Cochrane Eyes and Vision Group and also serves on the editorial board of several ophthalmic journals.

Laura Downie is a clinician scientist who has gained international recognition for research excellence in ocular disease, with awards, highly cited papers, international speaking engagements and appointments to key professional bodies. She is an Associate Professor in the Department of Optometry and Vision Sciences, Faculty of Medicine, Dentistry and Health Sciences, at the University of Melbourne, Victoria, Australia. In this role, she provides didactic and clinical training to eye care clinicians, leads the sub-specialty cornea clinic at University of Melbourne eyecare clinic and heads her own research laboratory, the 'Anterior Eye, Clinical Trials and Research Translation Unit.' Her research combines laboratory, clinical and implementation science as a foundation for improving patient outcomes, particular in the areas of anterior eye disease and age-related macular degeneration. She is a previous National Health and Medical Research Council (NHMRC) Translating Research Into Practice (TRIP) Fellow (2015-7), and has been awarded research funding from a diversity of sources, including the NHMRC, Macular Disease Foundation of Australia, Rebecca L Cooper Medical Foundation and industry. She graduated from the University of Melbourne with a Bachelor of Optometry in 2003, and completed her PhD, focusing on vascular, neuronal and glial cell changes in retinopathy of prematurity, at the same institution in 2008. She has undertaken post-graduate training in evidence synthesis and evidence-based medicine at the University of Oxford (UK), and completed the Women's Executive Leadership program at the Hass School of Business, UC Berkeley (US). Her research expertise spans across the fields of ocular biomarkers and diagnostics, evidence-based medicine, clinical trials, systematic reviews, critical appraisal and implementation science. She has authored, or co-authored, more than 75 peer-reviewed publications, and has achieved international recognition for her research achievements, including receiving the prestigious Irvin M and Beatrice Borish Award from the American Academy of Optometry in 2014. She has served as a member on several national and international expert panels, including the Tear Film and Ocular Surface Society International Dry Eye Workshop II (a global initiative to develop a consensus on dry eye care practices), standards committees, industry advisory boards and community and professional committees. She also serves on the Editorial Board of several journals, including Ophthalmology the leading journal in the discipline.

nutrients

MDPI

Editorial

Nutrition and Eye Health

John G. Lawrenson [1],* and Laura E. Downie [2]

[1]   Centre for Applied Vision Research, School of Health Sciences, City University of London,
     London EC1V 0HB, UK
[2]   Department of Optometry and Vision Sciences, The University of Melbourne, Parkville, Victoria 3010,
     Australia
*   Correspondence: j.g.lawrenson@city.ac.uk; Tel.: +44-(0)20-7040-4310

Received: 2 September 2019; Accepted: 4 September 2019; Published: 6 September 2019

Diet is a key lifestyle factor that can have long-term effects on ocular health. This Special Issue of Nutrients entitled 'Nutrition and Eye Health' contains 12 articles, including reviews and primary research studies, that report on a diverse range of topics relating to the role of nutrition in maintaining eye health, and the potential use of nutritional interventions for preventing or treating ocular disease. Collectively, these papers span a spectrum of ocular conditions, including corneal angiogenesis [1], cataract [2–4], diabetic retinopathy [5], age-related macular degeneration (AMD) [6,7], and experimental models of retinal disease [8–10]. In addition, clinically focussed papers report on the validation of a novel food frequency questionnaire for assessing long-chain omega-3 fatty acid intake in eye care practice [11], and evidence relating to the applicability of saffron for treating ocular disease [12].

Globally, approximately 250 million people suffer from varying degrees of vision loss [13]. Leading causes include several eye conditions considered in this Special Issue, such as cataract, AMD, glaucoma, and diabetic retinopathy. These conditions disproportionately affect older adults, and with an ageing population the number of affected individuals is predicted to increase exponentially [13]. Whilst the aetiology of age-related eye disease is complex and multifactorial, oxidative stress has been implicated as a common causative mechanism. The eye is particularly susceptible to oxidative stress as a result of its high oxygen consumption, high concentration of polyunsaturated fatty acids and cumulative exposure to high-energy visible light. This combination of factors leads to the generation of reactive oxygen species that can trigger oxidative damage to ocular tissues. Consequently, there has been significant research interest in the role of dietary antioxidants and the potential therapeutic benefits of antioxidant vitamin and mineral supplements as a simple and cost-effective strategy for disease prevention and/or control [14–17].

AMD is characterised by degenerative changes within the macula, the central area of the retina that is responsible for high-resolution vision, in people aged 55 years or older. AMD is a leading cause of severe vision impairment in European-derived populations. In the UK, the disease is responsible for over 50% of certifiable vision loss [18]. Although epidemiological studies have provided reasonably consistent evidence that diet is an important modifiable risk factor for AMD [19], concerns have been raised about the validity of findings from non-interventional studies due to the potential influence of confounding factors. For example, people with a particular dietary pattern may differ in other ways (e.g., the amount of exercise they undertake, their daily level of light exposure) and it is not typically possible to control for these differences [20].

In terms of primary research studies, the highest quality evidence to evaluate the efficacy and safety of therapeutic interventions derives from randomised controlled trials (RCTs). In RCTs, participants are randomly allocated to receive either the intervention or a comparator (typically placebo or no intervention), which minimises the potential for bias in the intervention assignment [21]. There is evidence from RCTs that prophylactic antioxidant vitamin or mineral supplementation does not prevent the development of AMD [22]. Five large RCTs have compared supplements containing vitamin E, beta-carotene, vitamin C, or antioxidant vitamin combinations with placebo in people from the general population. These trials randomised more than 75,000 people and followed their clinical outcomes between 4 to 10 years. People taking these supplements were found to have a similar risk of developing AMD to those not taking the supplements [22].

Other RCTs have investigated whether high-dose antioxidant vitamin and mineral supplements can slow the progression of AMD [23]. Most of these trials recruited small numbers of participants and were of relatively short duration, ranging from 9 months to 6 years. However, one large, multi-centre RCT conducted in the USA, the Age-Related Eye Disease Study (AREDS), randomised 3640 individuals with AMD to take supplement formulations containing combinations of vitamin C, E, beta-carotene, zinc, and copper, or a placebo, each day. A major conclusion from the AREDS was that daily, long-term supplementation with vitamin C (500 mg), vitamin E (400 international units (IU)), beta-carotene (15 mg), zinc (80 mg, as zinc oxide), and copper (2 mg, as cupric oxide) reduced the relative risk of progression to late-stage AMD from 28% (observed with placebo) to 20% at 5 years, in people with at least intermediate AMD. This means that for people with intermediate AMD, who are at the highest risk of progression to late AMD, 80 fewer cases would progress for every 1000 people taking the supplement. However, safety concerns were raised regarding high-dose supplementation of the carotenoid used in the original AREDS supplement, beta-carotene, in people who smoke [24]. In a follow-up study by the AREDS investigators, AREDS2, current smokers or those who had ceased smoking for less than 12 months before enrolment were not eligible to receive beta-carotene supplementation [25]. The primary analysis in AREDS2 demonstrated that adding lutein and zeaxanthin and/or omega-3 fatty acids to the AREDS formula was not associated with a significant reduction in the risk of progression to late-stage AMD compared with the original supplement. Lutein and zeaxanthin are carotenoids that are major components of macula pigment. They are proposed to have a protective role in the retina through their antioxidant properties and ability to act as a filter for blue light [26]. Exploratory analyses from AREDS2 suggested that lutein and zeaxanthin may be of value for reducing AMD progression when given without beta-carotene, but that more research was required to test this hypothesis [27].

Cataract is defined as any visible opacity within the otherwise clear crystalline lens of the eye. Cataract can be further classified as cortical, nuclear, or posterior sub-capsular, depending on the anatomical location of the opacity. Globally, over 60 million people are visually impaired due to cataract, however cataract-associated blindness shows significant geographical variation, accounting for less than 22% of blindness in high-income countries compared to more than 44% in South East Asia [13]. Age is the most significant risk factor for cataractogenesis. As the lens ages, conformational changes to lens proteins occur with subsequent aggregation, leading to a progressive loss of transparency and associated vision loss [28]. Oxidation reactions within the lens are thought to be a key factor in this process and there has been a significant amount of research on the role of antioxidant nutrients for preventing or slowing the progression of cataract [29]. Observational data suggest that the risk of cataracts can be reduced by a diet that contains optimal levels of vitamins C and E, the carotenoids lutein and zeaxanthin, and the daily use of multivitamin supplements [29]. However, RCTs that have compared antioxidant vitamin supplements (beta-carotene, vitamins C and E) to an inactive placebo or no supplement have been unable to detect any effect on the incidence or progression of cataract [30]. The lack of efficacy in these relatively short-term trials could suggest that a longer-term intake or a particular combination of antioxidants is required. A study of baseline factors that predicted cataract in the AREDS cohort found that the use of multivitamins supplements reduced the risk of developing nuclear cataracts over an approximate 10 year follow-up period [31].

In conclusion, age-related eye diseases, including cataract and AMD, are of global public health concern. Acquired vision loss associated with these conditions can be devastating to the individual through its detrimental impact on quality of life, and also impart substantial societal burden. Although the pathogenesis of these conditions is not fully understood, there is increasing evidence that their impact can, to some extent, be mitigated by targeting modifiable risk factors. Since diet and nutrition have been linked with the most common diseases affecting the elderly, dietary modification and nutritional supplementation for the prevention and treatment of these diseases has attracted a considerable amount of scientific attention. As evidenced by the quality and diversity of the contributions in this Special Issue, the role of nutrition in eye health remains a highly topical area, with scope for future research to enhance our understanding of the role of nutritional strategies for optimising eye health.

Author Contributions: J.G.L. and L.E.D. wrote the editorial.

Funding: This research received no external funding.

Conflicts of Interest: The authors declare no conflict of interest.

References

1. Estrella-Mendoza, M.F.; Jimenez-Gomez, F.; Lopez-Ornelas, A.; Perez-Gutierrez, R.M.; Flores-Estrada, J. Cucurbita argyrosperma Seed Extracts Attenuate Angiogenesis in a Corneal Chemical Burn Model. *Nutrients* **2019**, *11*, 1184. [CrossRef] [PubMed]

2. Braakhuis, A.J.; Donaldson, C.I.; Lim, J.C.; Donaldson, P.J. Nutritional Strategies to Prevent Lens Cataract: Current Status and Future Strategies. *Nutrients* **2019**, *11*, 1186. [CrossRef] [PubMed]

3. Lim, V.; Schneider, E.; Wu, H.; Pang, I.H. Cataract preventive role of isolated phytoconstituents: Findings from a decade of research. *Nutrients* **2018**, *10*, 1580. [CrossRef] [PubMed]

4. Zych, M.; Wojnar, W.; Dudek, S.; Kaczmarczyk-Sedlak, I. Rosmarinic and Sinapic Acids May Increase the Content of Reduced Glutathione in the Lenses of Estrogen-Deficient Rats. *Nutrients* **2019**, *11*, 803. [CrossRef] [PubMed]

5. Rossino, M.G.; Casini, G. Nutraceuticals for the treatment of diabetic retinopathy. *Nutrients* **2019**, *11*, 771. [CrossRef]

6. Lawrenson, J.G.; Evans, J.R.; Downie, L.E. A Critical Appraisal of National and International Clinical Practice Guidelines Reporting Nutritional Recommendations for Age-Related Macular Degeneration: Are Recommendations Evidence-Based? *Nutrients* **2019**, *11*, 823. [CrossRef] [PubMed]

7. Rinninella, E.; Mele, M.C.; Merendino, N.; Cintoni, M.; Anselmi, G.; Caporossi, A.; Gasbarrini, A.; Minnella, A.M. The role of diet, micronutrients and the gut microbiota in age-related macular degeneration: New perspectives from the gut(-)retina axis. *Nutrients* **2018**, *10*, 1677. [CrossRef]

8. Morita, Y.; Miwa, Y.; Jounai, K.; Fujiwara, D.; Kurihara, T.; Kanauchi, O. Lactobacillus paracasei KW3110 Prevents Blue Light-Induced Inflammation and Degeneration in the Retina. *Nutrients* **2018**, *10*, 1991. [CrossRef]

9. Kang, M.K.; Lee, E.J.; Kim, Y.H.; Kim, D.Y.; Oh, H.; Kim, S.I.; Kang, Y.H. Chrysin ameliorates malfunction of retinoid visual cycle through blocking activation of AGE-RAGE-ER stress in glucose-stimulated retinal pigment epithelial cells and diabetic eyes. *Nutrients* **2018**, *10*, 1046. [CrossRef]

10. Yu, M.; Yan, W.; Beight, C. Lutein and Zeaxanthin Isomers Protect against light-induced retinopathy via decreasing oxidative and endoplasmic reticulum stress in BALB/cJ Mice. *Nutrients* **2018**, *10*, 842. [CrossRef]

11. Zhang, A.C.; Downie, L.E. Preliminary Validation of a Food Frequency Questionnaire to Assess Long-Chain Omega-3 Fatty Acid Intake in Eye Care Practice. *Nutrients* **2019**, *11*, 817. [CrossRef] [PubMed]

12. Heitmar, R.; Brown, J.; Kyrou, I. Saffron (*Crocus sativus* L.) in Ocular Diseases: A Narrative Review of the Existing Evidence from Clinical Studies. *Nutrients* **2019**, *11*, 649. [CrossRef] [PubMed]

13. Flaxman, S.R.; Bourne, R.R.A.; Resnikoff, S.; Ackland, P.; Braithwaite, T.; Cicinelli, M.V.; Das, A.; Jonas, J.B.; Keeffe, J.; Kempen, J.H.; et al. Global causes of blindness and distance vision impairment 1990–2020: A systematic review and meta-analysis. *Lancet Glob. Health* **2017**, *5*, e1221–e1234. [CrossRef]

14. Weikel, K.A.; Chiu, C.J.; Taylor, A. Nutritional modulation of age-related macular degeneration. *Mol. Asp. Med.* **2012**, *33*, 318–375. [CrossRef]

15. Sideri, O.; Tsaousis, K.T.; Li, H.J.; Viskadouraki, M.; Tsinopoulos, I.T. The potential role of nutrition on lens pathology: A systematic review and meta-analysis. *Surv. Ophthalmol.* **2019**, *64*, 668–678. [CrossRef]

16. Loskutova, E.; O'Brien, C.; Loskutov, I.; Loughman, J. Nutritional supplementation in the treatment of glaucoma: A systematic review. *Surv. Ophthalmol.* **2019**, *64*, 195–216. [CrossRef] [PubMed]

17. Li, C.; Miao, X.; Li, F.; Wang, S.; Liu, Q.; Wang, Y.; Sun, J. Oxidative Stress-Related Mechanisms and Antioxidant Therapy in Diabetic Retinopathy. *Oxid. Med. Cell. Longev.* **2017**, *2017*, 9702820. [CrossRef]

18. Quartilho, A.; Simkiss, P.; Zekite, A.; Xing, W.; Wormald, R.; Bunce, C. Leading causes of certifiable visual loss in England and Wales during the year ending 31 March 2013. *Eye* **2016**, *30*, 602–607. [CrossRef]

19. Chiu, C.J.; Taylor, A. Nutritional antioxidants and age-related cataract and maculopathy. *Exp. Eye Res.* **2007**, *84*, 229–245. [CrossRef]

20. Downie, L.E.; Keller, P.R. Nutrition and age-related macular degeneration: Research evidence in practice. *Optom. Vis. Sci.* **2014**, *91*, 821–831. [CrossRef]

21. Ioannidis, J.P. Implausible results in human nutrition research. *BMJ* **2013**, *347*, f6698. [CrossRef] [PubMed]

22. Evans, J.R.; Lawrenson, J.G. Antioxidant vitamin and mineral supplements for preventing age-related macular degeneration. *Cochrane Database Syst. Rev.* **2017**, *7*, CD000253. [CrossRef] [PubMed]

23. Evans, J.R.; Lawrenson, J.G. Antioxidant vitamin and mineral supplements for slowing the progression of age-related macular degeneration. *Cochrane Database Syst. Rev.* **2017**, *7*, CD000254. [CrossRef] [PubMed]

24. Tanvetyanon, T.; Bepler, G. Beta-carotene in multivitamins and the possible risk of lung cancer among smokers versus former smokers: A meta-analysis and evaluation of national brands. *Cancer* **2008**, *113*, 150–157. [CrossRef] [PubMed]

25. Group, A.R.; Chew, E.Y.; Clemons, T.; SanGiovanni, J.P.; Danis, R.; Domalpally, A.; McBee, W.; Sperduto, R.; Ferris, F.L. The Age-Related Eye Disease Study 2 (AREDS2): Study design and baseline characteristics (AREDS2 report number 1). *Ophthalmology* **2012**, *119*, 2282–2289. [CrossRef]

26. Arunkumar, R.; Calvo, C.M.; Conrady, C.D.; Bernstein, P.S. What do we know about the macular pigment in AMD: The past, the present, and the future. *Eye* **2018**, *32*, 992–1004. [CrossRef] [PubMed]

27. Chew, E.Y.; Clemons, T.E.; Sangiovanni, J.P.; Danis, R.P.; Ferris, F.L., 3rd; Elman, M.J.; Antoszyk, A.N.; Ruby, A.J.; Orth, D.; Fish, G.E.; et al. Secondary analyses of the effects of lutein/zeaxanthin on age-related macular degeneration progression: AREDS2 report No. 3. *JAMA Ophthalmol.* **2014**, *132*, 142–149. [CrossRef]

28. Michael, R.; Bron, A.J. The ageing lens and cataract: A model of normal and pathological ageing. *Trans. R. Soc. B Biol. Sci.* **2011**, *366*, 1278–1292. [CrossRef]

29. Weikel, K.A.; Garber, C.; Baburins, A.; Taylor, A. Nutritional modulation of cataract. *Nutr. Rev.* **2014**, *72*, 30–47. [CrossRef]

30. Mathew, M.C.; Ervin, A.M.; Tao, J.; Davis, R.M. Antioxidant vitamin supplementation for preventing and slowing the progression of age-related cataract. *Cochrane Database Syst. Rev.* **2012**, *6*. [CrossRef]

31. Chang, J.R.; Koo, E.; Agron, E.; Hallak, J.; Clemons, T.; Azar, D.; Sperduto, R.D.; Ferris, F.L., 3rd; Chew, E.Y. Age-Related Eye Disease Study Group. Risk factors associated with incident cataracts and cataract surgery in the Age-related Eye Disease Study (AREDS): AREDS report number 32. *Ophthalmology* **2011**, *118*, 2113–2119. [CrossRef] [PubMed]

nutrients

MDPI

Article

Cucurbita argyrosperma Seed Extracts Attenuate Angiogenesis in a Corneal Chemical Burn Model

María Fernanda Estrella-Mendoza [1], Francisco Jiménez-Gómez [2], Adolfo López-Ornelas [2], Rosa Martha Pérez-Gutiérrez [1] and Javier Flores-Estrada [2,*]

[1] Laboratorio de Investigacíon de Productos Naturales, Escuela de Ingeniería Química e Industrias Extractivas, Instituto Politécnico Nacional, Unidad Profesional Adolfo López Mateos, Av. Instituto Politécnico Nacional S/N, Ciudad de México 07708, México; fernandaestrella29@hotmail.com (M.F.E.-M.); rmpg@prodigy.net.mx (R.M.P.-G.)

[2] División de Investigación, Hospital Juárez de México, Av. Instituto Politécnico Nacional 5160, Magdalena de las Salinas, Gustavo A. Madero, Ciudad de México 07760, México; microcirugiafco@hotmail.com (F.J.-G.); adolfolopezmd@gmail.com (A.L.-O.)

* Correspondence: javier_70_1999@yahoo.com; Tel.: +52-55-5747-7560 (ext.7476)

Received: 15 April 2019; Accepted: 17 May 2019; Published: 27 May 2019

Abstract: Severe corneal inflammation produces opacity or even perforation, scarring, and angiogenesis, resulting in blindness. In this study, we used the cornea to examine the effect of new anti-angiogenic chemopreventive agents. We researched the anti-angiogenic effect of two extracts, methanol (Met) and hexane (Hex), from the seed of *Cucurbita argyrosperma*, on inflamed corneas. The corneas of Wistar rats were alkali-injured and treated intragastrically for seven successive days. We evaluated: opacity score, corneal neovascularization (CNV) area, re-epithelialization percentage, and histological changes. Also, we assessed the inflammatory (cyclooxigenase-2, nuclear factor-kappaB, and interleukin-1β) and angiogenic (vascular endothelial growth factor A, VEGF-A; -receptor 1, VEGFR1; and -receptor 2, VEGFR2) markers. Levels of *Cox-2*, *Il-1β*, and *Vegf-a* mRNA were also determined. After treatment, we observed a reduction in corneal edema, with lower opacity scores and cell infiltration compared to untreated rats. Treatment also accelerated wound healing and decreased the CNV area. The staining of inflammatory and angiogenic factors was significantly decreased and related to a down-expression of *Cox-2*, *Il-1β*, and *Vegf*. These results suggest that intake of *C. argyrosperma* seed has the potential to attenuate the angiogenesis secondary to inflammation in corneal chemical damage.

Keywords: *Cucurbita argyrosperma*; corneal chemical burn; angiogenesis; corneal neovascularization (CNV); vascular endothelial growth factor (VEGF); interleukin-1β (IL-1β); cyclooxigenase-2 (COX-2); nuclear factor-kappaB (NF-κB)

1. Introduction

Angiogenesis, inflammation, and oxidative stress are important factors that predispose to and promote the progression of degenerative diseases, and corneal diseases are not an exception. Corneal neovascularization (CNV) caused by viral infections, autoimmune diseases, and chemical burns can progress to corneal ulceration and scaring, potentially leading to blindness. CNV is also associated with a high rate of corneal allograft rejection [1]. In this context, the nuclear factor-kappaB (NF-κB) signaling pathway in inflammatory, epithelial, and endothelial cells is a key step for the transcriptional overexpression of pro-inflammatory and proangiogenic factors, including interleukin-1β (IL-1β), cyclooxigenase-2 (COX-2), and vascular endothelial growth factor A (VEGF-A). The cyclooxigenase-2 enzyme increases the synthesis of prostaglandins to modulate cell proliferation, cell death, and tumor invasion in many types of cancer. Interleukin-1β, IL-6, and tumor necrosis factor alpha (TNF-α) regulate COX-2; therefore, they are overexpressed during inflammation [2–4].

In an alkali-burn corneal injury, VEGF-A is an important aspect of angiogenesis expressed in macrophages and epithelial cells. Vascular endothelial growth factor A binds to two main tyrosine kinase receptors, vascular endothelial growth factor receptor 1 (VEGFR1) and –receptor 2 (VEGFR2), on a vascular endothelial cell to promote migration, proliferation, and neovascularisation, along with monocyte/macrophage migration in the microenvironment injured [1,5,6]. Several therapeutic strategies to decrease CNV have been used, such as topical or subconjunctival corticosteroids and non-steroidal anti-inflammatory agents [7]. However, these have limited use and have a number of side effects [8,9]. Anti-VEGF therapy in chemically burned ocular tissues results in a substantial reduction of angiogenesis in both animal studies and clinical trials [10,11]. However, to establish its safety and efficacy, controlled and randomized trials to justify their continued use are required. Besides, systemic drug treatment is not recommended because of adverse effects. Therefore, it is important to search for new drugs for the systemic treatment of these disorders.

Current data in CNV models show that natural extracts from plants or bioactive compounds in plant extracts have angiogenic suppressing activity [12–14]. The genus *Cucurbita* (pumpkin) belongs to one of the 300 genera of the Cucurbitaceae family, and it is one of the most popular vegetables eaten in the world. Recently, pumpkin was recognized as a functional food, and *Cucurbita pepo*, *Cucurbita maxima*, *Cucurbita moschata*, *Cucurbita andreana*, and *Cucurbita ficifolia* are the most cultivated species [15]. Nutritionally, pumpkin seed has a high amount of polyunsaturated fatty acids as well as proteins, vitamins, several minerals, and other phytochemicals. The anti-diabetic, antioxidant, anti-carcinogenic, and anti-inflammatory properties of this seed are studied due to its high content of natural bioactive compounds, such as carotenoids, tocopherols, and sterols [16–19].

Cucurbita argyrosperma is an economically important species cultivated in Mesoamerica. Isozyme, morphological, and ecological analyses suggest that it was probably domesticated from the Mexican wild squash, *Cucurbita sororia* [20]. The seed is usually consumed as a snack or as an ingredient in traditional stews, although the scientific findings of its beneficial effects on human health have not been sufficiently evidenced, and the anti-neovascular effects of the secondary metabolites remain unknown. However, conceivably its phytochemical composition could be like the related species, showing anti-inflammatory effects as suggested. Besides, proangiogenic factors such as COX-2, IL-1β, and VEGF, including VEGFR1 and -R2, induced by the inflammatory agents have not been studied in these plants.

The aim of this study was to investigate the effect of the seed extracts from *C. argyrosperma* in the inflammatory and angiogenic process of attenuation. Herein, we show that hexanic and methanolic extracts from *C. argyrosperma* seed significantly attenuate the expression of proangiogenic factors during inflammation using a CNV model. Also, we observed by clinical manifestation that both extracts significantly diminish the CNV area. Significantly, corneal re-epithelialization was higher with hexane (Hex) extract treatment than methanol (Met) extract.

2. Materials and Methods

2.1. Extract Preparation

Pumpkins of *C. argyrosperma* were harvested in an agricultural field of Michoacán, México, and identified by a botanist in the herbarium of the National Polytechnic Institute (IPN). Voucher specimen number 4532 was deposited in the herbarium of the National School of Biological Sciences of IPN. One kilogram of seed was extracted with 3 L of hexane (50% *v/v*) and left to macerate for 8 days at room temperature. The crude extract was filtered for one hour in 8 μM-medium flow filter paper (Whatman®, Sigma-Aldrich, Inc., St. Louis, MO, USA), concentrated using a rotary vacuum evaporator and taken to dryness at 60 °C in a vacuum rotator until the complete removal of the solvent, obtaining a viscous residue (8.38 g/L). The same procedure was applied to the residue, using methanol for a sequential separation of the seed components. Each extract was stored in the dark at 4 °C until use.

2.2. Animal Model

Twenty-eight male Wistar rats weighing 200–250 g (12 weeks old) were used. Water and standard food were available ad libitum. The care and management of experimental animals were in accordance with the guidelines of the National Institutes of Health "Guide for the Care and Use of Laboratory Animals", the standards described by the Association for Research in Vision and Ophthalmology (ARVO), and Official Mexican Standard NOM-062-ZOO-1999. The study was approved by the Ethics Committee of the Hospital Juárez de México (HJM 2493/14-B).

2.3. Experimental Design

Animals were intraperitoneally anesthetized with pentobarbital sodium (0.5 mg/kg), inhaled sevoflurane, and one drop of ophthalmic tetracaine, to perform the chemical cauterization of the cornea. Central corneas from the right eyes were burned by applying 3-mm-diameter filter paper saturated with 1M NaOH solution for 30 s, with immediate washing with 10 mL of saline solution. To avoid infection, a drop of ophthalmic ciprofloxacin was applied every 24 h until the end of the study [21]. Each extract was dissolved in water and Tween-80 (20%) (Veh). Rats were randomly divided into four groups (*n* = 7 each) [21,22]: non-chemically burned healthy corneas (non-CB) treated only with the vehicle; chemically burned (CB) corneas treated with the vehicle (CB-Veh); CB corneas treated with hexanic extract (CB-Hex); and CB corneas treated with methanolic extract (CB-Met). All groups received 400 mg/kg/7 days of Hex/Met extracts or the vehicle in a single dose (0.5 mL) by oral gavage, at the same time daily (10:00 h). Animals were euthanized with an overdose of pentobarbital.

2.4. Clinical Manifestation

Corneal opacity, epithelial defects, and the CNV area were evaluated eight days after CB. Corneal opacity was scored using a scaling system from 0 to 4: 0 = no opacity, completely clear cornea; 1 = slightly hazy, iris and lens visible; 2 = moderately opaque, iris and lens still visible; 3 = severely opaque, iris and lens hardly visible; and 4 = completely opaque, with no visibility of the iris and lens [23].

The measurement of the CNV area (mm^2) was performed in vivo using a ruler under a microscope and photographed. The software program Image-Pro Plus version 6.0 software (by Media Cybernetics, Inc., Rockville, MD, USA) was used. Inferonasal quadrant was selected to calculate the neovascularized area, according to a previous report [24].

To evaluate corneal wound re-epithelialization, we used corneal fluorescein staining. Briefly, fluorescein sodium ophthalmic strip was instilled into the lower conjunctival sac and the cornea was examined using a slit lamp biomicroscope with cobalt blue light. Injured epithelial tissues retain the fluorescein staining, whereas the lack of stain indicates re-epithelialization. The re-epithelialization percentage was evaluated in the CB corneas, considering that a total area of approximately 7 mm^2 was 100 percent (a 3-mm disc saturated with 1M NaOH).

2.5. Histological Evaluation

Enucleated eyes (*n* = 4 per group) were immediately fixed in neutral formalin. Cut tissue slides (3–5 mm) were made (anteroposterior) and included the optic nerve. Slides were dehydrated in graded alcohols and embedded in paraffin. Histological sections of 2 μm were processed and stained with hematoxylin–eosin (HE). We measured the corneal thickness and cell infiltration in the peripheral region (500 μm beyond the limbus) using light microscopy (axioscope 2 plus, Carl Zeiss, Göttingen, Germany). The percentage of the infiltration was calculated in a masked fashion based on the density in the corneal stroma of the CB-Veh group.

Other 2-μm-sections were dewaxed and rehydrated with antigen recovery solution (ImmunoDNA Retriever 20× with Citrate; BioSB, Santa Barbara, CA, USA). Slides were then loaded into a Shandon Sequenza chamber (Thermo Shandon, Cheshire, United Kingdom). We used the procedure described for the polymer-based immunodetection system (PolyVue® mouse/rabbit 3,3'-Diaminobenzidine, DAB, detection system, Diagnostic BioSystems, Pleasanton, CA, USA). We applied 100 μL of IL-1β (Cat. No. sc-7884), NF-κB p65 (sc-8008), COX-2 (sc-1746), VEGF-A (sc-7269), and VEGFR1 (sc-31173). All antibodies were purchased from Santa Cruz Biotechnology, Inc. (Santa Cruz, CA, USA). The VEGFR2 antibody (MAB3571) was purchased from R&D Systems, Inc. (Minneapolis, MN, USA). All dilutions were at 1:200 and incubated overnight at 4 °C. Later, the enhancers Polyvue Plus and HRP were added and incubated with DAB plus/chromogen substrate and counterstained with hematoxylin. An Axio Imager.A2 microscope with an integrated camera, Axiocam ICc5, (Carl Zeiss Microscopy GmbH, Jena, Germany) was used for histological observation and image capture. Micrographs of the peripheral region of the cornea (three fields per side at 200× magnification) were taken to measure the mean staining intensity of these markers. Images were analyzed with the Image-Pro Plus software version 6.0 (Media Cybernetics, Inc., Rockville, MD, USA).

2.6. Quantitative Reverse-Transcription Polymerase Chain Reaction (qRT-PCR)

Total RNA was isolated from corneal tissue using TRIzol™ Reagent (Invitrogen, Boston, MA, USA) (n = 3 per group). One microgram of DNase I-treated RNA (Roche Applied Science, Mannheim, Germany) was reverse transcribed with SuperScript® II Reverse Transcriptase system (Life Technologies Corp., Carlsbad, CA, USA). Quantification of mRNA was carried out using qPCR with SYBR green and the following primers: *Cox-2* (5'-CTGAGGGGTTACCACTTCCA-3'; and 5'-CTTGAACACGGACTTGCTCA-3'); *Il-1β*(5'-AGGCTTCCTTGTGCAAGTGT-3' and 5'-TGAGTGACACTGCCTTCCTG-3'); *Vegf-a* (5'-GCCCATGAAGTGGTGAAGTT-3' and 5'-ACTCCAGGGCTTCATCATTG-3'); and *Gapdh* (5'-CTCATGACCACAGTCCATGC-3' and 5'-TTCAGCTCTGGGATGACCTT-3'). The cycling protocol was as follows: denaturation (95 °C for 10 min), 45 cycles of amplification (95 °C for 15 s, 59 °C for 15 s, and 72 °C for 20 s), and a final extension at 72 °C. A melting curve analysis was also performed to ascertain the specificity of the amplified product. The expression for each gene was normalized to *Gapdh*. Expression was quantified as fold-change using the ΔΔCt method.

2.7. Statistical Analysis

We used GraphPad Prism software (La Jolla, CA, USA) (version 5.0). Values are mean with standard deviation (mean ± SD). In all cases, we used unifactorial analysis of variance followed by Tukey's post-hoc analysis.

3. Results

3.1. Amelioration of Corneal Wound Repair

We evaluated corneal wound healing mediated by the extracts in the alkali-burn corneal model. Figure 1a shows that the treated groups had a significant reduction in corneal opacity score and CNV area compared to the CB-Veh ($p < 0.05$ and $p < 0.001$, respectively). However, CNV in the CB-Met treated group was lower than the CB-Hex group ($p < 0.05$) and furthermore did not show a significant increase in the percentage of re-epithelialization (Figure 1b).

Using HE-stained slides (Figure 1c), the non-CB group had an average corneal thickness of 336.7 ± 39.5 μm, with an intact epithelial and stromal layer. There were neither inflammatory cells nor blood vessels. Conversely, corneal integrity in the CB-Veh group was severely impaired, with a loss of the epithelial cell layers and disruption of stromal collagen fibers. The corneal thickness in this group was 592.3 ± 112.1 μm, with an average cell infiltration of 81 ± 11.9%. Corneal thickness in the CB-Hex group was 427.2 ± 113.7 μm, which was significantly reduced compared to the CB-Veh), with a reduction in cell infiltration (40 ± 8.1%). Similarly, the CB-Met group showed a decrease in cell infiltration (33.2 ±

5.9%) and corneal thickness (295.2 ± 62.67 µm) ($p < 0.001$) compared to CB-Veh. Corneal thickness in the CB-Met group was not significantly different from that of the non-CB group.

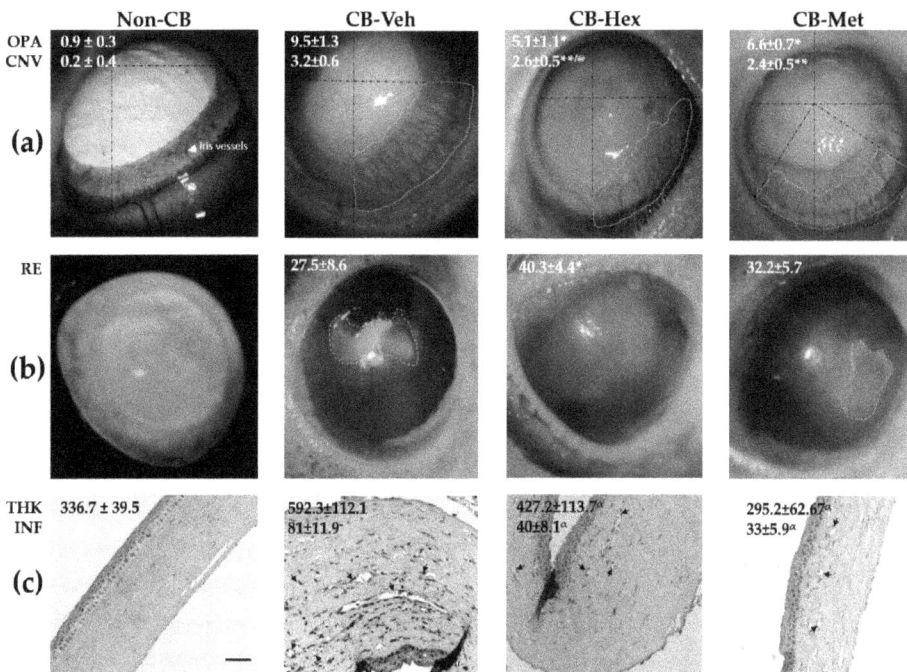

Figure 1. Methanol (Met) and hexane (Hex) extracts of *Cucurbita argyrosperma* seed in chemically burned corneas (CB) compared to the untreated group (CB corneas treated with the vehicle, CB-Veh). (**a**) Average of opacity score (OPA) and corneal neovascularization area (CNV) in mm². (**b**) Re-epithelialization percentage (RE). (**c**) Corneal thickness in microns (THK) and infiltration cell percentage (INF). Average value ± SD; * $p < 0.05$; ** $p < 0.01$, and $^\alpha$ $p < 0.001$ compared to CB-Veh. $^@$ $p < 0.05$ compared to CB-Met. Gray dotted lines show studied area, and black lines are the geometrical axis. CB-Hex: CB corneas treated with hexanic extract; CB-Met: CB corneas treated with methanolic extract. Arrows indicate the lumen of stromal blood vessels. (Scale bar = 100 µm).

3.2. Anti-inflammatory Effect

Several inflammatory cytokines implicated in alkali-induced corneal injury are regulated by the nuclear internalization of active NF-κB. Hence, we looked for the staining location of NF-κB in the corneal lesions. In the non-CB group, NF-κB was restricted to the cytoplasm of epithelial cells in the basal layer (Figure 2a). In the CB-Veh and both extract-treated groups, NF-κB was distributed in the nuclear compartment of endothelial and inflammatory cells. However, staining density decreased in the CB-Hex (41.13 ± 9.6) and the CB-Met (32.73 ± 8.1) compared to the CB-Veh (73.31 ± 10.4; $p < 0.0001$ both). Additionally, we performed measurements of the staining intensity for IL-1β (Figure 2b) and COX-2 (Figure 2c). In the non-CB, there was a low staining intensity for IL-1β (9.28 ± 2.6) compared to CB-Veh (75.95 ± 12.16; $p < 0.0001$), showing a distribution along with the corneal stroma as well as endothelial cells. Meanwhile, the intensities in the CB-Hex (42.16 ± 9.14) and the CB-Met (38.21 ± 7.9) were lower than the CB-Veh group ($p < 0.0001$). Staining intensity for IL-1β between the CB-Met and the CB-Hex showed no differences ($p > 0.05$). Likewise, the intensity for COX-2 was significantly different when comparing the CB-Veh (102.6 ± 13.08) to the CB-Hex (68.79 ± 10.73) and the CB-Met (37.15 ± 7.18) ($p < 0.0001$) groups. The non-CB had a detectable expression of 7.49 ± 3.48. Staining intensity for IL-1β and COX-2 in the cornea was also confirmed at the level of mRNA (Figure 3): *Il-1β* expression for the CB-Met (2.31 ± 0.30) and the CB-Hex (1.89 ± 0.11) groups was decreased compared to the CB-Veh group (3.03 ± 0.35; $p < 0.05$ and $p < 0.01$, respectively) (Figure 3a). The *Cox-2* in the CB-Hex (2.22 ± 0.10) and the CB-Met (1.91 ± 0.15) was also diminished compared to the Veh group (3.15 ± 0.31; $p < 0.01$) (Figure 3b). The *Il-1β* expression showed no difference between the CB-Met and the CB-Hex groups ($p > 0.05$); on the other hand, there were differences for *Cox-2* ($p < 0.05$).

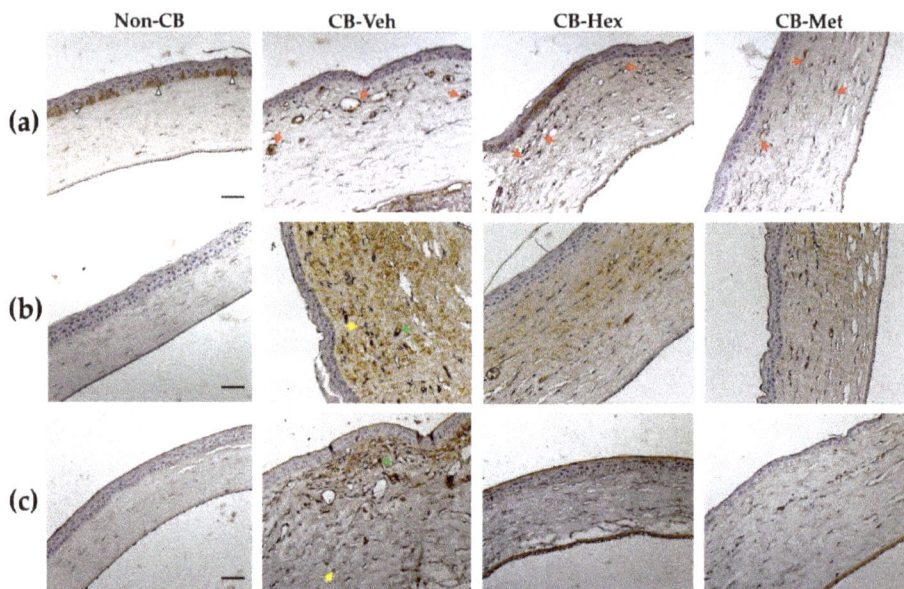

Figure 2. Micrograph of Met and Hex extracts of *C. argyrosperma* seed in the CB compared to the non-CB and CB-Veh groups. (**a**) Nuclei stained with anti-NF-κB p65 (red arrows). (**b**) Staining intensity for interleukin-1β (IL-1β) along with the corneal thickness. (**c**) Cyclooxigenase-2 (COX-2) staining in the studied groups. Arrowheads indicate the cytoplasmic distribution of NF-κB p65. Yellow and green arrows represent the minimum and maximum staining intensity, respectively, considered for software analysis. (Scale bar = 100 μm). CB: chemically burned corneas; CB-Veh: CB corneas treated with the vehicle; CB-Hex: CB corneas treated with hexanic extract; CB-Met: CB corneas treated with methanolic extract.

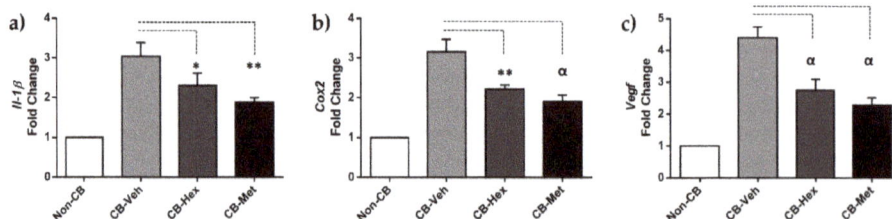

Figure 3. (**a**) qRT-PCR for *Il-1β*, (**b**) *Cox-2*, and (**c**) *Vegf-a*. Bars are the expression levels in each group. * $p \leq 0.05$, ** $p \leq 0.01$, and ᵅ $p \leq 0.001$ compared to the CB-Veh group. CB: chemically burned corneas; CB-Veh: CB corneas treated with the vehicle; CB-Hex: CB corneas treated with hexanic extract; CB-Met: CB corneas treated with methanolic extract.

3.3. Anti-Angiogenic Effect

Due to the anti-inflammatory effect in the CB-Hex and the CB-Met, we assessed whether it was related to an attenuation of CNV, by determining the staining intensity of VEGF-A and its receptors (Figure 4). In the CB-Veh, VEGF-A was in the cytoplasm and nucleus of the epithelial, endothelial, and other infiltrated cells with an intensity of 102.02 ± 14.04. A decrease in VEGF-A intensities was observed for the CB-Hex (71 ± 9.11) and CB-Met (61.3 ± 9.59) ($p < 0.001$) groups. Besides, staining intensity between VEGF-A in the CB-Met and CB-Hex groups showed differences ($p < 0.05$), and in the non-CB group it was about 16.25 ± 5.25 (Figure 4a). The *Vegf-a* expression was also confirmed (Figure 3c). The *Vegf-a* for the CB-Hex (2.74 ± 0.34) and CB-Met (2.29 ± 0.23) groups decreased compared to the CB-Veh (4.40 ± 0.34; $p < 0.05$ and $p < 0.01$, respectively).

Relevantly, staining for VEGFR1 was distributed in endothelial and inflammatory cells of the CB-Veh corneas (54.4 ± 6.8) and was higher when compared to the CB-Hex (33.13 ± 5.8; $p < 0.0001$) and CB-Met (25.47 ± 3.7; $p < 0.0001$) groups (Figure 4b). Vascular endothelial growth factor receptor 2 immunostaining was localized in the membrane region in endothelial cells, and in the CB-Veh (23.06 ± 3.5) it was higher compared to the CB-Hex (15.67 ± 2.6; $p < 0.0001$) and the CB-Met (10.95 ± 2.1; $p < 0.0001$) groups (Figure 4c).

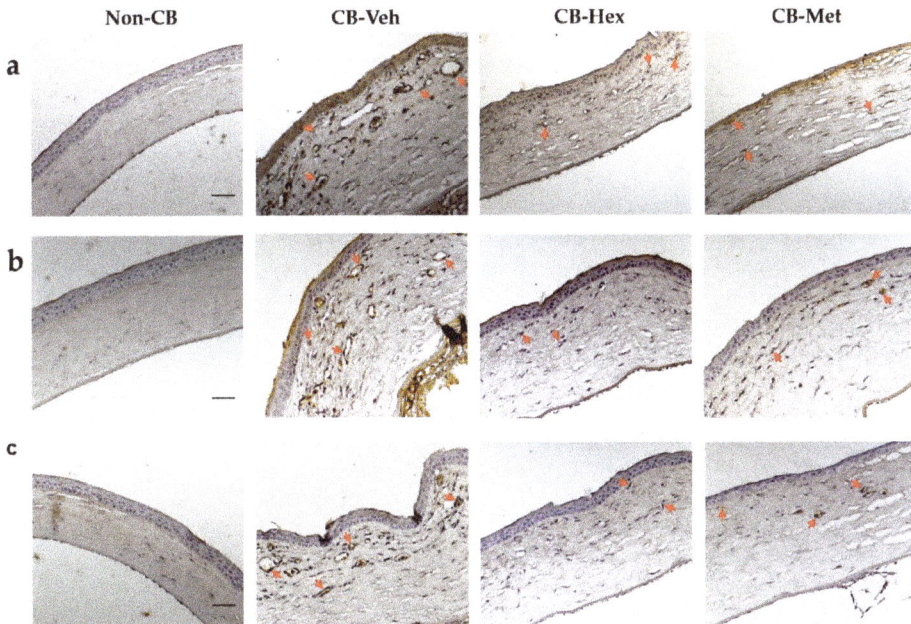

Figure 4. Immunolocalization of vascular endothelial growth factor A (VEGF-A) and its receptors, (vascular endothelial growth factor receptor 1, VEGFR1, and -receptor 2, VEGFR2), in the treated groups. (**a**) Staining intensities for VEGF-A. (**b**) and (**c**) Membranous staining in endothelial cells for VEGFR1 and VEGFR2, respectively (red arrows). Red arrows represent staining intensity considered for software analysis. (Scale bar = 100 μm). CB: chemically burned corneas; CB-Veh: CB corneas treated with the vehicle; CB-Hex: CB corneas treated with hexanic extract; CB-Met: CB corneas treated with methanolic extract.

4. Discussion

The cornea is a transparent, avascular, and immune-privileged tissue. However, the inflammatory response and growth of new vessels induced by infections, autoimmunity, and chemical burns may cause vision loss and a high rejection rate of corneal allografts if not treated effectively [1,25]. Recently, it has been demonstrated by histopathological and clinical observation in animal models that anti-neovascular topical treatments are successful at avoiding CNV [8,26]. To examine this phenomenon, experimental models combined with new therapeutic strategies have been used, primarily aimed at preserving corneal transparency through by attenuating the inflammatory and neovascularization responses.

Chemically burned corneas have long been used for this purpose because they are accompanied by the recruitment/migration of neutrophils and macrophages with resultant damage to the normal tissue structure. Furthermore, the release of oxidative derivatives, cytokines, chemokines, matrix metalloproteinases (MMPs), and growth factors including VEGF can influence corneal angiogenesis [21,27,28]. Topical treatments with monoclonal antibodies to VEGF-A or its receptor VEGFR2 suppress the mechanism of action of VEGF in endothelial cells, whereas dexamethasone inhibits CNV mediated by suppressing the activity of NF-κB, and decreasing the expression of IL-1β, COX-2, and VEGF [7,29]. In the same way, treatments with extracts of propolis and *Diospyros kaki*, as well as other purified phytochemicals (naringenin and (−)-Epigallocatechin gallate), have shown a decrease in CNV through down-regulation of VEGF-A, IL-1β, IL-6, and metalloproteases, promoting the healing of corneal wounds [12–14].

We aim to evidence the ability of the hexane and methanol extracts of *C. argyrosperma* seed to reduce CNV after an inflammatory stimulus induced by a chemical burn in rat cornea. To this end, we examined the nuclear localization of NF-κB p65, an important marker for the overexpression of VEGF-A, IL-1β, and COX-2 in damaged corneas. The receptor VEGFR2 in the endothelial cells of corneal stromal neovessels as an active marker of angiogenesis was studied. Angiogenesis is initiated when VEGFR2 is activated by tyrosine phosphorylation by VEGF-A binding. Consecutively, downstream pathways are activated, such as p38 MAPK and ERK1/2, producing a strong mitogenic and survival process [30,31]. In contrast, such a mitogenic signal is not equally induced by VEGFR1. Although VEGFR1 binds to VEGF-A with a higher affinity than VEGFR2, the induction of VEGFR1 phosphorylation is low and its downstream signaling is still poorly explored [32]. Vascular endothelial growth factor receptor 1 possesses anti-angiogenic activity by avoiding the union between VEGF-A and VEGR2. Corneal studies show that the proteolytic enzyme MMP14 can shave the extracellular domain of VEGFR1, converting it into a soluble receptor (sVEGFR) that acts as a decoy for VEGF-A [33]. Nevertheless, VEGFR1 indirectly induces angiogenesis by stimulating the migration of monocytes and macrophages directed towards the damaged microenvironment. For these reasons, this study evaluates both receptors because they are expressed predominantly in endothelial cells in the angiogenic environment: (1) VEGFR1 as a marker in the infiltration of monocytes and neovessels; and (2) VEGFR2 as an active marker of angiogenesis. Our results show that the staining intensity and expression of *Vegf-a* decreased along with the staining of VEGFR2 and VEGFR1 in endothelial cells of extract-treated corneas. Additionally, decreases in *Il-1β* and *Cox-2* expressions were also observed, suggesting that hexanic- and methanolic-extract components can attenuate these pro-angiogenic factors, likely through a lack of nuclear activation of NF-κB, which was also observed.

Treatment with hexanolic extract of *C. argyrosperma* seed shows a repairing of the corneal damage associated with a decrease in the expression of inflammatory and angiogenic factors. *Cucurbita pepo* seed extracts, containing mainly alpha-linoleic (ALA), linolenic acids (LA), tocopherols, and sterols, showed effective healing of skin wounds with a complete re-epithelialization, organization of collagen fibers, and absence of inflammatory cells [18]. Particularly, treatment with ALA in cultures of corneal epithelial cells and when it is topically applied to a dry eye animal model has anti-inflammatory activity, decreasing the release and expression of inflammatory factors (TNF-a, IL-6, IL-1β, and IL-8) regulated by the NF-κB pathway [34,35]. In addition, LA decreased corneal fluorescein staining and was associated with a significant decrease in the number of CD11b(+) cells [36].

Although we are still characterizing the bioactive compounds in the methanolic extract, we can speculate that the content of phytochemicals is similar to the *C. pepo* seed [16]. Some of these components include flavonoids such as quercetin, luteolin, and apigenin, which at low concentration intakes have protective anti-inflammatory effects on human retinal pigment epithelial damage by hypoxia, inhibiting VEGF and factors related to its activation [37]. For example, quercetin inhibits the production of inflammatory factors in VEGF-stimulated retinal photoreceptor cells, associated with inactivation of NF-κB as a consequence of the blockage of mitogen-activated protein kinases (MAPK) and protein kinase B (Akt) phosphorylation [38]. Nonetheless, *C. argyrosperma* may differ in the content and types of flavonoids from *C. pepo*. This would bring us closer to a better understanding of the mechanisms of attenuated angiogenesis by the phytochemicals contained in the methanol extract, which in turn could act synergistically, either directly or indirectly, in VEGF-A regulation.

Corneal inflammation eventually causes vision loss due to CNV. Corneal alkali-injury not only upregulated NF-κB, IL-1β, and COX-2 expression, it also significantly increased VEGF and its receptors VEGFR1 and VEGFR2 in endothelial cells. This work demonstrates for the first time, that methanolic or hexanic extracts of *C. argyrosperma* seed (400 mg/kg/7 days) improve the healing of corneal wounds caused by a chemical agent and may contribute to the anti-inflammatory properties of the phytochemicals in its composition. In addition, a significant reduction of the CNV was related to the attenuation of proangiogenic factors. Significantly, our results indicate that *C. argyrosperma* Hex extract is better than Met extract at reducing corneal re-epithelialization time, improving the healing process and thus preventing the entrance of microorganisms and inflammatory mediators into the deeper layers, probably through the inhibition of the NF-κB pathway for at least seven days after corneal alkali burn.

5. Conclusions

Intake of *Cucurbita argyrosperma* seed may be an option for preventing corneal angiogenesis. In addition, it might benefit wound healing or inhibit neovascularization in other degenerative pathologies. Further pharmacological and phytochemical studies are required to identify their constituents and accurately assess this activity.

Author Contributions: Conceptualization, R.M.P.-G.; Formal analysis, J.F.-E.; Investigation, M.F.E.-M., F.J.-G., R.M.P.-G. and J.F.-E.; Methodology, M.F.E.-M., F.J.-G., and R.M.P.-G.; Project administration, J.F.-E.; Resources, F.J.-G. and R.M.P.-G.; Supervision, J.F.-E.; Visualization, A.L.-O. and J.F.-E.; Writing—original draft, A.L.-O. and J.F.-E.; Writing—review and editing, A.L.-O. and J.F.-E.

Funding: This research was supported by the Hospital Juárez de México and School of Chemical Engineering and Extractive Industries from the National Polytechnic Institute.

Acknowledgments: We thank Julia D. Toscano-Garibay for her numerous comments on the manuscript.

Conflicts of Interest: The authors declare no conflict of interest.

References

1. Azar, D.T. Corneal angiogenic privilege: Angiogenic and antiangiogenic factors in corneal avascularity, vasculogenesis, and wound healing (an American Ophthalmological Society thesis). *Trans. Am. Ophthalmol. Soc.* **2006**, *104*, 264–302.
2. Kuwano, T.; Nakao, S.; Yamamoto, H.; Tsuneyoshi, M.; Yamamoto, T.; Kuwano, M.; Ono, M. Cyclooxygenase 2 is a key enzyme for inflammatory cytokine-induced angiogenesis. *FASEB J.* **2004**, *18*, 300–310. [CrossRef]
3. Nakao, S.; Kuwano, T.; Tsutsumi-Miyahara, C.; Ueda, S.; Kimura, Y.N.; Hamano, S.; Sonoda, K.H.; Saijo, Y.; Nukiwa, T.; Strieter, R.M.; et al. Infiltration of COX-2-expressing macrophages is a prerequisite for IL-1 beta-induced neovascularization and tumor growth. *J. Clin. Investig.* **2005**, *115*, 2979–2991. [CrossRef]
4. Sobolewski, C.; Cerella, C.; Dicato, M.; Ghibelli, L.; Diederich, M. The role of cyclooxygenase-2 in cell proliferation and cell death in human malignancies. *Int. J. Cell Biol.* **2010**, *2010*, 215158. [CrossRef]
5. Cursiefen, C.; Chen, L.; Borges, L.P.; Jackson, D.; Cao, J.; Radziejewski, C.; D'Amore, P.A.; Dana, M.R.; Wiegand, S.J.; Streilein, J.W. VEGF-A stimulates lymphangiogenesis and hemangiogenesis in inflammatory neovascularization via macrophage recruitment. *J. Clin. Investig.* **2004**, *113*, 1040–1050. [CrossRef] [PubMed]

6. Sivak, J.M.; Ostriker, A.C.; Woolfenden, A.; Demirs, J.; Cepeda, R.; Long, D.; Anderson, K.; Jaffee, B. Pharmacologic uncoupling of angiogenesis and inflammation during initiation of pathological corneal neovascularization. *J. Biol. Chem.* **2011**, *286*, 44965–44975. [CrossRef]

7. Nakao, S.; Hata, Y.; Miura, M.; Noda, K.; Kimura, Y.N.; Kawahara, S.; Kita, T.; Hisatomi, T.; Nakazawa, T.; Jin, Y.; et al. Dexamethasone inhibits interleukin-1beta-induced corneal neovascularization: Role of nuclear factor-kappaB-activated stromal cells in inflammatory angiogenesis. *Am. J. Pathol.* **2007**, *171*, 1058–1065. [CrossRef] [PubMed]

8. Gupta, D.; Illingworth, C. Treatments for corneal neovascularization: A review. *Cornea* **2011**, *30*, 927–938. [CrossRef]

9. Sheppard, J.D.; Comstock, T.L.; Cavet, M.E. Impact of the Topical Ophthalmic Corticosteroid Loteprednol Etabonate on Intraocular Pressure. *Adv. Ther.* **2016**, *33*, 532–552. [CrossRef] [PubMed]

10. Al-Debasi, T.; Al-Bekairy, A.; Al-Katheri, A.; Al Harbi, S.; Mansour, M. Topical versus subconjunctival anti-vascular endothelial growth factor therapy (Bevacizumab, Ranibizumab and Aflibercept) for treatment of corneal neovascularization. *Saudi J. Ophthalmol.* **2017**, *31*, 99–105. [CrossRef] [PubMed]

11. Zhou, C.; Robert, M.C.; Kapoulea, V.; Lei, F.; Stagner, A.M.; Jakobiec, F.A.; Dohlman, C.H.; Paschalis, E.I. Sustained Subconjunctival Delivery of Infliximab Protects the Cornea and Retina Following Alkali Burn to the Eye. *Investig. Ophthalmol. Vis. Sci.* **2017**, *58*, 96–105. [CrossRef]

12. Keshavarz, M.; Mostafaie, A.; Mansouri, K.; Shakiba, Y.; Motlagh, H.R. Inhibition of corneal neovascularization with propolis extract. *Arch. Med. Res.* **2009**, *40*, 59–61. [CrossRef] [PubMed]

13. Oguido, A.; Hohmann, M.S.N.; Pinho-Ribeiro, F.A.; Crespigio, J.; Domiciano, T.P.; Verri, W.A., Jr.; Casella, A.M.B. Naringenin Eye Drops Inhibit Corneal Neovascularization by Anti-Inflammatory and Antioxidant Mechanisms. *Investig. Ophthalmol. Vis. Sci.* **2017**, *58*, 5764–5776. [CrossRef] [PubMed]

14. Yang, S.J.; Jo, H.; Kim, K.A.; Ahn, H.R.; Kang, S.W.; Jung, S.H. Diospyros kaki Extract Inhibits Alkali Burn-Induced Corneal Neovascularization. *J. Med. Food* **2016**, *19*, 106–109. [CrossRef]

15. Kocyan, A.; Zhang, L.B.; Schaefer, H.; Renner, S.S. A multi-locus chloroplast phylogeny for the Cucurbitaceae and its implications for character evolution and classification. *Mol. Phylogenet. Evol.* **2007**, *44*, 553–577. [CrossRef]

16. Akomolafe, S.F.; Oboh, G.; Oyeleye, S.I.; Molehin, O.R.; Ogunsuyi, O.B. Phenolic Composition and Inhibitory Ability of Methanolic Extract from Pumpkin (*Cucurbita pepo* L) Seeds on Fe-induced Thiobarbituric acid reactive species in Albino Rat's Testicular Tissue In-Vitro. *J. Appl. Pharm. Sci.* **2016**, *6*, 115–120. [CrossRef]

17. Yadav, M.; Jain, S.; Tomar, R.; Prasad, G.B.; Yadav, H. Medicinal and biological potential of pumpkin: An updated review. *Nutr. Res. Rev.* **2010**, *23*, 184–190. [CrossRef]

18. Bardaa, S.; Ben Halima, N.; Aloui, F.; Ben Mansour, R.; Jabeur, H.; Bouaziz, M.; Sahnoun, Z. Oil from pumpkin (*Cucurbita pepo* L.) seeds: Evaluation of its functional properties on wound healing in rats. *Lipids Health Dis.* **2016**, *15*, 73. [CrossRef] [PubMed]

19. Medjakovic, S.; Hobiger, S.; Ardjomand-Woelkart, K.; Bucar, F.; Jungbauer, A. Pumpkin seed extract: Cell growth inhibition of hyperplastic and cancer cells, independent of steroid hormone receptors. *Fitoterapia* **2016**, *110*, 150–156. [CrossRef]

20. Sanchez-de la Vega, G.; Castellanos-Morales, G.; Gamez, N.; Hernandez-Rosales, H.S.; Vazquez-Lobo, A.; Aguirre-Planter, E.; Jaramillo-Correa, J.P.; Montes-Hernandez, S.; Lira-Saade, R.; Eguiarte, L.E. Genetic Resources in the "Calabaza Pipiana" Squash (*Cucurbita argyrosperma*) in Mexico: Genetic Diversity, Genetic Differentiation and Distribution Models. *Front. Plant. Sci.* **2018**, *9*, 400. [CrossRef]

21. Choi, H.; Phillips, C.; Oh, J.Y.; Stock, E.M.; Kim, D.K.; Won, J.K.; Fulcher, S. Comprehensive Modeling of Corneal Alkali Injury in the Rat Eye. *Curr. Eye Res.* **2017**, *42*, 1348–1357. [CrossRef]

22. Atiba, A.; Wasfy, T.; Abdo, W.; Ghoneim, A.; Kamal, T.; Shukry, M. Aloe vera gel facilitates re-epithelialization of corneal alkali burn in normal and diabetic rats. *Clin. Ophthalmol.* **2015**, *9*, 2019–2026. [CrossRef] [PubMed]

23. Yoeruek, E.; Ziemssen, F.; Henke-Fahle, S.; Tatar, O.; Tura, A.; Grisanti, S.; Bartz-Schmidt, K.U.; Szurman, P. Safety, penetration and efficacy of topically applied bevacizumab: Evaluation of eyedrops in corneal neovascularization after chemical burn. *Acta Ophthalmol.* **2008**, *86*, 322–328. [CrossRef] [PubMed]

24. Rogers, M.S.; Birsner, A.E.; D'Amato, R.J. The mouse cornea micropocket angiogenesis assay. *Nat. Protoc* **2007**, *2*, 2545–2550. [CrossRef] [PubMed]

25. Dana, M.R.; Streilein, J.W. Loss and restoration of immune privilege in eyes with corneal neovascularization. *Investig. Ophthalmol. Vis. Sci.* **1996**, *37*, 2485–2494.

26. Roshandel, D.; Eslani, M.; Baradaran-Rafii, A.; Cheung, A.Y.; Kurji, K.; Jabbehdari, S.; Maiz, A.; Jalali, S.; Djalilian, A.R.; Holland, E.J. Current and emerging therapies for corneal neovascularization. *Ocul. Surf.* **2018**, *16*, 398–414. [CrossRef] [PubMed]

27. Shakiba, Y.; Mansouri, K.; Arshadi, D.; Rezaei, N. Corneal neovascularization: Molecular events and therapeutic options. *Recent Pat. Inflamm. Allergy Drug Discov.* **2009**, *3*, 221–231. [CrossRef] [PubMed]

28. Edelman, J.L.; Castro, M.R.; Wen, Y. Correlation of VEGF expression by leukocytes with the growth and regression of blood vessels in the rat cornea. *Investig. Ophthalmol. Vis. Sci.* **1999**, *40*, 1112–1123.

29. Nakao, S.; Zandi, S.; Lara-Castillo, N.; Taher, M.; Ishibashi, T.; Hafezi-Moghadam, A. Larger therapeutic window for steroid versus VEGF-A inhibitor in inflammatory angiogenesis: Surprisingly similar impact on leukocyte infiltration. *Investig. Ophthalmol. Vis. Sci.* **2012**, *53*, 3296–3302. [CrossRef] [PubMed]

30. Kim, Y.M.; Hwang, S.; Kim, Y.M.; Pyun, B.J.; Kim, T.Y.; Lee, S.T.; Gho, Y.S.; Kwon, Y.G. Endostatin blocks vascular endothelial growth factor-mediated signaling via direct interaction with KDR/Flk-1. *J. Biol. Chem.* **2002**, *277*, 27872–27879. [CrossRef]

31. Gee, E.; Milkiewicz, M.; Haas, T.L. p38 MAPK activity is stimulated by vascular endothelial growth factor receptor 2 activation and is essential for shear stress-induced angiogenesis. *J. Cell Physiol.* **2010**, *222*, 120–126. [CrossRef]

32. Koch, S.; Claesson-Welsh, L. Signal transduction by vascular endothelial growth factor receptors. *Cold Spring Harb. Perspect. Med.* **2012**, *2*, a006502. [CrossRef]

33. Han, K.Y.; Chang, J.H.; Lee, H.; Azar, D.T. Proangiogenic Interactions of Vascular Endothelial MMP14 With VEGF Receptor 1 in VEGFA-Mediated Corneal Angiogenesis. *Invest. Ophthalmol. Vis. Sci.* **2016**, *57*, 3313–3322. [CrossRef]

34. Erdinest, N.; Shmueli, O.; Grossman, Y.; Ovadia, H.; Solomon, A. Anti-inflammatory effects of alpha linolenic acid on human corneal epithelial cells. *Investig. Ophthalmol. Vis. Sci.* **2012**, *53*, 4396–4406. [CrossRef]

35. Erdinest, N.; Shohat, N.; Moallem, E.; Yahalom, C.; Mechoulam, H.; Anteby, I.; Ovadia, H.; Solomon, A. Nitric oxide secretion in human conjunctival fibroblasts is inhibited by alpha linolenic acid. *J. Inflamm.* **2015**, *12*, 59. [CrossRef]

36. Rashid, S.; Jin, Y.; Ecoiffier, T.; Barabino, S.; Schaumberg, D.A.; Dana, M.R. Topical omega-3 and omega-6 fatty acids for treatment of dry eye. *Arch. Ophthalmol.* **2008**, *126*, 219–225. [CrossRef]

37. Chen, R.; Hollborn, M.; Grosche, A.; Reichenbach, A.; Wiedemann, P.; Bringmann, A.; Kohen, L. Effects of the vegetable polyphenols epigallocatechin-3-gallate, luteolin, apigenin, myricetin, quercetin, and cyanidin in primary cultures of human retinal pigment epithelial cells. *Mol. Vis.* **2014**, *20*, 242–258. [PubMed]

38. Lee, M.; Yun, S.; Lee, H.; Yang, J. Quercetin Mitigates Inflammatory Responses Induced by Vascular Endothelial Growth Factor in Mouse Retinal Photoreceptor Cells through Suppression of Nuclear Factor Kappa B. *Int. J. Mol. Sci.* **2017**, *18*, 2497. [CrossRef]

nutrients

MDPI

Article

A Critical Appraisal of National and International Clinical Practice Guidelines Reporting Nutritional Recommendations for Age-Related Macular Degeneration: Are Recommendations Evidence-Based?

John G. Lawrenson [1,*], Jennifer R. Evans [2] and Laura E. Downie [3]

1   Division of Optometry and Visual Science, City, University of London, Northampton Square, London EC1V OHB, UK
2   International Centre for Eye Health, Clinical Research Department, London School of Hygiene and Tropical Medicine, Keppel St, London WC1E 7HT, UK; Jennifer.Evans@lshtm.ac.uk
3   Department of Optometry and Vision Sciences, The University of Melbourne, Parkville, Victoria 3010, Australia; ldownie@unimelb.edu.au
*   Correspondence: j.g.lawrenson@city.ac.uk; Tel.: +44(0)20-7040-4310

Received: 26 March 2019; Accepted: 9 April 2019; Published: 11 April 2019

Abstract: Eye care professionals should have access to high quality clinical practice guidelines that ideally are underpinned by evidence from robust systematic reviews of relevant research. The aim of this study was to identify clinical guidelines with recommendations pertaining to dietary modification and/or nutritional supplementation for age-related macular degeneration (AMD), and to evaluate the overall quality of the guidelines using the Appraisal of Guidelines for Research and Evaluation II (AGREE II) instrument. We also mapped recommendations to existing systematic review evidence. A comprehensive search was undertaken using bibliographic databases and other electronic resources for eligible guidelines. Quality appraisal was undertaken to generate scores for each of the six AGREE II domains, and mapping of extracted nutritional recommendations was performed for systematic reviews published up to March 2017. We identified 13 national and international guidelines, developed or updated between 2004 and 2019. These varied substantially in quality. The lowest scoring AGREE II domains were for 'Rigour of Development', 'Applicability' (which measures implementation strategies to improve uptake of recommendations), and 'Editorial Independence'. Only four guidelines used evidence from systematic reviews to support their nutritional recommendations. In conclusion, there is significant scope for improving current Clinical Practice Guidelines for AMD, and guideline developers should use evidence from existing high quality systematic reviews to inform clinical recommendations.

Keywords: clinical practice guidelines; systematic reviews; age-related macular degeneration; nutritional supplements; diet; nutrition; AGREE II

1. Introduction

Age-related macular degeneration (AMD) is an ocular disease affecting the central area of the retina (the macula), which is responsible for high-resolution daytime vision [1]. In the early stages of AMD, the macula shows characteristic sub-retinal lipoprotein deposits, known as drusen, and the monolayer of pigmented cells beneath the retina (retinal pigment epithelium) shows areas of hypo- and hyper-pigmentation. As the disease progresses, the retinal pigment epithelium can become atrophic, with secondary dysfunction and a loss of retinal photoreceptors [2]. Less commonly, new blood vessels grow beneath, or within, the retina; these vessels have a tendency to leak, causing disruption to the macular architecture and ultimately scar formation (neovascular AMD). Early AMD is typically asymptomatic; however, later stages of AMD can have a significant negative impact on visual function and quality of life [3]. AMD is currently the leading cause of severe vision impairment among people aged 50 and over in high-income countries [4]. With increasing life expectancy these numbers are projected to increase substantially in the future.

AMD is a complex disorder with several non-modifiable and modifiable risk factors [5]. The retina is particularly susceptible to oxidative stress, as a result of its high oxygen consumption and exposure to light; animal and cell culture studies have identified oxidative stress as a contributory factor for the development of AMD [6]. A large body of observational and experimental research in humans has investigated the association between dietary antioxidants and AMD, in particular whether an increased intake of antioxidant vitamin and minerals or specific carotenoids can prevent or slow the progression of the disease [7,8]. The role of particular essential fatty acids has also been investigated, including exploiting the anti-inflammatory properties of long-chain omega-3 fatty acids.

Based on accumulating research evidence, dietary modification and/or nutritional supplementation has been proposed as a simple and potentially cost-effective strategy for modifying the risk of AMD. Although there is a large body of research in this area, the quality of studies is highly variable and conflicting results have been reported. These factors pose challenges for clinicians, when attempting to provide evidence-based recommendations to patients about the relative risks and benefits of nutrition-based interventions. Clinical practice guidelines can be used to aid clinical decision-making in such circumstances. These documents should contain recommendations that are informed by a systematic review of the available research evidence, and a consideration of the benefits and harms of alternative interventions [9]. In the process of guideline development, developers either elect to perform their own systematic review of the literature, or to instead integrate the results from previously published reviews. Systematic reviews involve a process of systematically identifying, appraising and synthesising findings from all relevant research studies relating to a particular research question, with the intent of minimising the potential for bias. The objectives of the present study were to:

(1) Identify clinical practice guidelines for AMD that contain nutritional recommendations, based upon a systematic literature search;

(2) Evaluate the quality of these guidelines using the Appraisal of Guidelines for Research and Evaluation II (AGREE II) tool; and

(3) Map clinical practice guideline recommendations, as related to nutrition and AMD, to relevant systematic reviews.

2. Materials and Methods

2.1. Eligibility Criteria

We included national and international clinical practice guidelines for AMD that contained nutritional recommendations, including dietary modification or nutritional supplementation, as strategies to prevent or slow the progression of the disease.

2.2. Search Strategy

We searched the Ovid MEDLINE and Embase bibliographic databases from January 1999 to January 2019 using search filters for identifying clinical guidelines that were developed by the Canadian Agency for Drugs and Technologies in Health (https://www.cadth.ca) (see Supplementary Materials). In addition to the bibliographic database searches, we also searched Guideline Central Summaries (https://www.guidelinecentral.com/summaries/) and the Turning Research into Practice (TRIP) database (https://www.tripdatabase.com/), and undertook a search of the webpages of national and international professional organisations for ophthalmology and optometry. The search was not limited by language and we used Google Translate to extract recommendations and appraise guidelines that were not written in English.

2.3. Study Selection and Data Extraction

Following removal of duplicates, two reviewers (JL/LD) independently screened the titles and abstracts identified from the bibliographic searches and resolved any discrepancies by discussion and consensus. We obtained full-text copies of potentially eligible guidelines and these were independently assessed by all three authors, working in pairs, to determine whether they met the inclusion criteria. Reasons for exclusion were documented at this stage. Two reviewers (all three authors working in pairs) independently extracted general characteristics of each guideline (e.g., author/organisation, country, year of publication, target audience) and details of the specific nutritional recommendations.

2.4. Appraisal of Clinical Guidelines

The quality of each clinical guideline was independently evaluated by two reviewers (all three authors working in pairs) using the Appraisal of Guidelines for Research and Evaluation (AGREE) II tool (https://www.agreetrust.org/). The AGREE II instrument evaluates the process of practice guideline development, and the quality of reporting, using 21 items organised into six key domains, as follows: (1) Scope and Purpose; (2) Stakeholder Involvement; (3) Rigour of Development; (4) Clarity of Presentation; (5) Applicability; and (6) Editorial Independence. Table 1 provides further details about the content of these domains. Each item within the AGREE II tool comprises a quality statement/concept, which is scored using a 7-point Likert rating scale. A score of '1' was given when there was no information that was relevant to the item or where the concept was very poorly reported. A score of '7' was given if the quality of reporting met the full criteria, as defined in the AGREE II User's Manual. The overall scores for each of the six domains were calculated by summing up all the scores of the individual items for that domain, and then scaling the total as a percentage of the maximum possible score for each domain. Therefore higher scores indicate higher guideline quality. An intra-class correlation coefficient (ICC) was used to evaluate overall inter-rater agreement.

Table 1. Clinical guideline quality domains used in the Appraisal of Guidelines for Research and Evaluation II (AGREE II) tool.

Domain	Content
Scope and Purpose (3 items)	Concerned with the overall aim of the guideline, the specific health questions, and the target population
Stakeholder Involvement (3 items)	Focuses on the extent to which the guideline was developed by the appropriate stakeholders and represents the views of its intended users
Rigour of Development (8 items)	Relates to the process used to gather and synthesize the evidence, the methods to formulate the recommendations, and to update them
Clarity of Presentation (3 items)	Deals with the language, structure, and format of the guideline
Applicability (4 items)	Pertains to the likely barriers and facilitators to implementation, strategies to improve uptake, and resource implications of applying the guideline
Editorial Independence (2 items)	Concerned with the formulation of recommendations not being unduly biased with competing interests

2.5. Mapping Clinical Guideline Recommendations to Systematic Review Evidence

Given the predominance of systematic reviews in the hierarchy of clinical evidence [10], we undertook an exercise to map the extracted guideline recommendations relating to diet or nutritional supplementation, in the context of AMD, to the evidence derived from systematic reviews. We also assessed whether recommendations were linked to particular citations (e.g., systematic reviews, individual randomised controlled trials (RCTs), or other study designs). To identify relevant systematic reviews on nutritional interventions for AMD we used two recently published studies [11,12] that identified and critically appraised systematic reviews of AMD interventions. Lindsley et al. [11]

identified 47 systematic reviews published between 2001 and 2014, of which 9 (19%) evaluated dietary supplements. A more recent study by Downie and colleagues [12] found 71 systematic reviews published between 2003 and 2017, with 10 (12%) relating to nutritional interventions in AMD. These studies identified 11 unique reviews [13–23], of which over 50% were published after 2014.

The process of mapping systematic reviews to guideline recommendations relating to nutrition was performed by a single reviewer and then independently checked by a second reviewer.

3. Results

3.1. Search Results

The searches identified 868 potentially relevant records. Following title and abstract screening, 838 were excluded. After a full-text review of 30 potentially eligible records, 17 were excluded as they did not contain recommendations for nutritional interventions. Thirteen guidelines were included in the final analysis. Figure 1 shows the flow diagram for guideline selection.

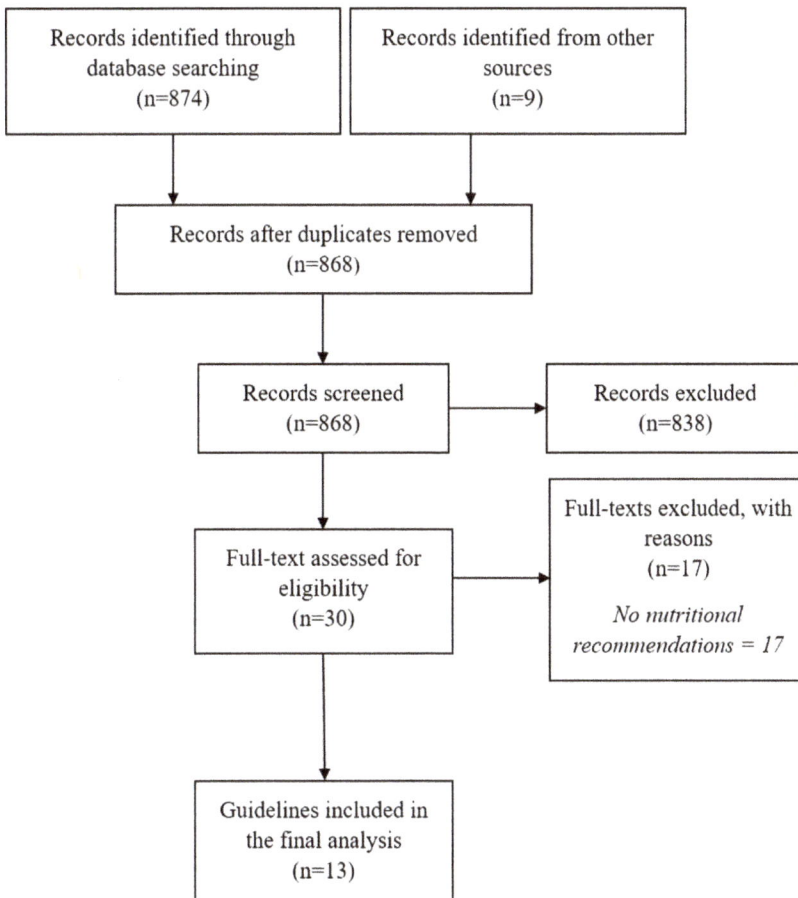

Figure 1. Selection process for identifying relevant clinical practice guidelines.

3.2. Characteristics of Included Guidelines and Nutritional Recommendations

Table 2 summarises the characteristics of 13 national guidelines from the United Kingdom [24,25], United States [26–28], Canada [29,30], Australia [31], New Zealand [32], Spain [33], Germany [34], and the Philippines [35], and an international ophthalmology guideline [36], that were published between 2004 and 2019. Three guidelines were produced by government organisations [24,25,32], while the remainder were developed by professional societies or associations for specific eye care professions.

Table 2. General characteristics of the included age-related macular degeneration (AMD) clinical practice guidelines.

Organisation	Reference	Year	Country	Target Audience
American Optical Association	[26]	2004	United States	Optometrists
International Council of Ophthalmology	[36]	2007	International	Optometrists
Spanish Retina and Vitreous Society (SERV)	[33]	2009	Spain	Ophthalmologists
Canadian Expert Consensus	[30]	2012	Canada	Ophthalmologists
German Ophthalmological Society (in German)	[34]	2014	Germany	Ophthalmologists
Eye Health Council of Ontario	[29]	2015	Canada	Health and eye care professionals
National Health Committee	[32]	2015	New Zealand	Healthcare professionals involved in the diagnosis and management of AMD
American Academy of Ophthalmology	[27]	2015	United States	Ophthalmologists
Vitreo-retina Society of the Philippines (VRSP)	[35]	2016	Philippines	Ophthalmologists
National Institute for Health and Care Excellence (NICE; Clinical Knowledge Summaries)	[24]	2016	United Kingdom	Primary care healthcare professionals
Clinical Advisory Committee [28]	[28]	2017	United States	Optometrists
NICE (NG82)	[25]	2018	United Kingdom	Healthcare professionals involved in the diagnosis and management of AMD
Optometry Australia	[31]	2019	Australia	Optometrists

Eight guidelines [24,26–29,31,34,35] provided nutritional recommendations (either positive or negative) for reducing the risk of developing AMD (primary prevention) (Table 3). Guidelines from Canada [29], Germany [34], and Australia [31] included recommendations on the value of dietary modification for primary prevention. With regard to supplements, a guideline produced by an expert panel of optometrists in the United States [28] recommended the use of xanthophil supplements (lutein, zeaxanthin, and mesozeaxanthin) in 'sub-clinical' AMD and another suggested that supplementation with antioxidants may be beneficial for those who are nutritionally deficient [26]. Five guidelines specifically stated that there was evidence that use of high- dose anti-oxidant vitamin and mineral supplements was not beneficial for primary prevention (Table 3) [24,27,31,34,35].

All of the included guidelines provided recommendations regarding nutritional strategies for secondary prevention (i.e., slowing the progression of AMD) (Table 4). Five guidelines [24,28,30,31,33] included dietary advice, consisting of encouraging people with AMD to eat a healthy diet and, in particular, to increase consumption of foods rich in the macular carotenoids (lutein and zeaxanthin) and/or eat more oily fish as a source of omega-3 essential fatty acids. With regard to nutritional supplements, all guidelines addressed the topic of antioxidant vitamin and mineral supplements; the vast majority referred to supplements containing the combinations of antioxidant vitamins, zinc and carotenoids used in the Age-Related Eye Disease Studies (AREDS) or AREDS2 studies [7,8]. Only the U.K. National Institute for Health and Care Excellence (NICE) guideline [25] did not advocate the use high-dose antioxidant vitamin and mineral supplements for the secondary prevention of AMD. The NICE guideline committee concluded that "the current clinical evidence was not able to demonstrate a clear treatment benefit of antioxidant vitamin and mineral supplement for people with early AMD and was insufficient to make a strong recommendation on the use of these supplements". The majority of guidelines acknowledged the risks as well as the benefits of antioxidant vitamin and mineral supplementation. Although acknowledging the lack of evidence, two guidelines [28,34] stated that omega-3 fatty acid supplements could be considered. By contrast, the NICE committee [25] concluded that "omega-3 fatty acid supplementation had no meaningful effect on AMD progression and visual acuity, and that therefore no recommendations could be made on this topic".

3.3. Clinical Guideline Quality Scores

The overall inter-rater agreement on scoring of the individual items was excellent (correlation coefficient: 0.81 (95% confidence interval (CI) 0.76 to 0.84)).

The quality of each guideline, across each of the AGREE II domains, is shown in Table 5. The median percentage scores (%) for different domains ranged from 13% to 75%, with substantial heterogeneity in quality between guidelines. Only two of six domains had a median score over 50% ('Scope and Purpose' and 'Clarity of presentation'). The lowest scoring domains were 'Applicability' (median 13% (range 0 to 98%)), 'Editorial Dependence' (median 20.8% (range 0 to 100%)), and Rigour of Development (median 20.4% (range 9.2 to 95.9%)). For the 'Rigour of Development' domain, only four of the 13 guidelines achieved a score ≥50%; this was largely due to a failure to demonstrate that systematic methods had been used to search for relevant evidence, no description of the criteria used for selecting and/or evaluating the evidence, and no information on the methods used by the developers to formulate recommendations.

3.4. Mapping Clinical Guideline Recommendations to Systematic Review Evidence

We extracted individual recommendations relating to nutritional strategies for both preventing and slowing the progression of AMD (see Tables 3 and 4 for an overview). For some guidelines, no evidence was provided to support nutritional recommendations [33,36]. For other guidelines, individual studies (i.e., RCTs or prospective cohort studies) were cited. Only four guidelines [24,25,30,34] included evidence from systematic reviews despite the availability of reliable systematic reviews for all of the extracted nutritional recommendations [13–23].

Table 3. Nutritional recommendations for primary prevention of age-related macular degeneration (AMD).

Clinical Guideline	Dietary Advice	Use of Antioxidant or Mineral Supplements	Use of Omega-3 Fatty Acid Supplements	Contraindications or Side Effects of Supplements	Systematic Review Cited with Recommendation
American Optical Association 2004	NR	✓[4]	NR	✓	None
International Council of Ophthalmology 2007	NR	NR	NR	N/A	N/A
Spanish Retina and Vitreous Society 2009	NR	NR	NR	N/A	N/A
Canadian Expert Consensus 2012	NR	NR	NR	N/A	N/A
German Ophthalmological Society 2014	✓[1]	✓[5]	NR	N/A	Yes
Eye Health Council of Ontario (Canada) 2015	✓[2]	✓[5]	NR	N/A	None
National Health Committee (New Zealand) 2015	NR	NR	NR	N/A	N/A
American Academy of Ophthalmology 2015	NR	✓[5]	NR	N/A	None
Vitreo-Retinal Society of the Philippines 2016	NR	✓[5]	NR	N/A	None
NICE Clinical Knowledge Summary (CKS) 2016	NR	✓[5]	NR	N/A	Yes
Clinical Advisory Committee (United States) 2017	NR	✓[6]	NR	NR	None
NICE Guideline (NG82) 2018	NR	NR	NR	N/A	N/A
Optometry Australia 2019	✓[3]	✓[7]	NR	NR	None

Abbreviations: ✓, recommendation included; NICE, National Institute for Health and Care Excellence; NR, no recommendation. Guideline recommendations: [1] Balanced diet; [2] A diet high in green, leafy vegetables (rich in antioxidants and carotenoids); [3] A diet high in macular carotenoids (zeaxanthin and lutein) and omega-3 fatty acids; [4] Antioxidant nutrient supplements (particularly for nutritionally deficient); [5] Evidence of no benefit for antioxidant vitamin and/or mineral supplements for primary prevention; [6] Supplement containing xanthophylls (lutein, zeaxanthin, meso-zeaxanthin) for 'sub-clinical' AMD. [7] Supplements not currently recommended for people with normal ageing changes.

Table 4. Nutritional recommendations for secondary prevention (i.e., slowing the progression) of age-related macular degeneration (AMD).

Clinical Guideline	Dietary Advice	Use of Antioxidant or Mineral Supplements	Use of Omega-3 Fatty Acid Supplements	Contraindications or Side Effects of Supplements	Systematic Review Cited with a Recommendation
American Optical Association 2004	NR	✓[6]	NR	✓	None
International Council of Ophthalmology 2007	NR	✓[7a]	NR	NR	None
Spanish Retina and Vitreous Society 2009	✓[1]	✓[7a]	NR	✓	None
Canadian Expert Consensus 2012	✓[2]	✓[7a]	NR	✓	Yes
German Ophthalmological Society 2014	NR	✓[7a]	✓[10]	✓	Yes
Eye Health Council of Ontario (Canada) 2015	✓[3]	✓[7a]	✓[10]	✓	None
National Health Committee (New Zealand) 2015	NR	✓[7a]	NR	NR	None
American Academy of Ophthalmology 2015	NR	✓[7a]	NR	✓	None
Vitreo-Retina Society of the Philippines 2016	NR	✓[7a]	NR	✓	None
NICE Clinical Knowledge Summary (CKS) 2016	✓[3]	✓[7a]	NR	✓	Yes
Clinical Advisory Committee (United States) 2017	✓[4]	✓[7a]	✓[9]	NR	None
NICE Guideline (NG82) 2018	NR	✓[8]	✓[10]	✓	Yes
Optometry Australia 2019	✓[5]	✓[7b]	NR	NR	None

Abbreviations: ✓, recommendation included; NICE, National Institute for Health and Care Excellence; NR, no recommendation. Guideline recommendations: [1] Diet rich in carotenoids (lutein and zeaxanthin) and omega-3 fatty acids; [2] Dietary intake of antioxidants, docosahexaenoic acid, and omega-3 fatty acids; [3] Diet high in fresh fruit and green, leafy vegetables (rich in antioxidants and carotenoids); [4] Consume oily fish rich in DHA and follow healthier eating styles e.g., Mediterranean diet; [5] A diet rich in green leafy vegetables, fish and antioxidants should be encouraged; [6] Antioxidant nutrient supplements; [7a] Age-Related Eye Disease Studies (AREDS) or AREDS2 supplement recommended; [7b] AREDS or AREDS2 supplement may be beneficial, and should be discussed in conjunction with the patient's general medical practitioner. [8] No evidence that benefit of AREDS supplement outweighs the risk; [9] Omega-3 fatty acid supplement recommended; [10] No evidence for benefit of omega-3 fatty acid supplements.

Table 5. Quality, measured in %, of the included AMD clinical guidelines, based on the six assessment domains of the Appraisal of Guidelines for Research and Evaluation II (AGREE II) appraisal tool. The scores for each of the six domains were calculated by summing up all the scores of the individual items for that domain, and then scaling the total as a percentage of the maximum possible score for each domain.

Clinical Guideline	AGREE II Domains (%)					
	Scope and Purpose	Stakeholder Involvement	Rigour of Development	Clarity of Presentation	Applicability	Editorial Independence
American Optical Association 2004	52.8	30.6	9.2	30.6	10.9	87.5
International Council of Ophthalmology 2007	58.3	30.6	10.2	58.3	2.2	0.0
Spanish Retina and Vitreous Society 2009	63.9	41.7	21.4	36.1	13.0	0.0
Canadian Expert Consensus 2012	75.0	47.2	36.7	75.0	28.3	20.8
German Ophthalmological Society 2014	30.6	25.0	19.4	52.8	0.0	0.0
Eye Health Council of Ontario (Canada) 2015	61.1	22.2	14.3	41.7	13.0	20.8
National Health Committee (New Zealand) 2015	75.0	13.9	20.4	55.6	30.4	0.0
American Academy of Ophthalmology 2015	83.3	63.9	72.4	94.4	17.4	75.0
Vitreo-Retina Society of the Philippines 2016	83.3	44.4	56.1	88.9	13.0	45.8
NICE Clinical Knowledge Summary (CKS) 2016	88.9	83.3	68.4	91.7	39.1	87.5
Clinical Advisory Committee (United States) 2017	52.8	22.2	9.2	44.4	4.3	0.0
NICE Guideline (NG82) 2018	94.4	88.9	95.9	100.0	97.8	100.0
Optometry Australia 2019	83.3	41.7	19.4	75.0	19.6	0.0
Median (range)	75.0 (range 30.6 to 94.4%)	41.7 (range 13.9 to 88.9%)	20.4 (range 9.2 to 95.9)	58.3 (range 30.6 to 100%)	13.0 (range 2.2 to 97.8%)	20.8 (range 0 to 100%)

4. Discussion

Nutritional supplements for 'eye health' are marketed at the general population, and are also widely recommended by optometrists and ophthalmologists for people who have clinical signs of AMD [37,38]. Clinical practice guidelines are a useful method for presenting evidence-based recommendations to health care professionals, and are intended to be used to inform clinical decision-making and reduce potential variations in practice. The aims of the current study were to identify all clinical guidelines for AMD that include recommendations relating to diet and/or nutrition; to evaluate the methodological quality of these guidelines using the AGREE II tool (the most widely used instrument for appraising clinical practice guidelines); and to investigate whether specific nutritional recommendations were underpinned by evidence from systematic reviews.

We identified 13 clinical practice guidelines meeting our inclusion criteria. The overall quality of these guidelines was judged to be low to moderate, with the median percentage scores for four of the six AGREE II quality domains being below 50%. The lowest scoring domains were 'Rigour of Development', 'Applicability', and 'Editorial Independence'. Low scores in these domains are of major concern, as they relate to: the process by which relevant research evidence is gathered and synthesised; the methods used to formulate the recommendations; the likely barriers and facilitators to implementation of the guideline recommendations; and how conflicts of interest are managed. It is important that guideline users are able to identify the evidence underpinning each recommendation, however many of the guidelines failed to make an explicit link to the evidence used to formulate recommendations. Transparency with respect to the reporting of conflicts of interest amongst guideline development panels is also essential, to avoid the potential for biased recommendations. Patient engagement in the development of the clinical practice guidelines was also generally poor, despite standards (including those defined by the World Health Organisation) [39] recommending the inclusion of patients on guideline development panels. The intent of this engagement is to ensure a focus on patient-centred guidelines that can enhance the quality of care [40].

Only one guideline (the National Institute for Health and Care Excellence (NICE) guideline NG82) achieved high scores in all of the AGREE II assessment domains. This is perhaps unsurprising, since the NICE guideline development manual states that NICE guidelines "are based on internationally accepted criteria of quality as detailed in the Appraisal of Guidelines of Research and Evaluation II (AGREE II) instrument". The development and implementation of high quality clinical practice guidelines requires substantial time, expertise and resources. Less formal guideline development groups, such as those produced by 'expert panels' or professional organisations, may lack the resources and/or methodological expertise to produce guidelines of the highest quality. In such circumstances, adapting existing high quality guidelines to local contexts may be an alternative to de novo development. The AMD Preferred Practice Patterns (PPP) Philippines (2016) [35] is an example of where this strategy has been adopted. This guideline was prepared by a panel from the Vitreo-Retina Society of the Philippines for the Philippine Academy of Ophthalmology, by adapting recommendations from the American Academy of Ophthalmology (AAO) PPP for AMD (updated in 2015).

Many robust systematic reviews have considered the merit of a variety of dietary and nutritional recommendations for AMD. In the present study, we used the systematic evidence searches performed by Lindsley et al. [11] and Downie et al. [12] to identify relevant systematic reviews, up to March 2017. These studies identified 11 unique systematic reviews on nutritional interventions for AMD, published between 2007 and 2016. Four of these reviews were published in the Cochrane Library. Mapping of clinical practice guideline recommendations to systematic review evidence showed that four of the included guidelines [24,25,30,34] made reference to systematic reviews to support their nutritional recommendations. Other guidelines included evidence from individual RCTs or non-randomised trials only. In some cases, recommendations were provided with no supporting citations. Although it is likely that a proportion of the systematic review evidence would not have been available to the panels developing the three earliest guidelines, published between 2004-2009 [26,33,36], this evidence would certainly have been available to later panels and could have been used to inform their recommendations.

Inconsistent uptake of evidence from systematic reviews by decision-makers and guideline developers is known to be related to several factors, including lack of awareness, lack of access and lack of familiarity [41]. The outcome of failing to use the best-available research evidence (i.e., systematic reviews) to inform practice guidelines is that this time-intensive, rigorous research effort is wasted. Furthermore, the entire rationale for the guideline to support evidence-based practice, and optimise patient outcomes is potentially not achieved.

Of the nine guidelines that considered nutritional approaches for the primary prevention of AMD, most considered the potential merit of dietary modification, although some also made recommendations in relation to nutritional supplementation. Several epidemiological studies have investigated whether specific dietary patterns and/or foods are associated with a reduced risk of developing AMD. A meta-analysis of prospective cohort, case-control and cross-sectional studies concluded that consumption of two or more servings of oily fish per week was beneficial in the primary prevention of AMD [13]. Another recent systematic review reported that a high consumption of vegetables rich in carotenoids and oily fish containing omega-3 fatty acids was beneficial for those at risk of AMD [42]. However, emphasising the potential difference in nutritional benefit(s) derived from whole foods versus supplementation, consuming anti-oxidant supplements does not prevent the primary onset of AMD [16]. It is therefore concerning that a recent (2017) U.S. 'guideline' [28], produced by a panel of optometrists, recommended prescribing xanthophil supplements (i.e., lutein, zeaxanthin and mesozeaxanthin) to patients with 'sub-clinical' AMD, as "it is better to prescribe a supplement than not to prescribe a supplement". Although termed a 'guideline', this document achieved the poorest quality scoring of all those included, with characteristics of a 'clinical viewpoint' rather than a 'guideline' per se.

Nutritional strategies for secondary prevention of AMD (i.e., slowing progression of the disease) were included in all of the clinical guidelines. Five guidelines [24,28,33,34] included dietary advice, consisting of recommendations for people with AMD to eat a healthy balanced diet and, specifically to increase the consumption of foods rich in the macular carotenoids and/or eat more fish (as a source of omega-3 fatty acids). A Mediterranean diet has been linked to a reduced risk of AMD progression [43]. Epidemiological studies suggest that a high dietary intake of omega-3 fatty acids is associated with a significant reduction in the risk of both intermediate [44,45] and late-stage AMD [46,47]. However, the best-available research evidence does not support long-chain omega-3 fatty acid supplementation for slowing disease progression [15]. Despite this, and acknowledging that there is a lack of evidence to substantiate such a position, two recent clinical guidelines [28,34] included a recommendation for clinicians to consider this approach.

With respect to high-dose anti-oxidant vitamin and mineral supplements for managing AMD, there were divergent recommendations. Notably, only the U.K. NICE guideline [25] did not advocate prescribing supplements containing the formulation of antioxidant vitamins, zinc and carotenoids investigated in the Age-Related Eye Disease Studies (AREDS [7] and AREDS2 [8]). This recommendation was based upon the committee's assessment of the limitations of the current, best-available evidence, and a need for "further research in this area". Indeed, there remain several key unanswered questions. For example, the minimum effective dose required for a given antioxidant to impart a potential retinoprotective effect remains unclear. Whether a single component, or a combination of components, represents the optimal formulation is also uncertain. The NICE guideline committee recommended "a well conducted randomised trial […] to provide an evidence base for the benefits and risks of individual components of the antioxidant supplements, and provide the ability to establish the treatment effect of antioxidant supplementation (the AREDS2 formula) on AMD progression by comparing AREDS2 formula with no treatment (for instance normal diet)". Although many of the guidelines considered the potential contraindications and/or side effects of high-dose antioxidant vitamin and mineral supplements, four of the guidelines did not articulate these risks [28,31,32,36]. Given that the decision to prescribe an intervention should balance the potential risks versus benefits of treatment, lack of this key information to guide clinical decision-making is a major shortcoming.

A key strength of the current study was the comprehensive search to identify eligible clinical practice guidelines. Data extraction and quality appraisal were conducted independently by two reviewers, using a recognised assessment tool (AGREE II), which has established metrics of reliability and validity [48]. Overall, the reporting of methodological details for clinical guideline development was poor and therefore it was not possible to make a judgement as to whether the guidelines had used an appropriate evidence-based approach in their development.

5. Conclusions

Despite the availability of robust systematic reviews evaluating the efficacy and safety of nutritional interventions for AMD, this study found evidence that these resources are infrequently used to support recommendations in AMD clinical practice guidelines. Consequently, guidelines often present conflicting recommendations that could lead to variations in clinical care. The AGREE II quality evaluations of the included guidelines identified several key areas that require improvement, particularly the rigour of development, managing potential conflicts of interests, and presenting strategies for implementing guideline recommendations into daily practice.

Supplementary Materials: The following are available online at http://www.mdpi.com/2072-6643/11/4/823/s1, File S1: Search Strategy.

Author Contributions: J.G.L., J.R.E. and L.E.D. were involved in all aspects of the study conception and design, data acquisition, analysis and interpretation. J.G.L. wrote the first draft of the manuscript and this was critically revised by J.R.E. and L.E.D. All authors approved the final version of the manuscript.

Conflicts of Interest: The authors declare no conflict of interest in relation to this work.

References

1. Coleman, H.R.; Chan, C.C.; Ferris, F.L., 3rd; Chew, E.Y. Age-related macular degeneration. *Lancet* **2008**, *372*, 1835–1845. [CrossRef]
2. Bhutto, I.; Lutty, G. Understanding age-related macular degeneration (AMD): Relationships between the photoreceptor/retinal pigment epithelium/Bruch's membrane/choriocapillaris complex. *Mol. Asp. Med.* **2012**, *33*, 295–317. [CrossRef]
3. Mitchell, J.; Bradley, C. Quality of life in age-related macular degeneration: A review of the literature. *Health Qual. Life Outcomes* **2006**, *4*, 97. [CrossRef] [PubMed]
4. Srinivasan, S.; Swaminathan, G.; Kulothungan, V.; Raman, R.; Sharma, T.; Medscape. Prevalence and the risk factors for visual impairment in age-related macular degeneration. *Eye* **2017**, *31*, 846–855. [CrossRef]
5. Downie, L.E.; Keller, P.R. Nutrition and age-related macular degeneration: Research evidence in practice. *Optom. Vis. Sci.* **2014**, *91*, 821–831. [CrossRef]
6. Gorusupudi, A.; Nelson, K.; Bernstein, P.S. The Age-Related Eye Disease 2 Study: Micronutrients in the Treatment of Macular Degeneration. *Adv. Nutr.* **2017**, *8*, 40–53. [CrossRef] [PubMed]
7. Age-Related Eye Disease Study Research Group. A randomized, placebo-controlled, clinical trial of high-dose supplementation with vitamins C and E, beta carotene, and zinc for age-related macular degeneration and vision loss: AREDS report no. 8. *Arch. Ophthalmol.* **2001**, *119*, 1417–1436. [CrossRef]
8. Age-Related Eye Disease Study 2 Research Group. Lutein + zeaxanthin and omega-3 fatty acids for age-related macular degeneration: The Age-Related Eye Disease Study 2 (AREDS2) randomized clinical trial. *JAMA* **2013**, *309*, 2005–2015. [CrossRef]
9. Institute of Medicine (US) Committee on Standards for Developing Trustworthy Clinical Practice Guidelines. *Clinical Practice Guidelines We Can Trust*; Graham, R.M.M., Wolman, D.M., Greenfield, S., Steinberg, E., Eds.; The National Academies Press: Washington, DC, USA, 2011; p. 290.
10. Wormald, R.; Evans, J. What Makes Systematic Reviews Systematic and Why are They the Highest Level of Evidence? *Ophthalmic Epidemiol.* **2018**, *25*, 27–30. [CrossRef]
11. Lindsley, K.; Li, T.; Ssemanda, E.; Virgili, G.; Dickersin, K. Interventions for Age-Related Macular Degeneration: Are Practice Guidelines Based on Systematic Reviews? *Ophthalmology* **2016**, *123*, 884–897. [CrossRef]

12. Downie, L.E.; Makrai, E.; Bonggotgetsakul, Y.; Dirito, L.J.; Kristo, K.; Pham, M.N.; You, M.; Verspoor, K.; Pianta, M.J. Appraising the Quality of Systematic Reviews for Age-Related Macular Degeneration Interventions: A Systematic Review. *JAMA Ophthalmol.* **2018**, *136*, 1051–1061. [CrossRef] [PubMed]

13. Chong, E.W.; Wong, T.Y.; Kreis, A.J.; Simpson, J.A.; Guymer, R.H. Dietary antioxidants and primary prevention of age related macular degeneration: Systematic review and meta-analysis. *BMJ* **2007**, *335*, 755. [CrossRef]

14. Evans, J. Antioxidant supplements to prevent or slow down the progression of AMD: A systematic review and meta-analysis. *Eye* **2008**, *22*, 751–760. [CrossRef] [PubMed]

15. Lawrenson, J.G.; Evans, J.R. Omega 3 fatty acids for preventing or slowing the progression of age-related macular degeneration. *Cochrane Database Syst. Rev.* **2012**, *11*, CD010015. [CrossRef] [PubMed]

16. Evans, J.R.; Lawrenson, J.G. Antioxidant vitamin and mineral supplements for preventing age-related macular degeneration. *Cochrane Database Syst. Rev.* **2012**, CD000253. [CrossRef]

17. Hodge, W.G.; Barnes, D.; Schachter, H.M.; Pan, Y.I.; Lowcock, E.C.; Zhang, L.; Sampson, M.; Morrison, A.; Tran, K.; Miguelez, M.; et al. Evidence for the effect of omega-3 fatty acids on progression of age-related macular degeneration: A systematic review. *Retina* **2007**, *27*, 216–221. [CrossRef]

18. Ma, L.; Liu, R.; Du, J.H.; Liu, T.; Wu, S.S.; Liu, X.H. Lutein, Zeaxanthin and Meso-zeaxanthin Supplementation Associated with Macular Pigment Optical Density. *Nutrients* **2016**, *8*, 426. [CrossRef] [PubMed]

19. Sin, H.P.; Liu, D.T.; Lam, D.S. Lifestyle modification, nutritional and vitamins supplements for age-related macular degeneration. *Acta Ophthalmol.* **2013**, *91*, 6–11. [CrossRef] [PubMed]

20. Vishwanathan, R.; Chung, M.; Johnson, E.J. A systematic review on zinc for the prevention and treatment of age-related macular degeneration. *Investig. Ophthalmol. Vis. Sci.* **2013**, *54*, 3985–3998. [CrossRef]

21. Wang, X.; Jiang, C.; Zhang, Y.; Gong, Y.; Chen, X.; Zhang, M. Role of lutein supplementation in the management of age-related macular degeneration: Meta-analysis of randomized controlled trials. *Ophthalmic Res.* **2014**, *52*, 198–205. [CrossRef]

22. Evans, J.R. Ginkgo biloba extract for age-related macular degeneration. *Cochrane Database Syst. Rev.* **2013**, CD001775. [CrossRef] [PubMed]

23. Evans, J.R.; Lawrenson, J.G. Antioxidant vitamin and mineral supplements for slowing the progression of age-related macular degeneration. *Cochrane Database Syst. Rev.* **2012**, *11*, CD000254. [CrossRef] [PubMed]

24. Anonymous. Clinical Knowledge Summaries. Macular Degeneration-Age-Related. Available online: https://cks.nice.org.uk/macular-degeneration-age-related (accessed on 25 March 2019).

25. Anonymous. Age-Related Macular Degeneration. NICE Guideline [NG82]. Available online: https://www.nice.org.uk/guidance/ng82 (accessed on 25 March 2019).

26. Anonymous; American Optical Association. Optometric Clinical Practice Guideline. Care of the Patient with Age-Related Macular Degeneration (CPG6). Available online: https://www.aoa.org/optometrists/tools-and-resources/clinical-care-publications/clinical-practice-guidelines (accessed on 25 March 2019).

27. Anonymous. American Academy of Ophthalmology Preferred Practice Pattern. Age-Related Macular Degeneration. Available online: https://www.aao.org/preferred-practice-pattern/age-related-macular-degeneration-ppp-2015 (accessed on 25 March 2019).

28. Anonymous. Practical Guidelines for the Treatment of AMD. Available online: https://www.reviewofoptometry.com/publications/ro1017-practical-guidelines-for-the-treatment-of-amd (accessed on 25 March 2019).

29. Anonymous. Guidelines for the Collaborative Management of Persons with Age-Related Macular Degeneration by Health- and Eye-Care Professionals. Available online: https://opto.ca/sites/default/files/resources/documents/cjo_journal_online_ehco_guidelines_eng_v2.pdf (accessed on 25 March 2019).

30. Cruess, A.F.; Berger, A.; Colleaux, K.; Greve, M.; Harvey, P.; Kertes, P.J.; Sheidow, T.; Tourville, E.; Williams, G.; Wong, D. Canadian expert consensus: Optimal treatment of neovascular age-related macular degeneration. *Can. J. Ophthalmol.* **2012**, *47*, 227–235. [CrossRef] [PubMed]

31. Optometry Australia. 2019 Clinical Practice Guide for the Diagnosis, Treatment and Management of Age-Related Macular Degeneration. Available online: http://www.optometry.org.au/media/1185775/amd_clinical_practice_guide_-_2019_final_designed_v5.pdf (accessed on 25 March 2019).

32. Anonymous; National Health Committee. Age-Related Macular Degeneration. Available online: http://www.moh.govt.nz/notebook/nbbooks.nsf/0/1169E93BA82E0E3BCC257F7F0007E835/$file/150620_age-related_macular_degeneration_t2_updated-june15.pdf (accessed on 25 March 2019).

33. Ruiz-Moreno, J.M.; Arias-Barquet, L.; Armada-Maresca, F.; Boixadera-Espax, A.; Garcia-Layana, A.; Gomez-Ulla-de-Irazazabal, F.; Mones-Carilla, J.; Pinero-Bustamante, A.; Suarez-de-Figueroa, M.; Sociedad Española de Retina yVítreo. Guidelines of clinical practice of the SERV: Treatment of exudative age-related macular degeneration (AMD). *Arch. Soc. Esp. Oftalmol.* **2009**, *84*, 333–344. [PubMed]

34. Deutsche Ophthalmologische, G. Dietary supplements in age-related macular degeneration. Current observations of the German Ophthalmological Society, the German Retina Society and the Professional Association of German Ophthalmologists (as of October 2014). *Klin. Mon. Augenheilkd.* **2015**, *232*, 196–201. [CrossRef]

35. Anonymous. Age-Related Macular Degeneration (AMD). Preferred Practice Patterns (PPP). Philippines. Available online: http://pao.org.ph/standard/Age%20Related%20Mac%20Degen.pdf (accessed on 25 March 2019).

36. Anonymous; International Council of Ophthalmology/International Federation of Ophthalmological Societies (ICO). International Clinical Guidelines. Age-Related Macular Degeneration (Management Recommendations). Available online: http://www.icoph.org/downloads/ICOARMDMa.pdf (accessed on 25 March 2019).

37. Downie, L.E.; Keller, P.R. The self-reported clinical practice behaviors of Australian optometrists as related to smoking, diet and nutritional supplementation. *PLoS ONE* **2015**, *10*, e0124533. [CrossRef]

38. Lawrenson, J.G.; Evans, J.R. Advice about diet and smoking for people with or at risk of age-related macular degeneration: A cross-sectional survey of eye care professionals in the UK. *BMC Public Health* **2013**, *13*, 564. [CrossRef] [PubMed]

39. Schünemann, H.J.; Fretheim, A.; Oxman, A.D. Improving the use of research evidence in guideline development: 10. Integrating values and consumer involvement. *Health Res. Policy Syst.* **2006**, *4*, 22. [CrossRef]

40. Armstrong, M.J.; Rueda, J.D.; Gronseth, G.S.; Mullins, C.D. Framework for enhancing clinical practice guidelines through continuous patient engagement. *Health Expect.* **2017**, *20*, 3–10. [CrossRef]

41. Wallace, J.; Nwosu, B.; Clarke, M. Barriers to the uptake of evidence from systematic reviews and meta-analyses: A systematic review of decision makers' perceptions. *BMJ Open* **2012**, *2*. [CrossRef]

42. Chapman, N.A.; Jacobs, R.J.; Braakhuis, A.J. Role of diet and food intake in age-related macular degeneration: A systematic review. *Clin. Exp. Ophthalmol.* **2019**, *47*, 106–127. [CrossRef] [PubMed]

43. Merle, B.M.J.; Colijn, J.M.; Cougnard-Gregoire, A.; de Koning-Backus, A.P.M.; Delyfer, M.N.; Kiefte-de Jong, J.C.; Meester-Smoor, M.; Feart, C.; Verzijden, T.; Samieri, C.; et al. Mediterranean Diet and Incidence of Advanced Age-Related Macular Degeneration: The EYE-RISK Consortium. *Ophthalmology* **2019**, *126*, 381–390. [CrossRef] [PubMed]

44. Christen, W.G.; Schaumberg, D.A.; Glynn, R.J.; Buring, J.E. Dietary omega-3 fatty acid and fish intake and incident age-related macular degeneration in women. *Arch. Ophthalmol.* **2011**, *129*, 921–929. [CrossRef]

45. Seddon, J.M.; George, S.; Rosner, B. Cigarette smoking, fish consumption, omega-3 fatty acid intake, and associations with age-related macular degeneration: The US Twin Study of Age-Related Macular Degeneration. *Arch. Ophthalmol.* **2006**, *124*, 995–1001. [CrossRef] [PubMed]

46. Chong, E.W.; Kreis, A.J.; Wong, T.Y.; Simpson, J.A.; Guymer, R.H. Dietary omega-3 fatty acid and fish intake in the primary prevention of age-related macular degeneration: A systematic review and meta-analysis. *Arch. Ophthalmol.* **2008**, *126*, 826–833. [CrossRef]

47. Sangiovanni, J.P.; Agron, E.; Meleth, A.D.; Reed, G.F.; Sperduto, R.D.; Clemons, T.E.; Chew, E.Y.; Age-Related Eye Disease Study Research Group. {omega}-3 Long-chain polyunsaturated fatty acid intake and 12-y incidence of neovascular age-related macular degeneration and central geographic atrophy: AREDS report 30, a prospective cohort study from the Age-Related Eye Disease Study. *Am. J. Clin. Nutr.* **2009**, *90*, 1601–1607. [CrossRef]

48. Brouwers, M.C.; Kho, M.E.; Browman, G.P.; Burgers, J.S.; Cluzeau, F.; Feder, G.; Fervers, B.; Graham, I.D.; Hanna, S.E.; Makarski, J.; et al. Development of the AGREE II, part 2: Assessment of validity of items and tools to support application. *CMAJ* **2010**, *182*, E472–E478. [CrossRef]

nutrients

MDPI

Article

Preliminary Validation of a Food Frequency Questionnaire to Assess Long-Chain Omega-3 Fatty Acid Intake in Eye Care Practice

Alexis Ceecee Zhang and Laura E. Downie *

Department of Optometry and Vision Sciences, The University of Melbourne, Parkville, 3010 Victoria, Australia; alexisz@student.unimelb.edu.au
* Correspondence: ldownie@unimelb.edu.au; Tel.: +61-3-9035-3043; Fax: +61-3-9035-9905

Received: 4 March 2019; Accepted: 8 April 2019; Published: 11 April 2019

Abstract: Clinical recommendations relating to dietary omega-3 essential fatty acids (EFAs) should consider an individual's baseline intake. The time, cost, and practicality constraints of current techniques for quantifying omega-3 levels limit the feasibility of applying these methods in some settings, such as eye care practice. This preliminary validation study, involving 40 adults, sought to assess the validity of a novel questionnaire, the Clinical Omega-3 Dietary Survey (CODS), for rapidly assessing long-chain omega-3 intake. Estimated dietary intakes of long-chain omega-3s from CODS correlated with the validated Dietary Questionnaire for Epidemiology Studies (DQES), Version 3.2, (Cancer Council Victoria, Melbourne, Australia) and quantitative assays from dried blood spot (DBS) testing. The 'method of triads' model was used to estimate a validity coefficient (ρ) for the relationship between the CODS and an estimated "true" intake of long-chain omega-3 EFAs. The CODS had high validity for estimating the ρ (95% Confidence Interval [CI]) for total long-chain omega-3 EFAs 0.77 (0.31–0.98), docosahexaenoic acid 0.86 (0.54–0.99) and docosapentaenoic acid 0.72 (0.14–0.97), and it had moderate validity for estimating eicosapentaenoic acid 0.57 (0.21–0.93). The total long-chain omega-3 EFAs estimated using the CODS correlated with the Omega-3 index ($r = 0.37$, $p = 0.018$) quantified using the DBS biomarker. The CODS is a novel tool that can be administered rapidly and easily, to estimate long-chain omega-3 sufficiency in clinical settings.

Keywords: omega-3; fatty acid; diet; dietary assessment; clinical survey; eye disease; dry eye; age-related macular degeneration; food frequency questionnaire; CODS

1. Introduction

Omega-3 polyunsaturated fatty acids (PUFAs) are essential fatty acids (EFAs) that cannot be synthesized de novo, and thus must be derived from food sources or dietary supplementation. The potential benefit of diets rich in omega-3 fatty acids has been shown in a variety of health conditions, such as hypercholesterolaemia and rheumatoid arthritis [1,2]. Omega-3 PUFAs exist in both short- and long-chain forms. The short-chain omega-3 fatty acids and alpha-linoleic acid (ALA) are derived from plant-based sources (e.g., flaxseed and walnuts) and they are a precursor to the more biologically potent long-chain omega-3 EFAs, docosahexaenoic acid (DHA) and eicosapentaenoic acid (EPA). Dietary long-chain omega-3 EFAs are found mostly in marine sources (e.g., oily fish and seafood). Once ingested, long-chain omega-3 PUFAs are incorporated into cellular membranes and play a role in cellular signaling, modulating systemic inflammation, and influencing immune function [3,4].

There is mounting evidence that diets rich in omega-3 EFAs may be beneficial for reducing the risk of development and/or progression of several ocular conditions, such as dry eye disease and age-related macular degeneration (AMD) [5–7]. The United States (US) Women's Health Study showed that a low dietary intake of omega-3 EFAs is associated with a higher incidence of dry eye disease in women [7]. The Blue Mountains Eye Study found that eating oily fish once per week, as compared with fewer than once per week, was associated with a lower risk of developing early-stage AMD [8]. A meta-analysis of prospective cohort, case-control, and cross-sectional studies suggested that consumption of two or more servings of oily fish per week was beneficial in the primary prevention of AMD [9]. An association has been found between higher dietary intake of omega-3 EFAs and significant risk reduction for developing more advanced, sight-threatening forms of AMD [10–12]. Omega-3 EFA supplementation may also lower intraocular pressure in adults [13].

Currently, the suggested dietary targets (SDT) for long-chain omega-3 EFA consumption, as recommended by the Australian National Health and Medical Research Council (NHMRC), in diets optimized to lower chronic disease risk, is 430 mg/day for women and 610 mg/day for men [14,15]. These recommendations are consistent with the National Heart Foundation position statement (2015) that recommends two to three servings of fish (serving size: 150–200 g), preferably oily fish, per week to achieve ~250–500 mg/day of combined EPA and DHA consumption [16]. The US Department of Health and Human Services Dietary Guidelines (2015–2020) recommend eight ounces of fish/seafood per week, which is approximately equivalent to 250 mg of EPA + DHA per day, and eight to 12 ounces in pregnancy [17]. These recommended values are similar to that of the European Food Safety Authority [18], while the French Agency for Food, Environmental and Occupational Health and Safety (ANSES) recommends 250 mg/day of each EPA and DHA. However, research suggests that approximately 80% of Australian adults do not meet this recommendation for daily intake [19]. Only ~15% of the French population meet the recommendation for daily intake of EPA, and 8% are estimated to meet the daily DHA recommendation. A study undertaken in the US estimated that low dietary intake of omega-3 fatty acids was a modifiable risk factor contributing to ~84,000 deaths in 2005 [20]. Therefore, there is a need for healthcare clinicians to actively enquire about their patients' diet, and preferably quantify their dietary intake of omega-3 EFAs, in order to identify individuals who are likely to benefit from dietary changes and/or supplementation to improve their health.

Various techniques can be used to quantify systemic omega-3 EFA levels, however, most of these methods are not ideally suited for direct application in all clinical settings [21]. Fatty acid levels present in subcutaneous adipose tissue are considered to be the most robust long-term marker of fatty acid intake [22]. Systemic EFA concentrations can be accurately estimated by assaying fatty acid levels in erythrocytes and plasma phospholipids [23]. Although both of these methods provide quantitative data relating to systemic fatty acid levels, the sample collection procedures are invasive, relatively costly, and time consuming (as they require off-site laboratory analyses), and therefore are not currently routinely performed in a clinical setting [24]. In addition, these tests are not readily accessible to all clinicians who may be in a position to provide relevant dietary advice regarding the potential benefits of omega-3 EFA intake to their patients (e.g., eye care clinicians, in relation to ocular health). Short-term dietary assessment, such as dietary records requiring self-monitored and detailed recordings over multiple days, and short-term recalls do not account for the day-to-day variation of a habitual diet [25]. Furthermore, both methods have been shown to underestimate average energy intake, particularly in populations with lower socioeconomic status, education, and literacy levels [26].

As an alternative, diet questionnaires provide a rapid, non-invasive method for estimating dietary fatty acid intake. Although nutrient intakes derived from several food frequency questionnaires (FFQs) have been shown to correlate with quantitative biological markers of omega-3 EFA intake [21], questionnaires designed for use in epidemiology research studies are often exhaustive and relatively time consuming, and not intended for use in routine clinical settings [27,28]. Several brief omega-3 dietary questionnaires have been developed for use in clinical settings [29], including for adults with psychological disorders [30,31], and to use for identifying older adults with inadequate omega-3 EFA intake in the context of cardiovascular disease risk [32]. The reliability estimates from different dietary surveys are dependent on the nutritional composition of food sources, which can vary across different geographical locations [33].

The aim of this study was to undertake preliminary validation of a novel, clinically applicable questionnaire, the Clinical Omega-3 Dietary Survey (CODS), for assessing dietary long-chain omega-3 EFA intake, as compared with an objective fatty acid blood biomarker and a previously validated survey designed to capture a full dietary profile. Information derived from this tool could then be used by clinicians to estimate whether an individual meets the recommended daily intake of long-chain omega-3 EFAs, or whether they might benefit from enhancing their dietary intake.

2. Materials and Methods

This cross-sectional research study was conducted in accordance with the tenets of the Declaration of Helsinki, and it was approved by the University of Melbourne Human Research Ethics Committee (HREC #1749830.1) and the St Vincent's Hospital Research Ethics Committee (HREC #HREC/17/SVHM/236).

2.1. Participants

Adult participants were recruited from the University of Melbourne and St Vincent's Hospital (Melbourne, Victoria, Australia) via advertisements. All participants provided written, informed consent to participate. Eligible participants were aged 18 years or older, and were not pregnant or breastfeeding. There were no eligibility restrictions with respect to general health conditions. Long-chain omega-3 EFA status was quantified using three methods, as detailed below: (a) the CODS; (b) the Dietary Questionnaire for Epidemiological Studies, Version 3.2 (DQES v3.2, Cancer Council Victoria, Australia); and (c) dried blood spot (DBS) testing (Xerion Pty Ltd, Victoria, Australia). For methods (a) and (b), daily long-chain omega EFA intake (mg/day) was calculated as the sum of EPA, DHA, and docosapentaenoic acid (DPA) intake. For method (c), the long-chain omega-3 EFAs present in erythrocytes (%) was calculated as the sum of EPA, DHA, and DPA concentrations (%) present in erythrocytes.

2.2. Clinical Omega-3 Dietary Survey

A clinically relevant FFQ, the CODS, was developed by the authors of this study to yield an estimate of long-chain omega-3 EFA intake, using the nutrient composition data available in the AUSNUT (2011–2013) food nutrient database (available online at: http://www. foodstandards.gov.au/science/monitoringnutrients/ausnut/foodnutrient/Pages/default.aspx). For natural foods containing the highest concentrations of total long-chain omega-3 fatty acids (ranging from 769.4 mg/100 g to 62 mg/100 g), the fatty acid composition was extracted from the nutritional database, and the food was considered for inclusion in the CODS.

The CODS (Figure S1) comprises three main sections, categorized according to the primary type of food as: (i) seafood, (ii) fish, and (iii) meat (including eggs). Seven species of seafood, 17 types of fish (including a generalized description for white-fleshed fish), and seven different types of meat/eggs are incorporated, and are used as a basis for estimating dietary long-chain omega-3 EFA intake. A final section, (iv), considers the consumption of long-chain omega-3 EFAs from dietary supplements. The study participants were required to provide details relating to the supplement formulation (brand, dose, form of omega-3 EFAs, and frequency of intake) and these data were incorporated into the analysis.

The CODS was administered by a single examiner (A.C.Z.), and the participants' responses were recorded on a paper-based survey. The first part of the survey asked participants to recall the average frequency (i.e., number of times per week or per month, as appropriate for the nominated frequency) of fish or seafood consumption (in any form, including fresh and canned) over the past three months. The method provided to the participants for estimating the approximate portion size (in amounts of 50 g, 100 g, 150 g and 200 g) required using the size of the examiner's palm as a reference for a 100 g portion size. Then, the participants were required to estimate the average portion size (in grams) of fish or seafood they typically consumed.

For each food source listed in sections (i), (ii), and (iii) of the CODS, participants were asked to estimate the average portion size per serving (g), and to note how often they consumed that type of food (i.e., average number of times per week or per month). In order to quantify egg intake, the average number of eggs per week or per month was reported. The participants were instructed to only include food sources that they consumed "regularly", defined as at least once per month over the past three months.

Quantification of the total long-chain omega-3 EFA intake was derived from the composition data available in the AUSNUT (2011–2013) food nutrient database [34], using a spreadsheet in Microsoft Excel 2017 (Microsoft Corp., Redmond, WA, USA).

2.3. Dietary Questionnaire for Epidemiology Studies

An online diet questionnaire, DQES v3.2, was purchased from an independent provider (Cancer Council Victoria, Australia) [35]. The DQES v3.2 was developed and validated for assessing intake of food and nutrients among Australian adults [36,37], and it had been previously utilized by the Melbourne Collaborative Cohort epidemiology study, involving a cohort of 810 participants.

The DQES v3.2 consisted of 142 food items, and it used extensive food compositional data derived from two Australian databases, AUSNUT 2007 and NUTTAB 2010 [38,39], to estimate dietary intake for a spectrum of 98 nutrients, including macronutrients (e.g., carbohydrates, proteins, fats) and micronutrients (e.g., vitamins, minerals). In addition, nutrient intakes from alcoholic beverages are surveyed, although these values were not incorporated into the analysis.

Nutritional information relating to EFAs was estimated from the questionnaire, and included a breakdown of individual fatty acids, derived from the NUTTAB 2010 food database [39]. The total long-chain omega-3 EFA intake (mg/day) was calculated as the sum of P205W3FD (EPA), P225W3FD (DPA), and P226W3FD (DHA). As the DQES is proprietary, we were unable to ascertain which specific food items were used to derive the estimate of omega-3 fatty acid intake for this survey.

Participants undertook the DQES online, using a self-administrated delivery method. Once completed, the nutritional analysis for each participant was undertaken by the survey supplier, with the results e-mailed to the study authors.

2.4. Dried Blood Spot Testing

The participants' fatty acid profiles were analyzed using dried DBS testing (PUFAcoat™ technology, Xerion Pty Ltd, Victoria, Australia). A single drop of capillary blood (~3 mL) was collected from each participant, using a sterile, single-use lancing device and spotted onto the proprietary PUFAcoat™ test cards, which are designed for long-term stabilization of long-chain PUFAs in dried blood samples. Prior to analysis, the DBS cards were air dried, placed in sealed cellophane bags, and stored in desiccants in a dark, temperature-controlled chamber. As per the manufacturer's instructions, all blood samples were sent for laboratory analyses within four weeks of collection.

The fatty acid analyses were performed in an independent laboratory (Waite Lipid Analysis Service (WLAS), University of Adelaide, South Australia, Australia), using established methods [40]. In brief, the fatty acid blood spots were transmethylated by mixing with H_2SO_4 (18M AR grade, BDH, Sussex, UK) in anhydrous methanol in a 5 mL sealed vial (Wheaton, Millville, USA), and then were heated for three hours at 70 °C. The fatty acid methyl esters were separated and quantified using gas chromatography (Hewlett-Packard 6890 gas chromatograph, equipped with a 50 m capillary column (0.32 mm internal diameter SGE, Victoria), coated with 70% cyanopropyl polysilphenylene-siloxane (BPX70) (0.25 μm film thickness) and fitted with a flame ionization detector. A helium carrier gas was used and the inlet split ratio was set to 20:1, with the injector temperature at 250 °C and detector temperature at 300 °C. Fatty acid methyl esters were identified by comparing the retention times and the peak area values of unknown samples to the standards using the ChemStation software (Hewlett Packard, CA, USA), and a normalized percentage was calculated based on the response factors.

The extracted long-chain omega-3 fatty acid output parameters were: 22:5n-3 (EPA), 22:5n-3 (DPA) and 22:6n-3 (DHA). The total long-chain omega-3 EFA erythrocyte concentration (%) was calculated as the sum of: C20:5n-3 (EPA) + C22:5n-3 (DPA) + C22:6n-3 (DHA). The Omega-3 index (%) was defined as the total percentage of EPA and DHA present in erythrocyte phospholipid membranes [40].

2.5. Statistical Analysis

On the basis of an estimated correlation coefficient of 0.45 between quantification methods, as previously reported in a similar study [41], a sample size of 36 participants was calculated to be required for 80% power at a 5% significance level.

Data normality testing was performed using the D'Agostino–Pearson omnibus normality test. The inter-group comparisons were analyzed using a Student's t-test or Mann–Whitney U test, as appropriate. The Spearman's correlation coefficient (r_s) was used to assess the relationship between outputs from the CODS, DQES, and DBS analyses, with respect to long-chain omega-3 EFA intake levels. The Spearman's correlation coefficients were interpreted with reference to similar studies, where $r_s \leq 0.35$ indicates a weak correlation, $r_s = 0.36–0.67$ indicates a moderate correlation, $r_s = 0.68–1$ indicates a good correlation and $r_s \geq 0.9$ indicating a very good correlation [28].

The 'method of triads' statistical model was used to derive a validity coefficient to estimate the relationship between a dietary measurement (e.g., the CODS) and an estimated "true" intake (Figure S2) [42,43]. Validity coefficients closer to 1 indicated a closer relationship between the estimated dietary score and the estimated "true" intake [27]. For this analysis, a validity coefficient <0.2 was considered to represent low validity, values between 0.2 and 0.6 were considered to represent moderate validity, and values >0.6 were regarded as indicative of high validity [43]. For a Heywood case, where a validity coefficient ≥ 1 was estimated, the validity coefficient was set to 1.

A bootstrap procedure was used to estimate the confidence interval (CI) for each validity coefficient [43,44]. For this method, each bootstrap drew 40 samples with replacement from the original sample, and the validity coefficient for each quantification method was calculated. A total of 1000 bootstrap samples were obtained to build a bootstrap distribution of validity coefficients using Microsoft Excel 2017 (Microsoft Corp., Redmond, WA, USA). The non-Heywood cases were used to calculate the 95 percentile CI for each validity coefficient.

The inter-method agreement, for the estimated daily total long-chain omega-3 EFA intake (mg/day) between the CODS and DQES, was examined using Bland–Altman analysis [45]. The mean difference (bias) and limits of agreement (LoA, defined as the bias ±1.96 standard deviations of the mean difference) were calculated. A regression analysis was used to analyze the potential relationship between the differences between the methods and the average of the two methods.

Group data are reported as the mean ± standard deviation (SD), unless otherwise specified.

3. Results

Forty participants were enrolled in the study, and they completed three independent methods for assessing long-chain omega-3 EFA intake. The mean (SD) age of participants was 45.7 (18.11) years. Among the 40 participants, 26 were female and 14 were male. The study cohort included individuals with diabetes mellitus (*n* = 19), however, there were no statistically significant differences between this subgroup of participants and the remaining subgroup, with respect to age, gender, or Omega-3 index measured using the DBS biomarker (*p* > 0.05 for each comparison).

3.1. Dietary Assessment of Long-Chain Omega-3 EFA Intake

The average time taken to conduct the CODS was three minutes and the average time taken to complete the DQES (v3.2) was 15 minutes.

Table 1 summarizes the estimated daily long-chain omega-3 EFA intake (mg/day), for each of DHA, DPA, and EPA, as assessed using the CODS and DQES (v3.2), for the study cohort (n = 40 participants). There was no significant inter-method difference for the estimated intake of omega-3 fatty acids (Table 1, $p > 0.05$ for all comparisons).

Table 1. Comparison of the median and inter-quartile range (IQR) values for long-chain omega-3 fatty acid intake, for the CODS and DQES v3.2 questionnaires, in the study cohort.

	CODS (mg/day)			DQES v3.2 (mg/day)			
	Median	IQR		Median	IQR		p-Value
		p25	p75		p25	p75	
Total LC omega-3 EFAs	443.7	204.1	650.0	362.6	200.7	577.9	0.83
EPA (20:5-n3)	140.5	55.9	209.2	91.4	49.9	168.0	0.37
DPA (22:5-n3)	75.2	39.3	102.2	77.8	41.3	116.2	0.41
DHA (22:6-n3)	221.7	95.8	320.2	184.9	98.2	353.7	0.98

Abbreviations: CODS, Clinical Omega-3 Dietary Survey; DHA, docosahexaenoic acid; DPA, docosapentaenoic acid; DQES, Dietary Survey for Epidemiology Studies v3.2; EFA, essential fatty acid; EPA, eicosapentaenoic acid; IQR, inter-quartile range; LC, long-chain; p25, 25th percentile; p75, 75th percentile.

3.2. Dried Blood Spot Analysis of Long-Chain Omega-3 EFAs Levels

Table 2 summarizes the estimated long-chain omega-3 EFAs present in erythrocytes, and the Omega-3 index as estimated using the dried blood spot test, for the study cohort.

Table 2. Concentration of long-chain omega-3 EFAs present in erythrocytes for the study cohort.

	Median	IQR	
		p25	p75
Total LC omega-3 EFAs (%)	4.15	3.58	5.30
EPA (20:5-n3) (%)	0.60	0.45	0.88
DPA (22:5-n3) (%)	1.25	1.08	1.66
DHA (22:6-n3) (%)	2.16	1.80	2.77
Omega-3 index	5.32	4.58	6.17

Abbreviations: EFA, essential fatty acid; EPA, eicosapentaenoic acid; DHA, docosahexaenoic acid; DPA, docosapentaenoic acid; IQR, inter-quartile range; LC, long-chain; total LC omega-3 EFAs, DHA + DPA + EPA; p25, 25th percentile; p75, 75th percentile.

3.3. Correlation between Quantification Methods

Table 3 summarizes the Spearman's correlation coefficients (r_s) for the association between the dietary survey methods (CODS and DQES v3.2, mg/day, Table 1) and DBS assays (%, Table 2) for quantifying DHA, DPA, EPA, and total long-chain omega-3 EFAs (DHA + DPA + EPA).

Table 3. Relationships between the long-chain omega-3 EFA quantification methods, calculated using the Spearman's correlation coefficient (r_s).

	CODS vs. DQES		DBS vs. DQES		DBS vs. CODS	
	r_s	p-Value	r_s	p-Value	r_s	p-Value
Total LC omega-3 EFAs	0.64	<0.0001	0.40	0.0096	0.38	0.017
EPA (20:5-n3)	0.60	<0.0001	0.32	0.042	0.18	0.28
DPA (22:5-n3)	0.53	0.0005	0.12	0.47	0.11	0.48
DHA (22:6-n3)	0.66	<0.0001	0.40	0.011	0.45	0.0038

Abbreviations: CODS, Clinical Omega-3 Dietary Survey; DBS, dried blood spot; DHA, docosahexaenoic acid; DPA, docosapentaenoic acid; DQES, Dietary Survey for Epidemiology Studies v3.2; EFA, essential fatty acid; EPA, eicosapentaenoic acid; LC, long-chain; total LC omega-3 EFAs, DHA + DPA + EPA.

Comparing the two dietary methods (CODS and DQES v3.2) there was a moderate-to-strong, positive correlation for each of total long-chain omega-3 EFAs ($r_s = 0.64$; $p < 0.0001$) (Figure 1A), EPA ($r_s = 0.60$; $p < 0.0001$) (Figure 1B), the DPA ($r_s = 0.53$; $p = 0.0005$) (Figure 1C) and DHA ($r_s = 0.66$; $p < 0.0001$) (Figure 1D).

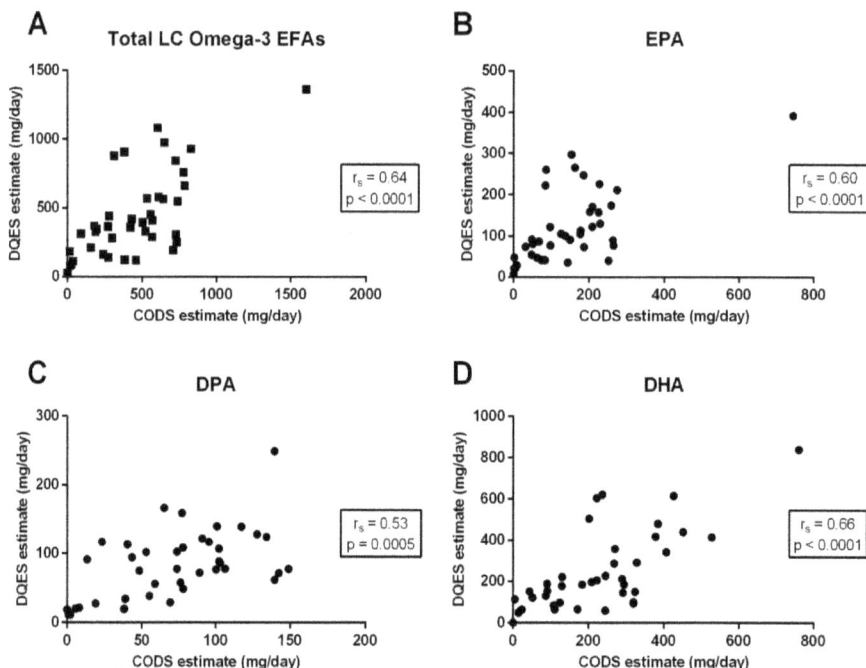

Figure 1. The relationship between calculated dietary long-chain (LC) omega-3 essential fatty acid (EFA) intake, quantified using the Spearman's correlation coefficients (r_s), and the Clinical Omega-3 Dietary Survey (CODS) versus the Dietary Questionnaire for Epidemiology Studies (DQES) v3.2, for (**A**) total LC omega-3 EFAs; (**B**) eicosapentaenoic acid (EPA); (**C**) docosapentaenoic acid (DPA); and (**D**) docosahexaenoic acid (DHA).

Comparing the DQES (v3.2) to the DBS assay, there was a moderate correlation for both total long-chain omega-3 EFAs ($r_s = 0.40$; $p = 0.0096$) (Figure 2A) and DHA ($r_s = 0.40$; $p = 0.011$, Figure 2D) and a weak, positive correlation for EPA ($r_s = 0.32$; $p = 0.042$) (Figure 2B). There was no significant relationship between the DQES (v3.2) and DPA levels measured using the DBS test ($r_s = 0.12$; $p = 0.47$) (Figure 2C).

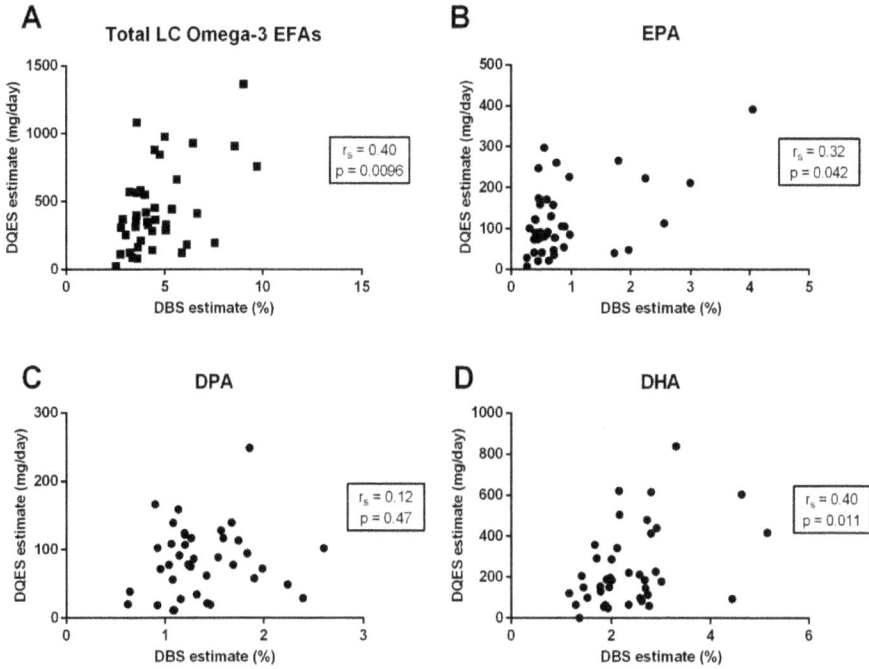

Figure 2. The relationship between total long-chain (LC) omega-3 essential fatty acids (EFAs) percentage (%) present in erythrocytes, measured using dried blood spot (DBS) biomarkers, and the dietary LC omega-3 EFA intake estimated using the Dietary Questionnaire for Epidemiology Studies (DQES) v3.2 for (**A**) total LC omega-3 EFAs, (**B**) eicosapentaenoic acid (EPA), (**C**) docosapentaenoic acid (DPA), and (**D**) docosahexaenoic acid (DHA). Correlation coefficients are calculated using the Spearman's correlation coefficients (r_s).

Comparing the CODS to the DBS fatty acid analysis, there was a moderate positive correlation between the two methods for total long-chain omega-3 EFAs (r_s = 0.38; p = 0.017) (Figure 3A) and DHA (r_s = 0.45; p = 0.0038) (Figure 3D), and a weak correlation for EPA (r_s = 0.18; p = 0.28) (Figure 3B), but no significant correlation for DPA (r_s = 0.11; p = 0.48) (Figure 3C).

For total long-chain omega-3 EFA intake, the relationship between both dietary methods (i.e., CODS and DQES) and the DBS Omega-3 index, which is a measure of the total EPA and DHA present in erythrocyte phospholipid membranes, was considered. There was a moderate, positive correlation between both the CODS (r_s = 0.37; p = 0.018) (Figure 4A) and DQES (r_s = 0.39; p = 0.013) (Figure 4B) questionnaires compared to the Omega-3 index estimated using the DBS assay.

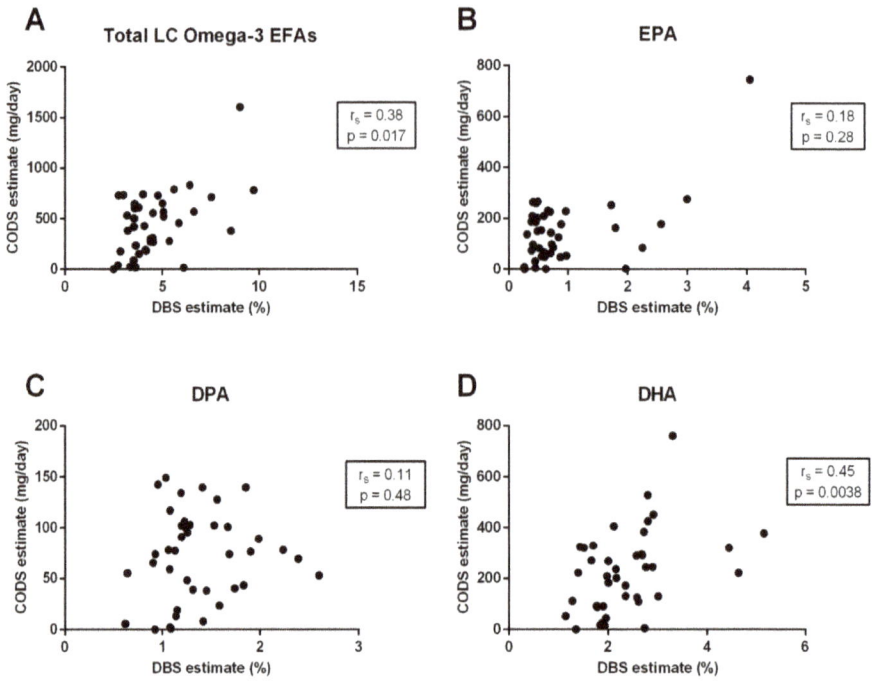

Figure 3. The relationship between total long-chain (LC) omega-3 essential fatty acids (EFAs) percentage (%) present in erythrocytes, measured using dried blood spot (DBS) biomarkers, and the dietary LC omega-3 EFA intake estimated using the Clinical Omega-3 Dietary Survey (CODS) for (**A**) total LC omega-3 EFAs, (**B**) eicosapentaenoic acid (EPA), (**C**) docosapentaenoic acid (DPA), and (**D**) docosahexaenoic acid (DHA). Correlation coefficients are calculated using the Spearman's correlation coefficients (r_s).

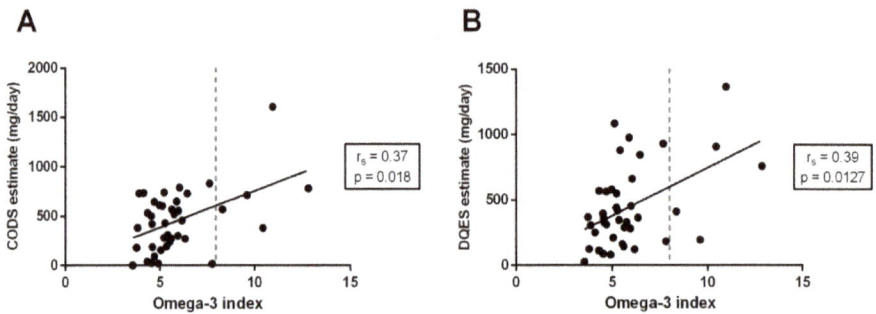

Figure 4. The relationship between the overall Omega-3 index, calculated as the total percentage (%) of EPA and DHA present in erythrocytes, and estimated using the dried blood spot (DBS) biomarker, and the dietary estimate of total long-chain omega-3 intake using (**A**) the Clinical Omega-3 Dietary Survey (CODS) and (**B**) the Dietary Questionnaire for Epidemiology Studies (DQES) v3.2. Correlation coefficients are calculated using the Spearman's correlation coefficients (r_s).

3.4. Method of Triads Analysis

Table 4 summarizes the validity coefficient (ρ) for each of the methods used to estimate the dietary omega-3 EFAs (CODS, DQES, and DBS), relative to an estimated true intake (T), calculated using the method of triads [43].

Table 4. Validity coefficients (ρ) calculated using the method of triads for each of the methods vs. the estimated true intake (T) for each of the long-chain omega-3 EFAs.

	CODS Validity Coefficient vs. T [ρQT] (95% CI)	DQES Validity Coefficient vs. T [ρRT] (95% CI)	DBS Validity Coefficient vs. T [ρBT] (95% CI)
Total LC omega-3 EFAs	0.77 (0.31–0.98)	0.83 (0.39–0.98)	0.49 (0.12–0.73)
EPA (20:5 n-3)	0.57 (0.21–0.93)	1.00 * (0.39–1.00)	0.31 (0.07–0.65)
DPA (22:5n-3)	0.72 (0.14–0.97)	0.73 (0.16–0.97)	0.16 (0.03–0.50)
DHA (22:6n-3)	0.86 (0.54–0.99)	0.77 (0.39–0.97)	0.52 (0.21–0.74)

* Validity coefficients >1 were set to 1.00 (Heywood cases). Abbreviations: CI, confidence interval; CODS, Clinical Omega-3 Dietary Survey; DBS, dried blood spot; DHA, docosahexaenoic acid; DPA, docosapentaenoic acid; DQES, Dietary Survey for Epidemiology Studies v3.2; EFA, essential fatty acid; EPA, eicosapentaenoic acid; LC, long-chain; total LC omega-3 EFAs, DHA + DPA + EPA; Q, questionnaire; R, reference method; T, true intake.

The CODS obtained high validity coefficients for estimating each of total long-chain omega-3 EFAs (ρ = 0.77, 95% CI: 0.31 to 0.98); DHA (ρ = 0.86, 95% CI: 0.54 to 0.99); and DPA (ρ = 0.72, 95% CI: 0.14 to 0.97). A moderate validity coefficient was obtained for estimating EPA (ρ = 0.57, 95% CI: 0.21 to 0.93).

The DQES obtained high validity coefficients for estimating total long-chain omega-3 EFAs (ρ = 0.83, 95% CI: 0.39 to 0.98); DHA (ρ = 0.77, 95% CI: 0.39 to 0.97); and DPA (ρ = 0.73, 95% CI: 0.16 to 0.97). A Heywood case was met for the validity coefficient estimate for EPA (ρ = 1.04), and thus the validity coefficient was set to 1.0 (95% CI: 0.39 to 1.00).

For the DBS assay, moderate validity coefficients (relative to the estimated true intake T) were obtained for estimating each of total long-chain omega-3 EFAs (ρ = 0.49, 95% CI: 0.12 to 0.73); DHA (ρ = 0.52, 95% CI: 0.21 to 0.74); and EPA (ρ = 0.31, 95% CI: 0.07 to 0.65). A low validity coefficient was obtained for estimating DPA (ρ = 0.16, 95% CI: 0.03 to 0.50).

3.5. Bland–Altman Analysis

The inter-method agreement was assessed using a Bland–Altman analysis (Figure 5) for the CODS- and the DQES-derived estimates for total daily long-chain omega-3 EFA intake (mg/day). Regression analysis indicated the absence of a significant inter-variable relationship across the spectrum of quantified values (*p* = 0.97). There was no significant global bias (mean difference) between these two assessment methods (mean ± standard error: 6.1 ± 38.6 mg/day, *p* > 0.05).

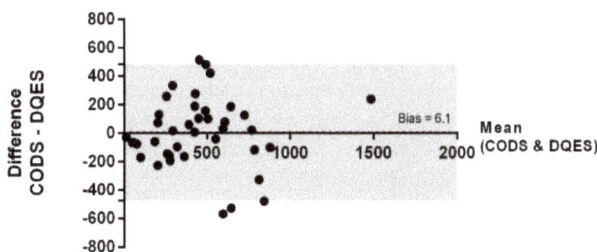

Figure 5. Bland–Altman plot comparing the estimated total daily intake of long-chain omega-3 EFAs (mg/day), measured using the CODS and DQES. The dotted line shows the bias (6.1 mg/day) for the comparison between the two methods, which was not statistically significant (95% CI: −32.5 to 44.7 mg/day). The grey shaded area highlights the limits of agreement (LoA).

4. Discussion

This preliminary validation study shows that a novel FFQ, the CODS, is a simple, valid tool for assessing long-chain omega-3 EFA intake in Australian adults. The CODS, which considers foods containing the highest concentrations of long-chain omega-3 EFAs, extracted from an Australian food compositional database (AUSNUT 2011–2013) [34], only requires a few minutes to complete.

The dietary estimates derived from the CODS correlated moderately well with long-chain omega-3 fatty acid intake quantified using the comprehensive, validated DQES (v3.2), as well as the systemic fatty acid profiles derived from dried blood spot analyses. Given that the CODS is straightforward to use, and can be rapidly completed in clinical settings, we propose that this survey can be applied clinically to estimate patients' long-chain omega-3 EFA intake, particularly in eye care settings where no similar tools currently exist. Therefore, this may provide information to inform clinical advice with respect to the sufficiency of a patient's estimated intake of dietary long-chain omega-3 EFAs, and thus any potential recommendations surrounding dietary modification and/or supplementation.

Knowledge of a patient's baseline omega-3 status is essential to develop informed, best-practice clinical recommendations relating to the relative appropriateness of dietary adjustment(s) and/or supplementation. Biomarker-based assessments of systemic omega-3 EFA levels, typically derived from subcutaneous adipose or blood assays, although accurate, are costly and time consuming, and therefore are not routinely applied in clinical settings. In addition, these types of biomarker assessments may not be readily available to all clinicians who require knowledge of patients' omega-3 fatty acid intake to provide informed clinical care. For example, as major providers of primary eye care, optometrists frequently provide clinical care to individuals who are at risk of, or who have existing, eye conditions where the natural history of the disease may be influenced by omega-3 EFA intake. Recent research demonstrates that eye care clinicians have identified a need for validated clinical tools to assess the degree of dietary omega-3 fatty acid sufficiency for their patients [46]. Furthermore, when attending eye examinations, there is an expectation among patients that the provision of comprehensive clinical care includes an assessment and relevant evidence-based advice surrounding dietary risk factors for ocular disease [47]. A major limitation with generic clinical recommendations regarding omega-3 EFA intake (e.g., to eat oily fish at least twice a week) is that such advice does not take into consideration the substantial variability of omega-3 fatty acid concentrations across different species of fish. For example, a 100 g serving of salmon provides approximately seven-fold more long-chain omega-3 EFAs than that of 100 g of lean fish, such as whiting [48]. In this regard, tailored questionnaires that are designed to specifically assess EFA intake may have the capacity to more accurately capture dietary intakes for specific nutrients (e.g., fatty acids) compared with generic questionnaires (i.e., questionnaires that assess total dietary intake but do not consider the specific fatty acid sources) [49].

Several FFQs have been developed to assess dietary omega-3 intake [21,50], which range from long [27] to short [29–31] in length. Questionnaires may perform differently in different geographic locations and patient populations, and as such, instruments should be validated by recruiting participants who are representative of the primary target population [51]. In terms of relatively short, clinically applicable FFQs, Sublette et al. developed and validated a 21-item questionnaire, which took approximately five minutes to complete, administered with a sample of 61 US adults with and without major depressive disorders. [31]. Dahl et al. developed and validated another brief, 10-minute questionnaire, administered with a sample of healthy Norwegian population [29]. A nine-item FFQ was developed based on the 2005 Canadian Nutrient File of Health Canada, conducted in a sample of women with low marine food intakes and with psychological stress [30]. It showed poor agreement with an Australian-based FFQ, when administered in an Australian population, however, estimates of the nutritional intake improved when nutritional composition data were replaced with Australian equivalents [32].

We validated the CODS relative to both an erythrocyte biomarker and a previously validated food frequency questionnaire (DQES v3.2). Nutritional information provided by the DQES differed from the CODS as it comprehensively considers all food groups available in Australia, in order to create a full dietary profile of consumed nutrients. In contrast, the CODS only assesses foods that are rich in omega-3 EFAs. Information relating to long-chain omega-3 EFA intake was extracted from the DQES based upon food compositional data from the NUTTAB (2010) database [39], whereas dietary estimates for the CODS were extracted from the independent AUSNUT (2011–2013) database [34].

Other potential dietary reference methods included 24-h food recalls, and short-term and long-term food diaries. These techniques provided an alternative, but not necessarily more accurate, method of assessing dietary intake [51]. The limitation of short-term food diaries is the day-to-day variation of foods consumed, and as such, a single observation may provide a poor measure of overall dietary intake [21]. Furthermore, although the food records provided a precise measure of dietary habits and portion sizes over the short monitoring period (e.g., seven days), bias may be induced as dietary behaviors of a participant may be influenced over the capture period [52].

The limitation of not capturing a baseline omega-3 status, which in practice may inform clinical recommendations, is similar to the confound of not quantifying this parameter in omega-3 EFA intervention trials, in order to consider this factor in the evaluation of therapeutic efficacy. For example, numerous clinical studies have sought to investigate the therapeutic efficacy of omega-3 EFA supplements for treating dry eye disease, with many reporting apparently differing results [53–56]. However, as recently reported in a Cochrane systematic review [57] of 34 randomized controlled trials assessing the effect(s) of oral omega-3 and/or omega-6 supplements on dry eye symptoms and signs, the vast majority of studies to date have not surveyed (through food questionnaires) or quantified (e.g., through blood testing) baseline systemic fatty acid levels. It is critical to have an understanding of basal levels, as individuals already achieving sufficient levels of PUFAs from food sources may not demonstrate the same response to a given dose of fatty acid supplementation as those with a diet deficient in some, or all, PUFAs [58,59]. The assessment of post-treatment omega-3 status is also a means of evaluating participant compliance (in addition to other measures such as returned capsule counts, compliance diaries, etc.), however, in this systematic review, it was only assessed in two of the included trials [54,55]. While the quantification of baseline omega-3 EFA levels is recommended as a standard for cardiovascular trials [58], currently, there is no such guidance in relation to ocular studies. If the cost of biological assays is prohibitive, the findings of our study suggest that the CODS may be a relevant surrogate marker for estimating dietary intake of long-chain omega-3 EFAs.

A potential advantage of using methods based on surveys, rather than biological biomarkers, to assess omega-3 EFA intake is the ability to disambiguate fatty acids consumed from food sources and nutritional supplements. This is of particular relevance to the clinical recommendations for omega-3 EFAs in the context of eye disease, where the source of omega-3 EFAs may influence clinical outcomes. For example, several epidemiological studies have shown that a high (food-sourced) dietary intake of omega-3 fatty acids is associated with a significantly reduced risk of developing age-related macular degeneration (AMD) [8,9], and a decreased risk of AMD progression in individuals with the established disease [10–12]. However, perhaps counter intuitively, a Cochrane collaboration systematic review of randomized controlled trial evidence reported that the use of omega-3 fatty acid nutritional supplements, for a follow-up period of up to five years, did not reduce the incidence of progression to late-stage AMD or the development of moderate-to-severe vision loss, compared to placebo supplementation [60]. There are currently no randomized controlled trials on dietary omega-3 EFA supplementation for the primary prevention of AMD. Therefore, whole food sources containing high levels of long-chain omega-3 EFAs have been shown to be retinoprotective, whereas, omega-3 supplementation does not appear to confer the same benefit. The specific mechanism(s) underlying the benefits of omega-3 EFAs in AMD have not been established, but likely derive from the anti-inflammatory and/or anti-oxidative effects of omega-3 EFAs [60], and the potential interaction of these fatty acids with other nutrients found in whole foods rich in these components.

The Omega-3 index is a measure of the relative concentration of EPA and DHA in erythrocyte membranes, and it is considered an acceptable marker for evaluating the risk of coronary heart disease [24]. An association has been reported between Omega-3 indices <4% and a high incidence of cardiovascular disease; shifting the Omega-3 index from 4% to 8% is also estimated to reduce the relative incidence of fatal coronary heart disease by 30% [61]. In the present study, we compared the DQES and CODS (survey-based) estimates of daily long-chain omega-3 EFA intake to the Omega-3 index quantified using the DBS biomarker assay. Total long-chain omega-3 EFA intake, estimated with both the CODS and DQES v3.2, showed a moderately strong, positive correlation with the Omega-3 index (CODS: r_s = 0.39, p = 0.013; DQES v3.2: r_s = 0.37, p = 0.018). Of note, on the basis of interpolation of the data shown in Figure 4, targeting a desired Omega-3 index of 8% [61] corresponds to an intake of ~600 mg/day of long-chain omega-3 EFAs, as estimated using the CODS. Consistently, this estimate agrees with NHMRC SDT recommendations for lowering chronic disease risk [15,19].

In general, the accuracy of dietary surveys depends on the recall ability of the participant and the accuracy and availability of nutritional composition data in the geographical region [33]. We acknowledge a limitation of the CODS is that it does not consider the conversion of ALA and short-chain omega-3 fatty acid to long-chain metabolites, or the consumption of omega-6 fatty acids, being the other major class of PUFAs. Foods containing ALA from nutrition composition databases were not considered due to the heterogeneity in ALA concentrations reported across foods in the same groups (e.g., the amount of ALA present in margarine varies from 1 g/100 to 8 g/100 g, depending on the brand and type i.e., monounsaturated or polyunsaturated) [34], and the additional time that would have been required to complete a questionnaire that incorporated these food sources. The decision to omit ALA estimates from the CODS was also based on the known efficiency of conversion of short- to long-chain omega-3 fatty acid forms *in vivo*, which has been reported to range from 4–20% for EPA and estimated at 5–9% for both DPA and DHA [62–64]. This conversion is further reduced in the presence of increasing circulating levels of omega-6 EFAs [62]. Long-chain omega-6 EFAs also competitively inhibit the incorporation of omega-3 EFAs into phospholipid membranes [65]. The poor validity coefficient between DPA intake estimated from survey methods and a biological biomarker has also been observed in other studies in the same region [27,41], and may be related to the selective uptake of DPA in tissues [66,67]. DPA, being an intermediate metabolite, is also highly interconvertible with EPA and less readily metabolized to DHA [66,67]. These factors may contribute to the lack of correlation between the DPA estimate using CODS and the DBS biomarker assay. Nevertheless, we observed a significant correlation between the CODS and the DBS biomarker assay, for total long-chain omega-3 fatty acids, DHA, and EPA. Further validation of the CODS, in particular within a larger population of patients in clinical practice, would be of value to confirm the generalizability of these findings.

In this preliminary investigation, we demonstrated that the CODS is a potentially useful tool for assessing long-chain omega-3 intake, validated in a population of Australian adults, factoring in both food sources and dietary supplementation. In addition, the CODS estimated daily intake of long-chain omega-3 EFAs was moderately well correlated to the Omega-3 index, which is a validated marker for cardiovascular disease risk. We propose that the CODS could provide a rapid, non-invasive tool (as a no-cost alternative application as compared with more costly investigations) for evaluating the relative sufficiency of a patient's dietary omega-3 EFA intake in a clinical setting. Future directions will include repeatability assessment and validation of the CODS against another well-controlled dietary reference method (e.g., a multiple-day food record), in a larger, homogenous population of participants. These additional investigations will be of value for strengthening the rigor of the CODS, and deriving further data to support its utility for implementation in eye care practice.

Supplementary Materials: The following are available online at http://www.mdpi.com/2072-6643/11/4/817/s1, Figure S1: The Clinical Omega-3 Dietary Survey, Figure S2: Diagram adapted from Ocke & Kaaks, to describe the method of triads [43].

Author Contributions: Conceptualization, A.C.Z. and L.E.D.; developing the methodology and collecting the data, A.C.Z.; supervised by L.E.D.; analysis was undertaken by A.C.Z. and L.E.D.; writing, including original draft preparation, review and editing, were undertaken by A.C.Z. and L.E.D.; funding acquisition, L.E.D.

Funding: This research was funded by the University of Melbourne—Melbourne Neuroscience Institute Interdisciplinary Seed Grant (LED, 2017).

Conflicts of Interest: The authors declare no conflict of interest. The funders had no role in the design of the study; in the collection, analyses, or interpretation of data; in the writing of the manuscript; or in the decision to publish the results.

References

1. Rajaei, E.; Mowla, K.; Ghorbani, A.; Bahadoram, S.; Bahadoram, M.; Dargahi-Malamir, M. The Effect of Omega-3 Fatty Acids in Patients with Active Rheumatoid Arthritis Receiving DMARDs Therapy: Double-Blind Randomized Controlled Trial. *Glob. J. Health Sci.* **2015**, *8*, 18–25. [CrossRef] [PubMed]

2. Yokoyama, M.; Origasa, H.; Matsuzaki, M.; Matsuzawa, Y.; Saito, Y.; Ishikawa, Y.; Oikawa, S.; Sasaki, J.; Hishida, H.; Itakura, H.; et al. Effects of eicosapentaenoic acid on major coronary events in hypercholesterolaemic patients (JELIS): A randomised open-label, blinded endpoint analysis. *Lancet* **2007**, *369*, 1090–1098. [CrossRef]

3. Calder, P.C. Omega-3 fatty acids and inflammatory processes: From molecules to man. *Biochem. Soc. Trans.* **2017**, *45*, 1105–1115. [CrossRef] [PubMed]

4. Simopoulos, A.P. Omega-3 fatty acids in inflammation and autoimmune diseases. *J. Am. Coll. Nutr.* **2002**, *21*, 495–505. [CrossRef] [PubMed]

5. Zhu, W.; Wu, Y.; Meng, Y.F.; Xing, Q.; Tao, J.J.; Lu, J. Fish Consumption and Age-Related Macular Degeneration Incidence: A Meta-Analysis and Systematic Review of Prospective Cohort Studies. *Nutrients* **2016**, *8*, 743. [CrossRef] [PubMed]

6. Downie, L.E.; Keller, P.R. Nutrition and age-related macular degeneration: Research evidence in practice. *Optom. Vis. Sci.* **2014**, *91*, 821–831. [CrossRef] [PubMed]

7. Miljanovic, B.; Trivedi, K.A.; Dana, M.R.; Gilbard, J.P.; Buring, J.E.; Schaumberg, D.A. Relation between dietary n-3 and n-6 fatty acids and clinically diagnosed dry eye syndrome in women. *Am. J. Clin. Nutr.* **2005**, *82*, 887–893. [CrossRef]

8. Tan, J.S.; Wang, J.J.; Flood, V.; Mitchell, P. Dietary fatty acids and the 10-year incidence of age-related macular degeneration: The Blue Mountains Eye Study. *Arch. Ophthalmol.* **2009**, *127*, 656–665. [CrossRef]

9. Chong, E.W.; Kreis, A.J.; Wong, T.Y.; Simpson, J.A.; Guymer, R.H. Dietary omega-3 fatty acid and fish intake in the primary prevention of age-related macular degeneration: A systematic review and meta-analysis. *Arch. Ophthalmol.* **2008**, *126*, 826–833. [CrossRef]

10. Sangiovanni, J.P.; Agrón, E.; Meleth, A.D.; Reed, G.F.; Sperduto, R.D.; Clemons, T.E.; Chew, E.Y. Age-Related Eye Disease Study Research G. {omega}-3 Long-chain polyunsaturated fatty acid intake and 12-y incidence of neovascular age-related macular degeneration and central geographic atrophy: AREDS report 30, a prospective cohort study from the Age-Related Eye Disease Study. *Am. J. Clin. Nutr.* **2009**, *90*, 1601–1607. [CrossRef]

11. Christen, W.G.; Schaumberg, D.A.; Glynn, R.J.; Buring, J.E. Dietary omega-3 fatty acid and fish intake and incident age-related macular degeneration in women. *Arch. Ophthalmol.* **2011**, *129*, 921–929. [CrossRef]

12. Seddon, J.M.; George, S.; Rosner, B. Cigarette Smoking, Fish Consumption, Omega-3 Fatty Acid Intake, and Associations with Age-Related Macular Degeneration: The US Twin Study of Age-Related Macular Degeneration. *Arch. Ophthalmol.* **2006**, *124*, 995–1001. [CrossRef]

13. Downie, L.E.; Vingrys, A.J. Oral Omega-3 Supplementation Lowers Intraocular Pressure in Normotensive Adults. *Transl. Vis. Sci. Technol.* **2018**, *7*, 1. [CrossRef]

14. Fats: Total Fat & Fatty Acids; Nutrient Reference Values for Australia and New Zealand. Available online: https://www.nrv.gov.au/nutrients/fats-total-fat-fatty-acids (accessed on 20 February 2019).

15. National Health and Medical Research Council. *Nutrient Reference Values for Australia and New Zealand including Recommended Dietary Intakes*; Australian Government Department of Health and Ageing: Canberra, Australia, 2006.

16. Nestel, P.; Clifton, P.; Colquhoun, D.; Noakes, M.; Mori, T.A.; Sullivan, D.; Thomas, B. Indications for Omega-3 Long Chain Polyunsaturated Fatty Acid in the Prevention and Treatment of Cardiovascular Disease. *Heart Lung Circ.* **2015**, *24*, 769–779. [CrossRef]

17. U.S. Department of Health and Human Services; U.S. Department of Agriculture. *2015–2020 Dietary Guidelines for Americans*, 8th ed.; Federal Government: Washington, DC, USA, 2015.

18. EFSA Panel on Dietetic Products, Nutrition, and Allergies (NDA). Scientific Opinion on Dietary Reference Values for fats, including saturated fatty acids, polyunsaturated fatty acids, monounsaturated fatty acids, trans fatty acids, and cholesterol. *EFSA J.* **2010**, *8*, 1461. [CrossRef]

19. Meyer, B.J. Australians are not Meeting the Recommended Intakes for Omega-3 Long Chain Polyunsaturated Fatty Acids: Results of an Analysis from the 2011-2012 National Nutrition and Physical Activity Survey. *Nutrients* **2016**, *8*, 111. [CrossRef]

20. Danaei, G.; Ding, E.L.; Mozaffarian, D.; Taylor, B.; Rehm, J.; Murray, C.J.; Ezzati, M. The preventable causes of death in the United States: Comparative risk assessment of dietary, lifestyle, and metabolic risk factors. *PLoS Med.* **2009**, *6*, e1000058. [CrossRef]

21. Overby, N.C.; Serra-Majem, L.; Andersen, L.F. Dietary assessment methods on n-3 fatty acid intake: A systematic review. *Br. J. Nutr.* **2009**, *102* (Suppl. 1), S56–S63. [CrossRef]

22. Baylin, A.; Kabagambe, E.K.; Siles, X.; Campos, H. Adipose tissue biomarkers of fatty acid intake. *Am. J. Clin. Nutr.* **2002**, *76*, 750–757. [CrossRef]

23. Saadatian-Elahi, M.; Slimani, N.; Chajes, V.; Jenab, M.; Goudable, J.; Biessy, C.; Ferrari, P.; Byrnes, G.; Autier, P.; Peeters, P.H.; et al. Plasma phospholipid fatty acid profiles and their association with food intakes: Results from a cross-sectional study within the European Prospective Investigation into Cancer and Nutrition. *Am. J. Clin. Nutr.* **2009**, *89*, 331–346. [CrossRef]

24. Harris, W.S. The omega-3 index as a risk factor for coronary heart disease. *Am. J. Clin. Nutr.* **2008**, *87*, 1997S–2002S. [CrossRef]

25. Wrieden, W.; Peace, H.; Armstrong, J.; Barton, K. A short review of dietary assessment methods used in national and Scottish research studies. In *Briefing Paper Prepared for the Working Group on Monitoring Scottish Dietary Targets Workshop*; Edinburgh University Press: Edinburgh, UK, 2003.

26. Institute of Medicine (US) Committee on Dietary Risk Assessment in the WIC Program. 5. Food-Based Assessment of Dietary Intake. In *Dietary Risk Assessment in the WIC Program*; National Academies Press (US): Washington, DC, USA, 2002.

27. Swierk, M.; Williams, P.G.; Wilcox, J.; Russell, K.G.; Meyer, B.J. Validation of an Australian electronic food frequency questionnaire to measure polyunsaturated fatty acid intake. *Nutrition* **2011**, *27*, 641–646. [CrossRef]

28. Rahmawaty, S.; Charlton, K.; Lyons-Wall, P.; Meyer, B.J. Development and validation of a food frequency questionnaire to assess omega-3 long chain polyunsaturated fatty acid intake in Australian children aged 9–13 years. *J. Hum. Nutr. Diet.* **2017**, *30*, 429–438. [CrossRef]

29. Dahl, L.; Maeland, C.A.; Bjorkkjaer, T. A short food frequency questionnaire to assess intake of seafood and n-3 supplements: Validation with biomarkers. *Open Nutr. J.* **2011**, *10*, 127. [CrossRef]

30. Lucas, M.; Asselin, G.; Mérette, C.; Poulin, M.-J.; Dodin, S. Validation of an FFQ for evaluation of EPA and DHA intake. *Public Health Nutr.* **2009**, *12*, 1783–1790. [CrossRef]

31. Sublette, M.E.; Segal-Isaacson, C.J.; Cooper, T.B.; Fekri, S.; Vanegas, N.; Galfalvy, H.C.; Oquendo, M.A.; Mann, J.J. Validation of a food frequency questionnaire to assess intake of n-3 polyunsaturated fatty acids in subjects with and without major depressive disorder. *J. Am. Diet. Assoc.* **2011**, *111*, 117–123. [CrossRef]

32. Dickinson, K.M.; Delaney, C.L.; Allan, R.; Spark, I.; Miller, M.D. Validation of a Brief Dietary Assessment Tool for Estimating Dietary EPA and DHA Intake in Australian Adults at Risk of Cardiovascular Disease. *J. Am. Coll. Nutr.* **2015**, *34*, 333–339. [CrossRef]

33. Stark, K.D.; Van Elswyk, M.E.; Higgins, M.R.; Weatherford, C.A.; Salem, N., Jr. Global survey of the omega-3 fatty acids, docosahexaenoic acid and eicosapentaenoic acid in the blood stream of healthy adults. *Prog. Lipid Res.* **2016**, *63*, 132–152. [CrossRef]

34. Food Standards Australia New Zealand. *AUSNUT 2011–13—Australian Food Composition Database*; FSANZ: Canberra, Australia, 2014. Available online: www.foodstandards.gov.au (accessed on 4 April 2018).

35. Giles, G.G.; Ireland, P.D. *Dietary Questionnaire for Epidemiological Studies (Version 3.2)*; Cancer Council Victoria: Melbourne, Australia, 1996.

36. Bassett, J.K.; English, D.R.; Fahey, M.T.; Forbes, A.B.; Gurrin, L.C.; Simpson, J.A.; Brinkman, M.T.; Giles, G.G.; Hodge, A.M. Validity and calibration of the FFQ used in the Melbourne Collaborative Cohort Study. *Public Health Nutr.* **2016**, *19*, 2357–2368. [CrossRef]

37. Hopkins, A.H.; Hodge, A.M.; Fletcher, A.S.; Bruinsma, F.J.; Bassett, J.K.; Popowski, L.V.; Brinkman, M.T.; English, D.R.; Giles, G.G.; Jayasekara, H.; et al. Cohort Profile: The Melbourne Collaborative Cohort Study (Health 2020). *Int. J. Epidemiol.* **2017**, *46*, 1757–1757i. [CrossRef]

38. Food Standards Australia New Zealand. *AUSNUT 2007—Australian Food Composition Tables*; FSANZ: Canberra, Australia, 2008.

39. Food Standards Australia New Zealand. *NUTTAB 2010—Australian Food Composition Tables*; FSANZ: Canberra, Australia, 2011.

40. Liu, G.; Muhlhausler, B.S.; Gibson, R.A. A method for long term stabilisation of long chain polyunsaturated fatty acids in dried blood spots and its clinical application. *Prostaglandins Leukot. Essent. Fatty Acids* **2014**, *91*, 251–260. [CrossRef] [PubMed]

41. Sullivan, B.L.; Williams, P.G.; Meyer, B.J. Biomarker validation of a long-chain omega-3 polyunsaturated fatty acid food frequency questionnaire. *Lipids* **2006**, *41*, 845–850. [CrossRef] [PubMed]

42. Da Silva, D.C.G.; Segheto, W.; de Lima, M.F.C.; Pessoa, M.C.; Pelúzio, M.C.G.; Marchioni, D.M.L.; Cunha, D.B.; Longo, G.Z. Using the method of triads in the validation of a food frequency questionnaire to assess the consumption of fatty acids in adults. *J. Hum. Nutr. Diet.* **2018**, *31*, 85–95. [CrossRef] [PubMed]

43. Ocke, M.C.; Kaaks, R.J. Biochemical markers as additional measurements in dietary validity studies: Application of the method of triads with examples from the European Prospective Investigation into Cancer and Nutrition. *Am. J. Clin. Nutr.* **1997**, *65*, 1240s–1245s. [CrossRef] [PubMed]

44. Efron, B.; Tibshirani, R. Bootstrap Methods for Standard Errors, Confidence Intervals, and Other Measures of Statistical Accuracy. *Stat. Sci.* **1986**, *1*, 54–75. [CrossRef]

45. Bland, J.M.; Altman, D.G. Measuring agreement in method comparison studies. *Stat. Methods Med. Res.* **1999**, *8*, 135–160. [CrossRef] [PubMed]

46. Downie, L.E.; Keller, P.R. The self-reported clinical practice behaviors of Australian optometrists as related to smoking, diet and nutritional supplementation. *PLoS ONE* **2015**, *10*, e0124533. [CrossRef] [PubMed]

47. Downie, L.E.; Douglass, A.; Guest, D.; Keller, P.R. What do patients think about the role of optometrists in providing advice about smoking and nutrition? *Ophthalmic Physiol. Opt.* **2017**, *37*, 202–211. [CrossRef]

48. Soltan, S.S.; Gibson, R.A. Levels of Omega 3 fatty acids in Australian seafood. *Asia Pac. J. Clin. Nutr.* **2008**, *17*, 385–390.

49. Meyer, B.J.; Swierk, M.; Russell, K.G. Assessing long-chain omega-3 polyunsaturated fatty acids: A tailored food-frequency questionnaire is better. *Nutrition* **2013**, *29*, 491–496. [CrossRef] [PubMed]

50. Serra-Majem, L.; Frost Andersen, L.; Henrique-Sanchez, P.; Doreste-Alonso, J.; Sanchez-Villegas, A.; Ortiz-Andrelluchi, A.; Negri, E.; La Vecchia, C. Evaluating the quality of dietary intake validation studies. *Br. J. Nutr.* **2009**, *102* (Suppl. 1), S3–S9. [CrossRef]

51. Cade, J.; Thompson, R.; Burley, V.; Warm, D. Development, validation and utilisation of food-frequency questionnaires—A review. *Public Health Nutr.* **2002**, *5*, 567–587. [CrossRef]

52. Thompson, F.E.; Byers, T. Dietary assessment resource manual. *J. Nutr.* **1994**, *124*, 2245s–2317s. [CrossRef]

53. Jones, L.; Downie, L.E.; Korb, D.; Benitez-Del-Castillo, J.M.; Dana, R.; Deng, S.X.; Dong, P.N.; Geerling, G.; Hida, R.Y.; Liu, Y.; et al. TFOS DEWS II Management and Therapy Report. *Ocul. Surf.* **2017**, *15*, 575–628. [CrossRef]

54. Deinema, L.A.; Vingrys, A.J.; Wong, C.Y.; Jackson, D.C.; Chinnery, H.R.; Downie, L.E. A Randomized, Double-Masked, Placebo-Controlled Clinical Trial of Two Forms of Omega-3 Supplements for Treating Dry Eye Disease. *Ophthalmology* **2017**, *124*, 43–52. [CrossRef]

55. Asbell, P.A.; Maguire, M.G.; Pistilli, M.; Ying, G.S.; Szczotka-Flynn, L.B.; Hardten, D.R.; Lin, M.C.; Shtein, R.M. n-3 Fatty Acid Supplementation for the Treatment of Dry Eye Disease. *N. Engl. J. Med.* **2018**, *378*, 1681–1690. [CrossRef]

56. Chinnery, H.R.; Naranjo Golborne, C.; Downie, L.E. Omega-3 supplementation is neuroprotective to corneal nerves in dry eye disease: A pilot study. *Ophthalmic Physiol. Opt.* **2017**, *37*, 473–481. [CrossRef]

57. Downie, L.E.; Ng, S.M.; Lindsley, K.; Akpek, E.K. Omega-3 and omega-6 polyunsaturated fatty acids for dry eye disease. *Cochrane Database Syst. Rev.* **2019**. In Press.

58. Rice, H.B.; Bernasconi, A.; Maki, K.C.; Harris, W.S.; von Schacky, C.; Calder, P.C. Conducting omega-3 clinical trials with cardiovascular outcomes: Proceedings of a workshop held at ISSFAL 2014. *Prostaglandins Leukot. Essent. Fatty Acids* **2016**, *107*, 30–42. [CrossRef]

59. Silva, V.; Singer, P. Membrane fatty acid composition of different target populations: Importance of baseline on supplementation. *Clin. Nutr. Exp.* **2015**, *1*, 1–9. [CrossRef]

60. Lawrenson, J.G.; Evans, J.R. Omega 3 fatty acids for preventing or slowing the progression of age-related macular degeneration. *Cochrane Database Syst. Rev.* **2015**, *4*, CD010015. [CrossRef]

61. Harris, W.S.; Del Gobbo, L.; Tintle, N.L. The Omega-3 Index and relative risk for coronary heart disease mortality: Estimation from 10 cohort studies. *Atherosclerosis* **2017**, *262*, 51–54. [CrossRef]

62. Gerster, H. Can adults adequately convert alpha-linolenic acid (18:3n-3) to eicosapentaenoic (20:5n-3) and docosahexaenoic acid (22:6n-3)? International journal for vitamin and nutrition research. *Int. J. Vitam. Nutr. Res.* **1998**, *68*, 159–173.

63. Burdge, G.C.; Wootton, S.A. Conversion of alpha-linolenic acid to eicosapentaenoic, docosapentaenoic and docosahexaenoic acids in young women. *Br. J. Nutr.* **2002**, *88*, 411–420. [CrossRef]

64. Brenna, J.T. Efficiency of conversion of alpha-linolenic acid to long chain n-3 fatty acids in man. *Curr. Opin. Clin. Nutr.* **2002**, *5*, 127–132. [CrossRef]

65. Simopoulos, A.P. The importance of the omega-6/omega-3 fatty acid ratio in cardiovascular disease and other chronic diseases. *Exp. Biol. Med.* **2008**, *233*, 674–688. [CrossRef]

66. Kaur, G.; Cameron-Smith, D.; Garg, M.; Sinclair, A.J. Docosapentaenoic acid (22:5n-3): A review of its biological effects. *Prog. Lipid Res.* **2011**, *50*, 28–34. [CrossRef]

67. Miller, E.; Kaur, G.; Larsen, A.; Loh, S.P.; Linderborg, K.; Weisinger, H.S.; Turchini, G.M.; Cameron-Smith, D.; Sinclair, A.J. A short-term n-3 DPA supplementation study in humans. *Eur. J. Nutr.* **2013**, *52*, 895–904. [CrossRef]

nutrients

MDPI

Article

Rosmarinic and Sinapic Acids May Increase the Content of Reduced Glutathione in the Lenses of Estrogen-Deficient Rats

Maria Zych *, Weronika Wojnar, Sławomir Dudek and Ilona Kaczmarczyk-Sedlak

Department of Pharmacognosy and Phytochemistry, School of Pharmacy with the Division of Laboratory Medicine in Sosnowiec, Medical University of Silesia, Katowice, Jagiellońska 4, 41-200 Sosnowiec, Poland; wwojnar@sum.edu.pl (W.W.); sdudek@sum.edu.pl (S.D.); isedlak@sum.edu.pl (I.K.-S.)
* Correspondence: mzych@sum.edu.pl; Tel.: +48-32-364-15-25

Received: 28 February 2019; Accepted: 5 April 2019; Published: 9 April 2019

Abstract: Oxidative stress is believed to be associated with both postmenopausal disorders and cataract development. Previously, we have demonstrated that rosmarinic and sinapic acids, which are diet-derived antioxidative phenolic acids, counteracted some disorders induced by estrogen deficiency. Other studies have shown that some phenolic acids may reduce cataract development in various animal models. However, there is no data on the effect of phenolic acids on oxidative stress markers in the lenses of estrogen-deficient rats. The study aimed to investigate whether administration of rosmarinic acid and sinapic acid affects the antioxidative abilities and oxidative damage parameters in the lenses of estrogen-deficient rats. The study was conducted on three-month-old female Wistar rats. The ovariectomized rats were orally treated with rosmarinic acid at doses of 10 and 50 mg/kg or sinapic acid at doses of 5 and 25 mg/kg, for 4 weeks. The content of reduced glutathione (GSH), oxidized glutathione and amyloid β_{1-42}, as well as products of protein and lipid oxidation, were assessed. Moreover, the activities of superoxide dismutase, catalase, and some glutathione-related enzymes in the lenses were determined. Rosmarinic and sinapic acids in both doses resulted in an increase in the GSH content and glutathione reductase activity. They also improved parameters connected with protein oxidation. Since GSH plays an important role in maintaining the lens transparency, the increase in GSH content in lenses after the use of rosmarinic and sinapic acids seems to be beneficial. Therefore, both the investigated dietary compounds may be helpful in preventing cataract.

Keywords: rosmarinic acid; sinapic acid; lenses; estrogen-deficient rats; oxidative stress; reduced glutathione

---

1. Introduction

Cataract, a visual impairment characterized by opacification of the lens, may be classified as an age-related disorder. Population-based studies indicate that lens opacities occur more often in women than in men [1,2]. Although estrogen deficiency occurs commonly in elderly women, which might suggest a link between this condition and cataract development, the data on the effect of estrogen on the opacity of the lens is contradictory. On the one hand, the meta-analysis from 2013 [3] showed that hormone replacement therapy reduces the risk of this disease and in vitro studies demonstrated the protective effect of estradiol against oxidative stress in the epithelial cells of the lens [4,5]. On the other hand, although there was a report suggesting that, in the experimental animals exposed to radiation, administration of estradiol may protect against lens opacity [6], other reports showed that treatment with estradiol may induce cataract [7,8]. The role of estrogens in cataract development and cataract dependence on gender has been presented by Zetterberg and Celojevic in a comprehensive review [9].

It is assumed that the development of disorders associated with estrogen deficiency in postmenopausal women, such as vasomotor symptoms, cardiovascular diseases or osteoporosis, is connected with oxidative stress [10–12]. Oxidative stress is also considered to be one of the causes of lens opacity [13,14]. Increased production of reactive oxygen species (ROS) and weakened antioxidant system leads to oxidative lens damage, which results in protein aggregation and lens turbidity [14,15]. It is believed that the use of antioxidants in the form of dietary components may be helpful to prevent disorders resulting from post-menopausal oxidative stress [16].

Antioxidants include, among others, phenolic acids, which are components of food products and medicinal plants [17]. The examples of phenolic acids are rosmarinic acid and sinapic acid, which are hydroxycinnamic acid derivatives. Rosmarinic acid is found mainly in plants of the Lamiaceae family, which are widely used as spices and medicinal plants, such as rosemary, spearmint, and lemon balm [18], while sinapic acid occurs in vegetables (especially from the Brassicaceae family, like tronchuda cabbage or broccoli), and fruits (e.g., strawberries or citruses) [19].

Our previous studies showed that both rosmarinic acid and sinapic acid had a positive effect on parameters related to glucose and lipid metabolism, as well as on some parameters of oxidative stress in the serum of ovariectomized rats in the early phase of estrogen deficiency [20,21]. Based on various experimental in vitro and in vivo animal studies, there are also suggestions on the possibility of using phenolic acids, including rosmarinic acid, to reduce cataract development [22–25]. However, there is still no data on the effects of plant-derived antioxidants, including rosmarinic and sinapic acid on oxidative stress parameters in the lenses exposed to estrogen deficiency. Based on literature data and our previous results, we hypothesized that both rosmarinic acid and sinapic acid may also show a protective antioxidative effect in the lenses of estrogen-deficient rats. Therefore, the study aimed to investigate the effect of rosmarinic and sinapic acids on the antioxidative abilities and oxidative damage parameters in the lenses of ovariectomized rats in the early phase of estrogen deficiency.

2. Materials and Methods

2.1. Animals and Drugs

The experiment was carried out on three-month-old female Wistar rats. The experiment was conducted under the approval of the Local Ethics Committee in Katowice (permission numbers: 38/2015, 148/2015, and 66/2016). The rats were purchased at the Center of Experimental Medicine, Medical University of Silesia (Katowice, Poland).

In the course of the experiment the following drugs were administered orally to the rats: rosmarinic acid (Sigma-Aldrich, St. Louis, MO, USA), sinapic acid (Sigma-Aldrich, St. Louis, MO, USA) and estradiol hemihydrate (Estrofem, Novo Nordisk A/S, Bagsvard, Denmark). As anesthetics ketamine (Ketamina 10%, Biowet Puławy, Puławy, Poland) and xylazine (Xylapan, Vetoquinol Biowet, Gorzów Wlkp., Poland) were used.

2.2. Experimental Design

During the acclimation period (13 days) and during the experiment, the animals had unlimited access to standard feed (Labofeed B, Wytwórnia Pasz "Morawski", Kcynia, Poland) and drinking water. The rats were divided into 7 groups: (n = 10):

- sham-operated control rats (SHAM);
- ovariectomized control rats (OVX);
- ovariectomized rats treated with estradiol at a dose of 0.2 mg/kg (OVX+ESTR);
- ovariectomized rats treated with rosmarinic acid at a dose of 10 mg/kg (OVX+RA10);
- ovariectomized rats treated with rosmarinic acid at a dose of 50 mg/kg (OVX+RA50);
- ovariectomized rats treated with sinapic acid at a dose of 5 mg/kg (OVX+SA5);
- ovariectomized rats treated with sinapic acid at a dose of 25 mg/kg (OVX+SA25).

The OVX+ESTR group of rats was used as a positive control.

As previously described [20,21], rats from the SHAM group underwent a sham surgery, and in the other animals, bilateral ovariectomy was carried out. The sham and ovariectomy surgery were performed under general anesthesia by intraperitoneal (i.p.) administration of the mixture of ketamine and xylazine (87.5 and 12.5 mg/kg i.p., respectively).

Seven days after ovariectomy and sham surgery, the administration of rosmarinic acid, sinapic acid or estradiol to rats started. Phenolic acids and estradiol were administered orally (p.o.) using an intragastric tube once a day for 4 weeks in the form of water solution or suspension, both prepared with the addition of Tween 20 (maximum 1 µL of Tween 20 per 1 mL of water). The sham-operated and ovariectomized control rats were vehicle treated with water containing the same amount of Tween 20, in the same volume of 2 mL/kg p.o. To adjust the volume of administered substances, the rats were weighed twice a week. On the next day after the last administration of drugs and overnight fasting, the animals were sacrificed under general anesthesia (ketamine and xylazine) by cardiac exsanguinations and then the uterus, thymus, liver, right kidney, and eyeballs were removed. Serum obtained from the blood was used to determine biochemical parameters and parameters of oxidative stress, which were previously presented together with body mass and masses of selected organs [20,21]. The lenses were isolated from the eyes, weighed, and homogenized in a glass homogenizer in ice-cold 10 mM phosphate-buffered saline pH 7.4, giving 10% homogenates (w/v). Part of the total homogenate was frozen, and then used to determine TBARS (thiobarbituric acid reactive substances) and amyloid β_{1-42}. The rest was centrifuged at $10,000 \times g$ at 4 °C for 15 min. The supernatant was frozen and used to determine the remaining biochemical parameters. All spectrophotometric measurements were carried out with the use of a Tecan Infinite M200 PRO plate reader with Magellan 7.2 software (Tecan Austria, Grödig, Austria).

2.3. Determination of Soluble Protein in the Lenses

Determination of soluble protein was conducted according to Lowry's method [26]. BSA was used to prepare the calibration curve, and the protein content was expressed in milligram per gram of the lens.

2.4. Determination of Superoxide Dismutase and Catalase Activities and Oxidative Damage Products Content in the Lenses

To determine the activities of the following antioxidant enzymes: superoxide dismutase (SOD) and catalase (CAT), Cayman kits (Cayman Chemical MI, USA) were used. The activities of SOD and CAT were expressed in U or nanomole/min, respectively, per milligram of protein.

The method of Ohkawa et al. [27] was used to determine the content of TBARS (thiobarbituric acid reactive substances) in the total homogenate of the lenses. This method is based on the reaction between lipid peroxidation products and thiobarbituric acid. TBARS content is expressed in nanomole per gram of the lens. The intensity of the obtained color was determined spectrophotometrically at the wavelength of 535 nm. To establish a standard curve, 1,1,3,3-tetraethoxypropane (Sigma-Aldrich, St. Louis, MO, USA) was used.

The concentration of advanced oxidation protein products (AOPP) in the lens homogenate was determined using spectrophotometric method described by Witko-Sarsat et al. [28]. The calibration curve was established using chloramine T (Sigma-Aldrich, St. Louis, MO, USA), while the absorbance was measured at the wavelength of 340 nm. The content of AOPP was expressed in nanomole chloramine T equivalents per milligram of protein.

2.5. Determination of Glutathione-related Enzymes Activities in the Lenses

Glutathione peroxidase (GPx) and glutathione reductase (GR) activities were determined using Cayman kits. The activities of GPx and GR were expressed in nanomole of reduced nicotinamide adenine dinucleotide phosphate (NADPH) oxidized during 1 min per milligram of protein.

Activity of glucose-6-phosphate dehydrogenase (G6PD) was measured with Pointe Sci. Kit (Pointe Scientific, Canton, MI, USA), while to determine the activity of γ-glutamyl transpeptidase (GGT), the BioSystems kit was used (Costa Brava, Barcelona, Spain). The activity of G6PD was expressed in nanomole of $NADP^+$ reduced during 1 min per milligram of protein and the activity of GGT was expressed in nanomole of 3-carboxy-4-nitroaniline formed during 1 min per milligram of protein.

2.6. Determination of Glutathione in the Lenses

The concentration of total glutathione (TotGSH) and the concentration of oxidized glutathione (GSSG) in the lens homogenate was determined by Cayman kit (Cayman Chemical MI, USA). The concentration of reduced glutathione (GSH) was calculated according to the formula: GSH = TotGSH - 2×GSSG (nmol/mL), and then the GSH/GSSG ratio was determined. The content of GSH and GSSG in the lenses is expressed in nanomole per milligram of protein.

2.7. Determination of Amyloid β$_{1-42}$ Content in the Lenses

ELISA kit (Bioassay Technology Laboratory, Shanghai, Yangpu, China) was used to determine the content of amyloid β$_{1-42}$. Following the manufacturer's instructions, total homogenates were centrifuged at 2500 RPM for 20 min, and amyloid β$_{1-42}$ was determined in the obtained supernatants. The content of amyloid β$_{1-42}$ was expressed in nanogram per gram of the lens.

2.8. Statistical Analysis

The results are presented as the arithmetic mean ± SEM. One-way ANOVA followed by Duncan's post-hoc test were applied to assess statistical significance of the results (Statistica 12 software, StatSoft Polska, Kraków, Poland). The results were assumed statistically significant if $p \le 0.05$.

3. Results

3.1. Effect of Rosmarinic Acid and Sinapic Acid on the Lens Mass and Lens Soluble Protein Content

The average mass of the lens, as well as the soluble protein content in the lenses of the ovariectomized control rats, did not change statistically as compared to the lenses in the sham-operated rats. The administration of rosmarinic acid or sinapic acid in both doses did not lead to any changes in the average mass of the lens or in the content of soluble protein of the lenses compared to ovariectomized control rats. Similarly, administration of estradiol to the ovariectomized rats did not cause any changes in these parameters (Table 1).

Table 1. Effects of rosmarinic acid and sinapic acid on the average lens mass and lens soluble protein content in ovariectomized rats.

Parameter/Group	SHAM	OVX	OVX + ESTR	OVX + RA10	OVX + RA50	OVX + SA5	OVX + SA25
Average mass of the lens (g)	0.059 ± 0.002	0.058 ± 0.003	0.055 ± 0.002	0.055 ± 0.001	0.056 ± 0.002	0.056 ± 0.002	0.055 ± 0.001
Soluble protein (mg/g of the lens)	280.6 ± 6.8	302.5 ± 7.8	295.8 ± 5.8	290.4 ± 3.2	291.7 ± 7.2	291.9 ± 5.2	286.7 ± 4.2

Rosmarinic acid at doses of 10 mg/kg (OVX+RA10) and 50 mg/kg (OVX+RA50), sinapic acid at doses of 5 mg/kg (OVX+SA5) and 25 mg/kg (OVX+SA25) or estradiol at a dose 0.2 mg/kg (OVX+ESTR) were administered orally to ovariectomized rats, once daily for 28 days. SHAM: sham-operated control rats; OVX: ovariectomized control rats. Results are presented as the mean ± SEM. No statistically significant differences in results for both parameters were demonstrated by ANOVA.

3.2. Effect of Rosmarinic Acid and Sinapic Acid on Superoxide Dismutase and Catalase Activities and on Oxidative Damage Products Content in the Lenses

In the lenses of the ovariectomized rats, no significant changes in the SOD and CAT activities were observed compared to the sham-operated rats. The administration of estradiol and phenolic acids did not cause any significant changes in the activities of these enzymes when compared to the ovariectomized control rats (Table 2). Estrogen deficiency in the ovariectomized rats did not affect the content of AOPP and TBARS in the lenses as compared to the sham-operated rats. The use of rosmarinic acid at doses of 10 and 50 mg/kg and sinapic acid at doses of 5 and 25 mg/kg p.o. led to a decrease of the AOPP content in the lenses in comparison to the ovariectomized control rats, whereas estradiol did not exert such an effect. The administration of estradiol and phenolic acids did not significantly change the content of TBARS in the lenses as compared to the ovariectomized rats (Figure 1).

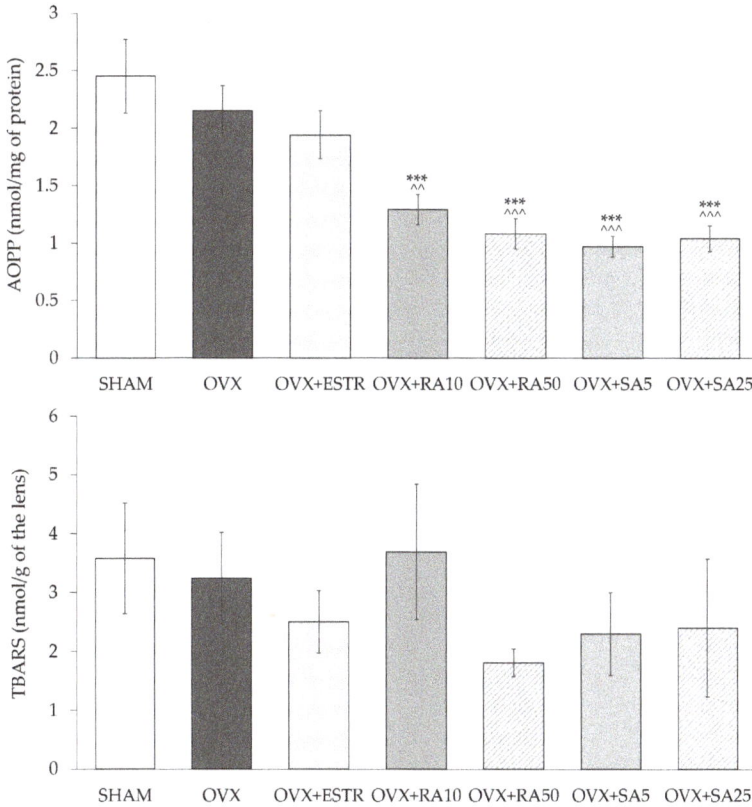

Figure 1. Effect of rosmarinic acid and sinapic acid on the AOPP and TBARS content in the lenses of ovariectomized rats. Rosmarinic acid at doses of 10 mg/kg (OVX + RA10) and 50 mg/kg (OVX + RA50), sinapic acid at doses of 5 mg/kg (OVX + SA5) and 25 mg/kg (OVX + SA25) or estradiol at a dose 0.2 mg/kg (OVX + ESTR) were administered orally to ovariectomized rats, once daily for 28 days. SHAM: sham-operated control rats; OVX: ovariectomized control rats; TBARS: thiobarbituric acid reactive substances; AOPP: advanced oxidation protein products. Results are presented as the mean ± SEM. One-way ANOVA followed by Duncan's test were used for evaluation of the significance of the results. *** $p < 0.001$: significant differences with regard to the SHAM control rats. ^^ $p < 0.01$, ^^^ $p < 0.001$—significant differences with regard to the OVX control rats. No statistically significant differences in results for TBARS were demonstrated by ANOVA.

Table 2. Effect of rosmarinic acid and sinapic acid on the superoxide dismutase (SOD) and catalase (CAT) activities in the lenses of ovariectomized rats.

Parameter/Group	SHAM	OVX	OVX + ESTR	OVX + RA10	OVX + RA50	OVX + SA5	OVX + SA25
SOD (U/mg of protein)	0.194 ± 0.017	0.156 ± 0.003	0.170 ± 0.007	0.164 ± 0.005	0.174 ± 0.006	0.171 ± 0.012	0.167 ± 0.002
CAT (nmol/min/mg of protein)	0.085 ± 0.012	0.033 ± 0.009	0.052 ± 0.014	0.075 ± 0.014	0.076 ± 0.020	0.060 ± 0.018	0.063 ± 0.010

Rosmarinic acid at doses of 10 mg/kg (OVX + RA10) and 50 mg/kg (OVX + RA50), sinapic acid at doses of 5 mg/kg (OVX + SA5) and 25 mg/kg (OVX + SA25) or estradiol at a dose 0.2 mg/kg (OVX + ESTR) were administered orally to ovariectomized rats, once daily for 28 days. SHAM: sham-operated control rats; OVX: ovariectomized control rats; SOD: superoxide dismutase (1 U of SOD determines the amount of enzyme required to exhibit 50% dismutation of the superoxide radical); CAT: catalase. Results are presented as the mean ± SEM. No statistically significant differences in results for SOD and CAT were demonstrated by ANOVA.

3.3. Effect of Rosmarinic Acid and Sinapic Acid on Glutathione-Related Enzymes Activities in the Lenses

The GR, G6PD, and GGT activities in the lenses were decreased in the ovariectomized control rats in a statistically significant manner, whereas the GPx activity showed no statistically significant difference in comparison to the sham-operated control rats. The administration of estradiol to the estrogen-deficient rats did not affect the activities of the examined glutathione-related enzymes. A statistically significant increase in the GR activity was observed after the administration of rosmarinic acid at doses of 10 and 50 mg/kg and sinapic acid at doses of 5 and 25 mg/kg. After administration rosmarinic acid at 50 mg/kg, there was a tendency to increase ($p = 0.058$) in the G6PD activity, whereas the administration of 25 mg/kg of sinapic acid significantly increased the activity of this enzyme when compared to the ovariectomized control rats. Rosmarinic acid and sinapic acid had no effect on the activities of GPx and GGT (Table 3).

Table 3. Effects of rosmarinic acid and sinapic acid on the glutathione-related enzymes activities in the lenses of ovariectomized rats.

Parameter/Group	SHAM	OVX	OVX + ESTR	OVX + RA10	OVX + RA50	OVX + SA5	OVX + SA25
GPx (nmol/min/mg of protein)	2.31 ± 0.08	2.15 ± 0.08	2.04 ± 0.07	2.26 ± 0.11	2.21 ± 0.06	2.24 ± 0.08	2.28 ± 0.05
GR (nmol/min/mg of protein)	0.367 ± 0.047	0.220 ± 0.036 **	0.291 ± 0.034	0.357 ± 0.027 ^^	0.352 ± 0.026 ^^	0.323 ± 0.026 ^	0.346 ± 0.021 ^
G6PD (nmol/min/mg of protein)	1.99 ± 0.10	1.30 ± 0.14*	1.39 ± 0.17 *	1.52 ± 0.19	1.83 ± 0.16	1.47 ± 0.13	2.00 ± 0.27 ^
GGT (nmol/min/mg of protein)	0.039 ± 0.005	0.024 ± 0.004*	0.028 ± 0.003	0.024 ± 0.004 *	0.020 ± 0.002 **	0.025 ± 0.005 *	0.031 ± 0.002

Rosmarinic acid at doses of 10 mg/kg (OVX + RA10) and 50 mg/kg (OVX + RA50), sinapic acid at doses of 5 mg/kg (OVX + SA5) and 25 mg/kg (OVX + SA25) or estradiol at a dose 0.2 mg/kg (OVX + ESTR) were administered orally to ovariectomized rats, once daily for 28 days. SHAM: sham-operated control rats; OVX: ovariectomized control rats; GPx: glutathione peroxidase; GR: glutathione reductase, G6PD: glucose-6-phosphate dehydrogenase. Results are presented as the mean ± SEM. One-way ANOVA followed by Duncan's test was used for evaluation of the significance of the results. * $p \le 0.05$, ** $p < 0.01$: significant differences with regard to the SHAM control rats. ^ $p \le 0.05$, ^^ $p < 0.01$: significant differences with regard to the OVX control rats. No statistically significant differences in results for GPx were demonstrated by ANOVA.

3.4. Effect of Rosmarinic Acid and Sinapic Acid on Glutathione Content in the Lenses

A statistically significant decrease in the content of the reduced glutathione (GSH) in the lenses was observed while the content of the oxidized glutathione (GSSG) and the GSH/GSSG ratio did not change in the ovariectomized control rats compared to the sham-operated rats. The administration of estradiol did not change the content of GSH and GSSG or GSH/GSSG ratio, whereas the administration of phenolic acids (rosmarinic acid at 10 and 50 mg/kg and sinapic acid at dose 5 and 25 mg/kg) resulted in the statistically significant increase in the GSH content in the lenses, without impact on the GSSG content when compared to the ovariectomized control rats. The use of rosmarinic acid in the estrogen-deficient rats at both doses did not affect the GSH/GSSG ratio, while the use of sinapic acid at both doses caused a significant increase in the GSH/GSSG ratio, compared to the ovariectomized control rats (Figure 2).

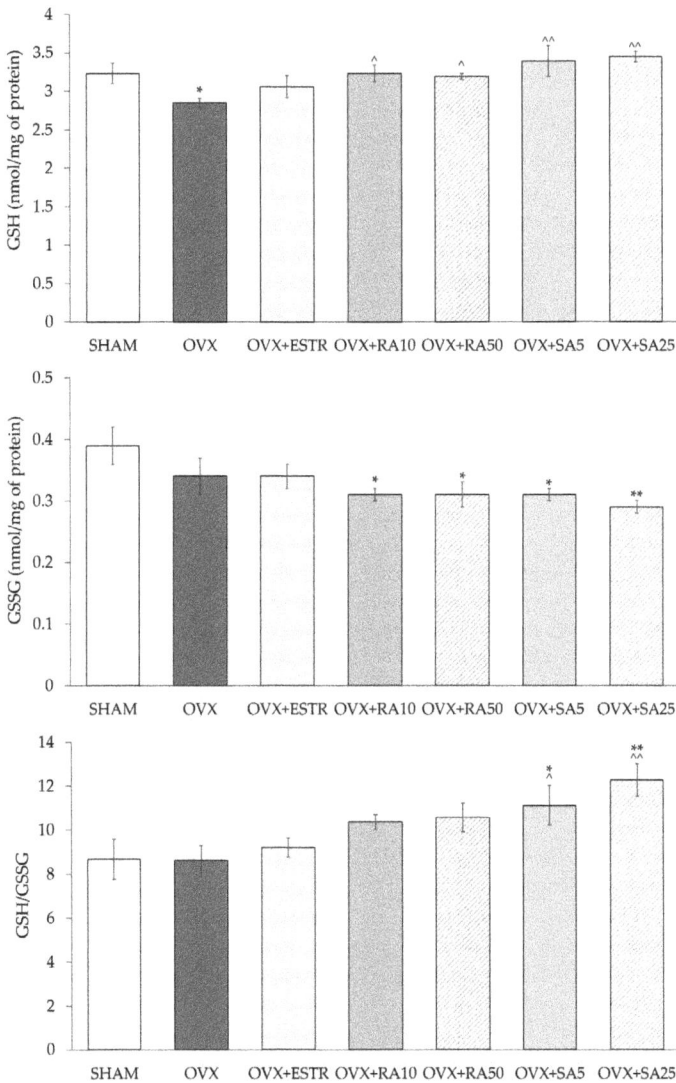

Figure 2. Effect of rosmarinic acid and sinapic acid on the GSH and GSSG content and on the GSH/GSSG ratio in the lenses of ovariectomized rats. Rosmarinic acid at doses of 10 mg/kg (OVX + RA10) and 50 mg/kg (OVX + RA50), sinapic acid at doses of 5 mg/kg (OVX + SA5) and 25 mg/kg (OVX + SA25) or estradiol at a dose 0.2 mg/kg (OVX+ESTR) were administered orally to ovariectomized rats, once daily for 28 days. SHAM: sham-operated control rats; OVX: ovariectomized control rats; GSH: reduced glutathione; GSSG: oxidized glutathione. Results are presented as the mean ± SEM. One-way ANOVA followed by Duncan's test were used for evaluation of the significance of the results. * $p \leq 0.05$, ** $p < 0.01$: significant differences with regard to the SHAM control rats. ^ $p \leq 0.05$, ^^ $p < 0.01$: significant differences with regard to the OVX control rats.

3.5. Effect of Rosmarinic Acid and Sinapic Acid on Amyloid β_{1-42} Content in the Lenses

The content of amyloid β_{1-42} in the lenses of the ovariectomized rats significantly decreased as compared to the sham-operated rats. The administration of rosmarinic acid and sinapic acid did not result in any statistically significant changes in the content of amyloid β_{1-42}. Likewise, treatment with estradiol did not affect the amyloid β_{1-42} content in the lenses of ovariectomized rats (Figure 3).

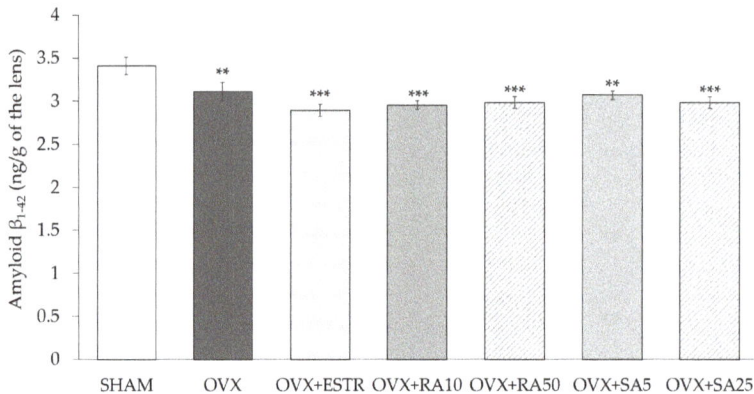

Figure 3. Effect of rosmarinic acid and sinapic acid on the amyloid β_{1-42} content and on the GSH/GSSG ratio in the lenses of ovariectomized rats. Rosmarinic acid at doses of 10 mg/kg (OVX + RA10) and 50 mg/kg (OVX + RA50), sinapic acid at doses of 5 mg/kg (OVX + SA5) and 25 mg/kg (OVX + SA25) or estradiol at a dose 0.2 mg/kg (OVX+ESTR) were administered orally to ovariectomized rats, once daily for 28 days. SHAM: sham-operated control rats; OVX: ovariectomized control rats. Results are presented as the mean ± SEM. One-way ANOVA followed by Duncan's test were used for evaluation of the significance of the results. ** $p < 0.01$, *** $p < 0.001$: significant differences with regard to the SHAM control rats.

4. Discussion

The lens is a transparent structure located in the front part of the eye. It is the most important part of the optical system of the eye, which projects a reduced, inverted, and exceptionally clear image on the retina. The lack of cell nuclei and other light-scattering organelles contributes to the transparency of the lens. Light scattering is also minimized due to the close apposition of the lens fiber cells [29]. It has recently been pointed out that the lens is not a passive optical component, but an active tissue (which may, for example, protect the anterior segment of the eye from oxygen or its metabolites, as well as can release GSH and adenosine triphosphate (ATP) to other eye tissues), the removal of which can contribute to the development of other eye diseases [30]. The artificial lens is not capable of performing metabolic functions but only serves as an optical element. Therefore, although contemporary cataract surgery is safe, it is still recommended to avoid removing lenses and to put more emphasis on preventing cataract formation.

Scientific reports based on observational studies indicate that a well-balanced diet rich in vegetables and fruits, containing about 150 g of protein, high intake of vitamin C, vitamin E, and a reduced amount of simple sugars, as well as supplementation with other vitamins or carotenoids, may contribute to delaying the cataract progression [31]. According to the in vitro and in vivo experiments, dietary components derived from medicinal plants, such as flavonoids, phenolic acids, terpenes, carotenoids or phytosterols, seem to be effective in preventing opacity of the lenses [22,23,31]. They can act through various mechanisms, of which the most important is anti-oxidative and anti-glycating activities [23].

The aim of the presented study was to investigate the effect of dietary components: rosmarinic acid and sinapic acid on antioxidative abilities parameters (GSH and enzymes associated with GSH and SOD, CAT), as well as products of oxidative damage from lipids and proteins (TBARS and AOPP, respectively) in the lenses obtained from the rats 5 weeks after ovariectomy. Rosmarinic acid was administered to animals at doses of 10 and 50 mg/kg, while sinapic acid at doses of 5 and 25 mg/kg. As we discussed before [20,21] the doses of phenolic acids have been selected so that the smaller ones (10 mg/kg rosmarinic acid and 5 mg/kg sinapic acid) correspond to the amount that can be consumed in the diet. Five times higher doses were used to determine whether they exert a stronger therapeutic effect than achievable dietary doses. The doses used in this experiment could be considered safe as acute toxicity tests conducted for rosmarinic acid and sinapic acid revealed that both the acids are non-toxic even at a dose of 2000 mg/kg when administered orally to rats [32,33].

Estrogen deficiency in ovariectomized animals was manifested by a decrease in estradiol and progesterone concentration in the serum, a decrease in uterus mass, and enhanced body mass gain, as well as the changes in parameters related to glucose and lipid metabolism [20,21]. We have previously reported that the use of estradiol, sinapic acid [21], and rosmarinic acid [20] in these rats had a positive effect on the serum parameters associated with glucose and lipid metabolism and also increased serum GSH concentration.

The increase in GSH content in the lenses, which was reduced by estrogen deficiency, was also observed in the present study in estrogen-deficient rats administered with rosmarinic acid and sinapic acid. GSH plays an important role in maintaining the lens transparency, and simultaneously in the regulation of the lens redox state [34–36]. GSH may form reversible disulfide bonds with protein thiol groups. Therefore, it protects proteins from permanent oxidation and, in result, from their aggregation and loss of function [37]. Moreover, GSH is a cofactor for numerous enzymes, such as thioltransferase (TT-ase), which uses GSH to dethiolate protein-thiol disulfides [36,38]. GSH can also be used by other antioxidative enzymes, such as glutathione peroxidase (GPx), to neutralize H_2O_2 [36] or glutathione S transferases (GST). GSTs of many classes (such as pi or mu) use GSH as a substrate to neutralize electrophilic xenobiotics [39], regulate pro- and antiapoptotic pathways in many tissues [40–42] and polymorphism in genes encoding GSTs may be an important risk factor in cataractogenesis [43–45]. GSH synthesis takes place in the lens epithelium and the outer part of the lens cortex. The required amino acids are supplied from the aqueous humor and from the decomposition of GSH in the gamma-glutamyl cycle in which the γ-glutamyl transferase (GGT) plays a major role. The role of GGT is to break down extracellular GSH, GSSG, and S-glutathione conjugates, thus, providing cells with amino acids which are necessary for intracellular GSH synthesis [34]. GSH, as a complete tripeptide, may also be transferred to the lens from aqueous humor. The content of GSH in the lenses decreases with age. It is believed that this is a result of, for example, a decreased glutamate cysteine ligase (GCL) activity, and, hence, a reduction in the GSH de novo synthesis, but also weakening of the GSH regeneration system from the oxidized form, which includes GR and G6PD [34].

In this study, the reduced content of GSH in the lenses of the ovariectomized control rats compared to the lenses of the sham-operated rats was observed simultaneously with decreased GR, G6PD, and GGT activities. GR is necessary to reduce GSSG using NADPH, while G6PD catalyzes the first phase of the pentose phosphate pathway, during which glucose-6 phosphate is transformed into 6-phosphoglucono-δ lactone, and $NADP^+$ is reduced to NADPH [34,46]. NADPH is essential for the activity of many enzymes, including GR. Lowering of the GSH content may, therefore, be a result of a weakened regeneration from GSSG, but there is also a possibility that the decrease of GSH content may be an effect of the reduced synthesis resulting from decreased GCL activity. Based on the results obtained in this study, the determination of the mechanisms responsible for lowering the GSH content in estrogen-deficient rats is not entirely possible. It seems to be quite surprising that even though GR, G6PD, and GGT activities are lowered in the lenses of the rats which underwent ovariectomy, the GSSG content is not elevated and GSH/GSSG ratio remained unchanged when compared to the sham-operated animals. In the report of Umapathy et al. [47], the authors suggest that the excess of GSSG is exported from the lens to the neighboring structures as an early response to oxidative stress to minimize the possible damage and maintain lenticular GSH redox state [47]. A decreased GSH content in the lenses is well documented in various rat cataract models [48–52]. However, there are only few reports describing the GSH content or antioxidative abilities in the lenses of the laboratory animals with estrogen deficiency [53,54]. A study conducted on ovariectomized mice showed that GSH content in the lenses did not change when compared with control animals [53]. Acer et al. [54] examined total non-enzymatic antioxidant content in the lenses of ovariectomized rats conducting total antioxidant capacity (TAC) test and noted that TAC in the lenses of estrogen-deficient rats was significantly lower than in the lenses of control rats [54]. The results for GSH content in the lenses obtained in our study overlap with these for TAC presented by Acer et al. [54], possibly due to the fact that GSH is a predominant non-enzymatic antioxidant in the lenses [47].

The increase in the GSH content in the lenses of ovariectomized rats treated with rosmarinic acid and sinapic acid in both doses was accompanied by an increase in the GR activity. There was also a tendency to increase and significant increase of G6PG activity after the use of 50 mg/kg rosmarinic acid and 25 mg/kg sinapic acid, respectively. No effect on GPx and GGT activities was noted after administration of both phenolic acids regardless of the used dose. Therefore, it seems that the increase in the GSH content in the lenses after the use of phenolic acids may partly result from increased regeneration from GSSG. It is also probable that the increase in the GSH content is due to its increased synthesis, as rosmarinic acid is described to up-regulate the catalytic subunits of GCL in hepatic stellate cells [55]. There are reports indicating that the use rosmarinic acid [56,57] and sinapic acid [58], in different experimental rodent models, led to the increased content of GSH in various tissues and organs, such as kidneys and liver [56–58]. An important indicator of cellular redox status as well as for the redox state in tissues is the GSH/GSSG ratio [59]. In the present study, there was a statistically significant increase in this ratio in the lenses of the estrogen-deficient rats after treatment with both doses of sinapic acid.

A period of 35 days after ovariectomy in rats corresponds to approximately 3.3 years in postmenopausal women [60]. In the early postmenopausal period, some changes in the organism are not very pronounced. This may explain why in the present study there were no statistically significant changes in the activities of antioxidant enzymes or the content of oxidative damage parameters, such as TBARS or AOPP. Estradiol and phenolic acids in the present study did not affect the activities of SOD, CAT, and the content of TBARS, but both the rosmarinic acid and sinapic acid reduced the AOPP content in the lenses. Although the content of AOPP did not increase in ovariectomized rats as compared to the sham-operated rats, the reduction in the AOPP content seems to be a favorable change, since AOPP may promote ROS formation via the receptor for advanced glycation end products (RAGE)-dependent pathway [61]. Reduction in the AOPP content was also observed as a result of using the plant-derived antioxidants, such as diosmin, naringenin or resveratrol, in the lenses of rats with experimentally induced diabetes [62–64].

One of the parameters which depict changes occurring in the lens during cataract development is the content of amyloid β_{1-42}. In rats, 5 weeks after performing ovariectomy, a reduction in the amyloid β_{1-42} content was observed, which is consistent with the results indicating that its expression is reduced in the early and middle stages of age-related cataract development in human lens epithelial cells [65]. In addition, in Upjohn Pharmaceutical Limited (UPL) rats (a dominant hereditary cataract model derived from Sprague-Dawley rats), there was no increase in the content of amyloid β_{1-42} in the lenses until complete opacity occurred [66]. There was no effect of rosmarinic acid and sinapic acid on the amyloid β_{1-42} content in the lenses.

In our previous study, we observed that even though estradiol administered orally to ovariectomized rats caused an increase in the uterus mass and a decrease in the thymus mass (estrogenic activity), it did not increase the estradiol level in the serum [21]. In the presented study it was noted that oral administration of estradiol did not affect oxidative stress-related parameters in the lenses of ovariectomized rats, including the GSH content. Similar findings were made in another study, in which estradiol administered to estrogen-deficient rats revealed no effect on apoptosis rate in the lens epithelial cells, which was increased in the ovariectomized control rats [67]. Unlike the treatment with estradiol, administration of sinapic acid at both doses caused a significant increase in the estradiol concentration in the serum of the ovariectomized rats [21]. What is more, after treatment with rosmarinic acid at the higher dose (50 mg/kg) there was a trend to an increase in the estradiol concentration in the serum of ovariectomized rats [20]. Therefore, although, it appears that the action of rosmarinic and sinapic acids on GSH content in the lenses of ovariectomized rats results rather from their anti-oxidative activity, there is also a possibility that the mechanism of their action could be somehow estrogen-dependent.

Our research has some limitations. First of all, further studies are required to determine unequivocally whether mechanism underlying changes in GSH content in the lenses after administration of the phenolic acids is connected with direct effects of these compounds on anti-oxidative status or rather with their phytoestrogenic activity. Moreover, the present study did not assess the effect of rosmarinic acid and sinapic acid on the cytoplasmatic expression of antioxidant enzymes. To affirm our results, further molecular studies using Western blot or real-time PCR could be helpful. Since there is a possibility of simultaneous consumption of rosmarinic acid and sinapic acid, it would also be interesting to investigate the effects of these phenolic acids combined.

5. Conclusions

Rosmarinic and sinapic acids contributed to the increase in GSH content in the lenses of rats in the early phase of estrogen deficiency. Due to the important role of GSH in maintaining the transparency of the lenses, it seems that these phenolic acids may exert a beneficial effect on the redox status in the eye lenses of ovariectomized rats, and, thus, may be a supporting factors in the prevention of cataract formation.

Author Contributions: Conceptualization, M.Z.; methodology, M.Z. and W.W.; formal analysis, M.Z., I.K.-S., W.W. and S.D.; investigation, M.Z and W.W.; writing—original draft preparation, M.Z.; writing—review and editing, M.Z., I.K.-S., W.W. and S.D.; visualization, W.W.; supervision, I.K.-S.

Funding: This research was funded by the Medical University of Silesia grant number KNW-1-087/N/8/0.

Acknowledgments: The authors thank Ms. Anna Bońka, MSc for her technical assistance.

Conflicts of Interest: The authors declare no conflict of interest.

References

1. Klein, B.E.K.; Klein, R.; Linton, K.L.P. Prevalence of age-related lens opacities in a population. The Beaver Dam Eye Study. *Ophthalmology* **1992**, *99*, 546–552. [CrossRef]
2. Zhang, J.S.; Xu, L.; Wang, Y.X.; You, Q.S.; Da Wang, J.; Jonas, J.B. Five-year incidence of age-related cataract and cataract surgery in the adult population of Greater Beijing. The Beijing Eye Study. *Ophthalmology* **2011**, *118*, 711–718. [CrossRef]
3. Lai, K.; Cui, J.; Ni, S.; Zhang, Y.; He, J.; Yao, K. The effects of postmenopausal hormone use on cataract: A meta-analysis. *PLoS ONE* **2013**, *8*, e78647. [CrossRef]
4. Celojevic, D.; Petersen, A.; Karlsson, J.-O.; Behndig, A.; Zetterberg, M. Effects of 17β-estradiol on proliferation, cell viability and intracellular redox status in native human lens epithelial cells. *Mol. Vis.* **2011**, *17*, 1987–1996. [PubMed]
5. Gottipati, S.; Cammarata, P.R. Mitochondrial superoxide dismutase activation with 17 β-estradiol-treated human lens epithelial cells. *Mol. Vis.* **2008**, *14*, 898–905. [PubMed]
6. Dynlacht, J.R.; Valluri, S.; Lopez, J.; Greer, F.; DesRosiers, C.; Caperell-Grant, A.; Mendonca, M.S.; Bigsby, R.M. Estrogen protects against radiation-induced cataractogenesis. *Radiat. Res.* **2008**, *170*, 758–764. [CrossRef] [PubMed]
7. Dynlacht, J.R.; Tyree, C.; Valluri, S.; Desrosiers, C.; Mendonca, M.S.; Timmerman, R.; Bigsby, R.M.; Mendonca, M.S.; Desrosiers, C.; Tyree, C.; et al. Effect of estrogen on radiation-induced cataractogenesis. *Radiat. Res.* **2006**, *165*, 9–15. [CrossRef] [PubMed]

8. Bigsby, R.M.; Valluri, S.; Lopez, J.; Mendonca, M.S.; Caperell-Grant, A.; Desrosiers, C.; Dynlacht, J.R. Ovarian hormone modulation of radiation-induced cataractogenesis: Dose-response studies. *Invest. Ophthalmol. Vis. Sci.* **2009**, *50*, 3304–3310. [CrossRef] [PubMed]

9. Zetterberg, M.; Celojevic, D. Gender and cataract—The role of estrogen. *Curr. Eye Res.* **2015**, *40*, 176–190. [CrossRef]

10. Cagnacci, A.; Cannoletta, M.; Palma, F.; Bellafronte, M.; Romani, C.; Palmieri, B. Relation between oxidative stress and climacteric symptoms in early postmenopausal women. *Climacteric* **2015**, *18*, 631–636. [CrossRef]

11. Manolagas, S.C. From estrogen-centric to aging and oxidative stress: A revised perspective of the pathogenesis of osteoporosis. *Endocr. Rev.* **2010**, *31*, 266–300. [CrossRef]

12. Cervellati, C.; Bergamini, C.M. Oxidative damage and the pathogenesis of menopause related disturbances and diseases. *Clin. Chem. Lab. Med.* **2016**, *54*, 739–753. [CrossRef] [PubMed]

13. Liu, Y.-C.; Wilkins, M.; Kim, T.; Malyugin, B.; Mehta, J.S. Cataracts. *Lancet* **2017**, *390*, 600–612. [CrossRef]

14. Nita, M.B.; Grzybowski, A. The role of the reactive oxygen species and oxidative stress in the pathomechanism of the age-related ocular diseases and other pathologies of the anterior and posterior eye segments in adults. *Oxid. Med. Cell. Longev.* **2016**, *2016*, 3164734. [CrossRef]

15. Babizhayev, M.A.; Yegorov, Y.E. Reactive oxygen species and the aging eye: Specific role of metabolically active mitochondria in maintaining lens function and in the initiation of the oxidation-induced maturity onset cataract—A novel platform of mitochondria-targeted antioxidants with broad therapeutic potential for redox regulation and detoxification of oxidants in eye diseases. *Am. J. Ther.* **2016**, *23*, 98–117.

16. Miquel, J.; Ramírez-Boscá, A.; Ramírez-Bosca, J.V.; Alperi, J.D. Menopause: A review on the role of oxygen stress and favorable effects of dietary antioxidants. *Arch. Gerontol. Geriatr.* **2006**, *42*, 289–306. [CrossRef]

17. Zhang, H.; Tsao, R. Dietary polyphenols, oxidative stress and antioxidant and anti-inflammatory effects. *Curr. Opin. Food Sci.* **2016**, *8*, 33–42. [CrossRef]

18. Amoah, S.K.S.; Sandjo, L.P.; Kratz, J.M.; Biavatti, M.W. Rosmarinic acid—Pharmaceutical and clinical aspects. *Planta Med.* **2016**, *82*, 388–406. [CrossRef] [PubMed]

19. Nićiforović, N.; Abramovič, H. Sinapic acid and its derivatives: Natural sources and bioactivity. *Compr. Rev. Food Sci. Food Saf.* **2014**, *13*, 34–51. [CrossRef]

20. Zych, M.; Kaczmarczyk-Sedlak, I.; Wojnar, W.; Folwarczna, J. Effect of rosmarinic acid on the serum parameters of glucose and lipid metabolism and oxidative stress in estrogen-deficient rats. *Nutrients* **2019**, *11*, 267. [CrossRef]

21. Zych, M.; Kaczmarczyk-Sedlak, I.; Wojnar, W.; Folwarczna, J. The effects of sinapic acid on the development of metabolic disorders induced by estrogen deficiency in rats. *Oxid. Med. Cell. Longev.* **2018**, *2018*, 9274246. [CrossRef]

22. Lim, V.; Schneider, E.; Wu, H.; Pang, I.-H. Cataract preventive role of isolated phytoconstituents: Findings from a decade of research. *Nutrients* **2018**, *10*, 1580. [CrossRef]

23. Kaur, A.; Gupta, V.; Francis, A.; Ahmad, M.; Bansal, P. Nutraceuticals in prevention of cataract—An evidence based approach. *Saudi J. Ophthalmol.* **2017**, *31*, 30–37. [CrossRef]

24. Sunkireddy, P.; Jha, S.N.; Kanwar, J.R.; Yadav, S.C. Natural antioxidant biomolecules promises future nanomedicine based therapy for cataract. *Colloids Surf. B Biointerfaces* **2013**, *112*, 554–562. [CrossRef]

25. Chemerovski-Glikman, M.; Mimouni, M.; Dagan, Y.; Haj, E.; Vainer, I.; Allon, R.; Blumenthal, E.Z.; Adler-abramovich, L.; Segal, D.; Gazit, E. Rosmarinic acid restores complete transparency of sonicated human cataract ex vivo and delays cataract formation in vivo. *Sci. Rep.* **2018**, *8*, 9341. [CrossRef]

26. Lowry, O.H.; Rosebrough, N.J.; Farr, A.L.; Randall, R.J. Protein measurement with the Folin phenol reagent. *J. Biol. Chem.* **1951**, *193*, 265–275.

27. Ohkawa, H.; Ohishi, N.; Yagi, K. Assay for lipid peroxides in animal tissues thiobarbituric acid reaction. *Anal. Biochem.* **1979**, *95*, 351–358. [CrossRef]

28. Witko-Sarsat, V.; Friedlander, M.; Capeillère-Blandin, C.; Nguyen-Khoa, T.; Nguyen, A.T.; Zingraff, J.; Jungers, P.; Descamps-Latscha, B. Advanced oxidation protein products as a novel marker of oxidative stress in uremia. *Kidney Int.* **1996**, *49*, 1304–1313. [CrossRef]

29. Bassnett, S.; Shi, Y.; Vrensen, G.F.J.M. Biological glass: Structural determinants of eye lens transparency. *Philos. Trans. R. Soc. B* **2011**, *366*, 1250–1264. [CrossRef]

30. Lim, J.C.; Umapathy, A.; Grey, A.C.; Vaghefi, E.; Donaldson, P.J. Novel roles for the lens in preserving overall ocular health. *Exp. Eye Res.* **2017**, *156*, 117–123. [CrossRef]

31. Sella, R.; Afshari, N.A. Nutritional effect on age-related cataract formation and progression. *Curr. Opin. Ophthalmol.* **2019**, *30*, 63–69. [CrossRef] [PubMed]

32. Jayanthy, G.; Subramanian, S. Rosmarinic acid, a polyphenol, ameliorates hyperglycemia by regulating the key enzymes of carbohydrate metabolism in high fat diet—STZ induced experimental diabetes mellitus. *Biomed. Prev. Nutr.* **2014**, *4*, 431–437. [CrossRef]

33. Bais, S.; Kumari, R.; Prashar, Y. Therapeutic effect of sinapic acid in aluminium chloride induced dementia of Alzheimer's type in rats. *J. Acute Dis.* **2017**, *6*, 154–162. [CrossRef]

34. Fan, X.; Monnier, V.M.; Whitson, J. Lens glutathione homeostasis: Discrepancies and gaps in knowledge standing in the way of novel therapeutic approaches. *Exp. Eye Res.* **2017**, *156*, 103–111. [CrossRef]

35. Nye-Wood, M.G.; Spraggins, J.M.; Caprioli, R.M.; Schey, K.L.; Donaldson, P.J.; Grey, A.C. Spatial distributions of glutathione and its endogenous conjugates in normal bovine lens and a model of lens aging. *Exp. Eye Res.* **2017**, *154*, 70–78. [CrossRef]

36. Lou, M.F. Redox regulation in the lens. *Prog. Retin. Eye Res.* **2003**, *22*, 657–682. [CrossRef]

37. Dalle-Donne, I.; Milzani, A.; Gagliano, N.; Colombo, R.; Giustarini, D.; Rossi, R. Molecular mechanisms and potential clinical significance of S-glutathionylation. *Antioxid. Redox Signal.* **2008**, *10*, 445–473. [CrossRef]

38. Xing, K.-Y.; Lou, M.F. Effect of age on the thioltransferase (glutaredoxin) and thioredoxin systems in the human lens. *Investig. Ophthalmol. Vis. Sci.* **2010**, *51*, 6598–6604. [CrossRef]

39. Dasari, S.; Ganjayi, M.S.; Yellanurkonda, P.; Basha, S.; Meriga, B. Role of glutathione S-transferases in detoxification of a polycyclic aromatic hydrocarbon, methylcholanthrene. *Chem. Biol. Interact.* **2018**, *294*, 81–90. [CrossRef]

40. La Russa, D.; Brunelli, E.; Pellegrino, D. Oxidative imbalance and kidney damage in spontaneously hypertensive rats: Activation of extrinsic apoptotic pathways. *Clin. Sci.* **2017**, *131*, 1419–1428. [CrossRef]

41. Simic, T.; Savic-Radojevic, A.; Pljesa-Ercegovac, M.; Matic, M.; Mimic-Oka, J. Glutathione S-transferases in kidney and urinary bladder tumors. *Nat. Rev. Urol.* **2009**, *6*, 281–289. [CrossRef]

42. Chauhan, A.K.; Mittra, N.; Singh, B.K.; Singh, C. Inhibition of glutathione S-transferase-pi triggers c-jun N-terminal kinase-dependent neuronal death in Zn-induced Parkinsonism. *Mol. Cell. Biochem.* **2019**, *452*, 95–104. [CrossRef]

43. Sun, W.; Su, L.; Sheng, Y.; Shen, Y.; Chen, G. Is there association between Glutathione S Transferases polymorphisms and cataract risk: A meta-analysis? *BMC Ophthalmol.* **2015**, *15*, 84. [CrossRef]

44. Qi, R.; Gu, Z.; Zhou, L. The effect of GSTT1, GSTM1 and GSTP1 gene polymorphisms on the susceptibility of age-related cataract in Chinese Han population. *Int. J. Clin. Exp. Med.* **2015**, *8*, 19448–19453.

45. Sireesha, R.; Laxmi, S.G.B.; Mamata, M.; Reddy, P.Y.; Goud, P.U.; Rao, P.V.; Reddy, G.B.; Vishnupriya, S.; Padma, T. Total activity of glutathione-S-transferase (GST) and polymorphisms of GSTM1 and GSTT1 genes conferring risk for the development of age related cataracts. *Exp. Eye Res.* **2012**, *98*, 67–74. [CrossRef]

46. Ganea, E.; Harding, J.J. Glutathione-related enzymes and the eye. *Curr. Eye Res.* **2006**, *31*, 1–11. [CrossRef]

47. Umapathy, A.; Li, B.; Donaldson, P.J.; Lim, J.C. Functional characterisation of glutathione export from the rat lens. *Exp. Eye Res.* **2018**, *166*, 151–159. [CrossRef]

48. Su, S.; Leng, F.; Guan, L.; Zhang, L.; Ge, J.; Wang, C.; Chen, S.; Liu, P. Differential proteomic analyses of cataracts from rat models of type 1 and 2 diabetes. *Investig. Ophthalmol. Vis. Sci.* **2014**, *55*, 7848–7861. [CrossRef]

49. Arnal, E.; Miranda, M.; Almansa, I.; Muriach, M.; Barcia, J.M.; Romero, F.J.; Diaz-Llopis, M.; Bosch-Morell, F. Lutein prevents cataract development and progression in diabetic rats. *Graefe's Arch. Clin. Exp. Ophthalmol.* **2009**, *247*, 115–120. [CrossRef]

50. Kim, J.; Choung, S. *Pinus densiflora* bark extract prevents selenite-induced cataract formation in the lens of Sprague Dawley rat pups. *Mol. Vis.* **2017**, *23*, 638–648.

51. Cao, S.; Gao, M.; Wang, N.; Liu, N.; Du, G.; Lu, J. Prevention of selenite-induced cataratogenesis by *Ginkgo biloba* extract (Egb761) in Wistar rats. *Curr. Eye Res.* **2015**, *40*, 1028–1033. [CrossRef]

52. Ozkol, H.U.; Koyuncu, I.; Tuluce, Y.; Dilsiz, N.; Soral, S.; Ozkol, H. Anthocyanin-rich extract from *Hibiscus sabdariffa* calyx counteracts UVC-caused impairments in rats. *Pharm. Biol.* **2015**, *53*, 1435–1441. [CrossRef]

53. Nazıroğlu, M.; Güler, M.; Küçükayaz, M.; Övey, İ.S.; Özgül, C. Apple cider vinegar supplementation modulates lipid peroxidation and glutathione peroxidase values in lens of ovariectomized mice. *Cell Membr. Free Radic. Res.* **2012**, *3*, 209–214.

54. Acer, S.; Pekel, G.; Küçükatay, V.; Karabulut, A.; Yağcı, R.; Çetin, E.N.; Akyer, Ş.P.; Şahın, B. Oxidative stress of crystalline lens in rat menopausal model. *Arq. Bras. Oftalmol.* **2016**, *79*, 222–225. [CrossRef]

55. Lu, C.; Zou, Y.; Liu, Y.; Niu, Y. Rosmarinic acid counteracts activation of hepatic stellate cells via inhibiting the ROS-dependent MMP-2 activity: Involvement of Nrf2 antioxidant system. *Toxicol. Appl. Pharmacol.* **2017**, *318*, 69–78. [CrossRef]

56. Tavafi, M.; Ahmadvand, H. Effect of rosmarinic acid on inhibition of gentamicin induced nephrotoxicity in rats. *Tissue Cell* **2011**, *43*, 392–397. [CrossRef]

57. Vanithadevi, B.; Anuradha, C.V. Effect of rosmarinic acid on insulin sensitivity, glyoxalase system and oxidative events in liver of fructose-fed mice. *Int. J. Diabetes Metab.* **2008**, *16*, 35–44.

58. Silambarasan, T.; Manivannan, J.; Raja, B.; Chatterjee, S. Prevention of cardiac dysfunction, kidney fibrosis and lipid metabolic alterations in L-NAME hypertensive rats by sinapic acid—Role of HMG-CoA reductase. *Eur. J. Pharmacol.* **2016**, *777*, 113–123. [CrossRef]

59. Enns, G.M.; Cowan, T.M. Glutathione as a redox biomarker in mitochondrial disease—Implications for therapy. *J. Clin. Med.* **2017**, *6*, 50. [CrossRef]

60. Sengupta, P. The laboratory rat: Relating its age with human's. *Int. J. Prev. Med.* **2013**, *4*, 624–630.

61. Guo, Z.J.; Niu, H.X.; Hou, F.F.; Zhang, L.; Fu, N.; Nagai, R.; Lu, X.; Chen, B.H.; Shan, Y.X.; Tian, J.W.; et al. Advanced oxidation protein products activate vascular endothelial cells via a RAGE-mediated signaling pathway. *Antioxid. Redox Signal.* **2008**, *10*, 1699–1712. [CrossRef]

62. Wojnar, W.; Kaczmarczyk-Sedlak, I.; Zych, M. Diosmin ameliorates the effects of oxidative stress in lenses of streptozotocin-induced type 1 diabetic rats. *Pharmacol. Rep.* **2017**, *69*, 995–1000. [CrossRef]

63. Wojnar, W.; Zych, M.; Kaczmarczyk-Sedlak, I. Antioxidative effect of flavonoid naringenin in the lenses of type 1 diabetic rats. *Biomed. Pharmacother.* **2018**, *108*, 974–984. [CrossRef]

64. Sedlak, L.; Wojnar, W.; Zych, M.; Mrukwa-Kominek, E.; Kaczmarczyk-Sedlak, I. Effect of resveratrol, a dietary-derived polyphenol, on the oxidative stress and polyol pathway in the lens of rats with streptozotocin-induced diabetes. *Nutrients* **2018**, *10*, 1423. [CrossRef]

65. Xu, J.; Li, D.; Zheng, T.; Lu, Y. β-amyloid expression in age-related cataract lens epithelia and the effect of β-amyloid on oxidative damage in human lens epithelial cells. *Mol. Vis.* **2017**, *23*, 1015–1028.

66. Nagai, N.; Ito, Y. Excessive hydrogen peroxide enhances the attachment of amyloid β1-42 in the lens epithelium of UPL rats, a hereditary model for cataracts. *Toxicology* **2014**, *315*, 55–64. [CrossRef]

67. Özcura, F.; Dündar, S.O.; Çetin, E.D.; Beder, N.; Dündar, M. Effect of estrogen replacement therapy on lens epithelial cell apoptosis in an experimental rat model. *Int. Ophthalmol.* **2010**, *30*, 279–284. [CrossRef]

nutrients

MDPI

Article

Lactobacillus paracasei KW3110 Prevents Blue Light-Induced Inflammation and Degeneration in the Retina

Yuji Morita [1,*,†], Yukihiro Miwa [2,3,†], Kenta Jounai [4], Daisuke Fujiwara [1], Toshihide Kurihara [2,3] and Osamu Kanauchi [1]

[1] Research Laboratories for Health Science & Food Technologies, Kirin Company, Ltd., 1-13-5, Fukuura Kanazawa-ku, Yokohama-shi 236-0004, Japan; d-fujiwara@kirin.co.jp (D.F.); kanauchio@kirin.co.jp (O.K.)
[2] Laboratory of Photobiology, Keio University School of Medicine, Tokyo 160-8582, Japan; yukihiro226@gmail.com (Y.M.); kurihara@z8.keio.jp (T.K.)
[3] Department of Ophthalmology, Keio University School of Medicine, Tokyo 160-8582, Japan
[4] Technical Development Center, Koiwai Dairy Products Co Ltd. Sayama, Saitama 350-1321, Japan; k_jounai@koiwai-dairy.co.jp
* Correspondence: Yuji_Morita@kirin.co.jp; Tel.: +81-801-002-4546
† Contributed equally.

Received: 12 November 2018; Accepted: 13 December 2018; Published: 15 December 2018

Abstract: Age-related macular degeneration and retinitis pigmentosa are leading causes of blindness and share a pathological feature, which is photoreceptor degeneration. To date, the lack of a potential treatment to prevent such diseases has raised great concern. Photoreceptor degeneration can be accelerated by excessive light exposure via an inflammatory response; therefore, anti-inflammatory agents would be candidates to prevent the progress of photoreceptor degeneration. We previously reported that a lactic acid bacterium, *Lactobacillus paracasei* KW3110 (*L. paracasei* KW3110), activated macrophages suppressing inflammation in mice and humans. Recently, we also showed that intake of *L. paracasei* KW3110 could mitigate visual display terminal (VDT) load-induced ocular disorders in humans. However, the biological mechanism of *L. paracasei* KW3110 to retain visual function remains unclear. In this study, we found that *L. paracasei* KW3110 activated M2 macrophages inducing anti-inflammatory cytokine interleukin-10 (IL-10) production in vitro using bone marrow-derived M2 macrophages. We also show that IL-10 gene expression was significantly increased in the intestinal immune tissues 6 h after oral administration of *L. paracasei* KW3110 in vivo. Furthermore, we demonstrated that intake of *L. paracasei* KW3110 suppressed inflammation and photoreceptor degeneration in a murine model of light-induced retinopathy. These results suggest that *L. paracasei* KW3110 may have a preventive effect against degrative retinal diseases.

Keywords: *Lactobacillus paracasei* KW3110; retina; light; macrophage

1. Introduction

In recent years, blue light has been used in several visual display terminals (VDTs), including computers, smart phones, and tablet devices; thus, opportunities of human exposure to blue light have increased. Excessive exposure to blue light can cause photoreceptor degeneration in the retina [1] and may be related to age-related macular degeneration (AMD) [2,3] and retinitis pigmentosa [4]. AMD and retinitis pigmentosa are the leading causes of blindness in the elderly population [5]. Recently, natural compounds in foods have attracted worldwide attention in an attempt to treat light-induced ocular problems, in particular, antioxidants in foods [6–8]. However, the mechanism of light-induced retinal damage has not been completely elucidated.

Although multiple factors such as oxidative stress and hypoxia have been reported to have a critical role in photoreceptor degeneration [9], retinal inflammation is also believed to be associated with the progression of photoreceptor degeneration [10–12]. In a previous report, recruitment and polarization of macrophages were shown to be involved in the pathogenesis of light-induced retinal degeneration in vivo [10]. Macrophages can be grouped into at least two subgroups, the classically activated inflammatory M1 phenotype and the alternatively activated M2 phenotype [13,14]. M1 macrophages produce several inflammatory cytokines including interleukin-1β (IL-1β) and cause inflammatory reactions. In contrast, M2 macrophages are associated with anti-inflammatory reactions including tissue remodeling through the production of anti-inflammatory cytokines such as interleukin-10 (IL-10) [15,16]. In the retina, the polarization of M2 macrophages is also thought to promote retinal cell survival in several mouse models [10,17].

Dietary nutrients or constituents have been reported to have potential in protecting photo-stressed retina [18,19]. However, the mechanism of action especially related to retinal inflammation is not well understood. Lactic acid bacteria have been widely used as sources of probiotics and paraprobiotics to enhance gut barrier function and improve the immune system. Some strains have been reported to attenuate several inflammatory phenomena including diarrhea, allergies, and metabolic disorders.

Our group has previously reported that *Lactobacillus paracasei* KW3110 (*L. paracasei* KW3110) suppressed excessive inflammation including dermatitis in mice [20–22] and humans [23]. In addition, we have shown that the intake of *L. paracasei* KW3110 mitigated VDT load-induced ocular disorders, including eye fatigue, in Japanese healthy adults [24]. In this study, we have investigated the ability of *L. paracasei* KW3110 to activate M2 macrophages in vitro and in vivo to attenuate blue light-induced retinal degeneration. We then examined the protective effects of *L. paracasei* KW3110 on retinal functions.

2. Materials and Methods

2.1. Animals

Four-week-old mice (BALB/c, male) were purchased (Charles River Japan, Kanagawa, Japan) and acclimatized for 1 week with free access to water and a basic diet AIN93G (Oriental Yeast, Tokyo, Japan) before all experiments were performed.

All animal procedures and experiments were performed in accordance with the Association for Research in Vision and Ophthalmology Statement for the Use of Animals in Ophthalmic and Vision Research. Animal procedures and experiments were also approved by the Laboratory Animal Care Committee for Experimental Animals of our institute: the approval ID was AN10134-Z00. All efforts were made to minimize animal suffering.

2.2. Preparation of Bone Marrow-Derived Macrophages

Bone marrow-derived M2 macrophages were generated as previously described [25,26]. Briefly, bone marrow cells were extracted from BALB/c mice, and erythrocytes were generated and harvested after brief exposure to 0.168 M NH_4Cl. Cells were then cultured at a density of 5×10^5 cells/mL for 7 days in RPMI1640 medium supplemented with 10% fetal calf serum (FCS) and 5,000 U/mL of macrophage colony stimulating-factor (M-CSF; R and D Systems, Minneapolis, MN, USA). Lipo-teichoic acid (LTA; Invitrogen, Carlsbad, CA, USA) was added at a concentration of 10 ng/mL and *L. paracasei* KW3110 was added at concentrations of 0.1, 1, and 10 µg/mL. The cultures were continued for 24 h. *L. paracasei* KW3110 was prepared as described in a previous study [23].

2.3. Enzyme-Linked Immunosorbent Assay (ELISA)

The concentrations of cytokines in cell culture supernatants were measured using a mouse IL-10 ELISA kit (BD Biosciences, San Jose, CA, USA).

2.4. Oral Administration and Sample Collection

The mice were orally administered saline (Otsuka Pharmaceutical) containing 50 mg of heat-killed *L. paracasei* KW3110. Mesenteric lymph nodes (MLNs) were removed at 0, 2, 6, 10, and 24 h after treatments (BALB/c, male, *n* = 8/each time point). The tissues were soaked in RNAlater RNA Stabilization Reagent (Qiagen, Hilden, Germany) and kept at −80 °C until RNA extraction.

2.5. RNA Preparation and Quantitative RT-PCR from Tissues

Total RNA was extracted from MLNs using the RNeasy Mini kit (Qiagen), and cDNAs were prepared using an iScript cDNA synthesis kit (BioRad, Hercules, CA, USA) according to the manufacturer's instructions. The resulting products were subjected to quantitative RT-PCR using SYBR Premix Ex Taq (Takara Bio, Otsu, Japan) and a LightCycler PCR system (Roche Diagnostics, Basel, Switzerland). The relative expression levels of the gene were normalized to glyceraldehyde-3-phosphate dehydrogenase (*Gapdh*). The primers used for PCR were as follows: *Gapdh* forward (F) (AACGACCCCTTCATTGAC) and *Gapdh* reverse (R) (TCCACGACATACTCAGCAC), *Il10* F (CAGAGCCACATGCTCCTAGA) and *Il10* R (TGTCCAGCTGGTCCTTTGTT).

2.6. Light Exposure

After acclimatization, the mice (BALB/c, male) were divided by equal average weights into three groups (*n* = 6). The non-light exposure control mice group and the light exposure mice group were maintained on AIN93G purified rodent diet (Zeigler, Gardners, PA, USA). In addition, the light exposure *L. paracasei* KW3110 mice group was fed the AIN93G diet containing approximately 1 mg heat-killed *L. paracasei* KW3110/day/mouse. All mice were housed in specific pathogen-free conditions under a 12-h light-dark photo cycle and had ad libitum access to water and the diet. Two weeks later, light exposure experiments were performed. Mice were exposed to blue light as previously described with slight modifications [18]. Briefly, the mice were dark-adapted for 12 h before light exposure. The mice were then exposed to 5000 lux of blue light (CCS Inc., Kyoto, Japan, peak at 470 nm) for 3 h, starting at 9:00 a.m., in exposure boxes maintained at 23 °C. After light exposure, the mice were maintained under a dim cyclic light (5 lux, 12 h on/off).

2.7. Retinal Cell Preparations

Three days after the start of light exposure, the retinas were digested with 1 mg/mL collagenase II (Worthington, Lakewood, NJ, USA) for 40 min at 37 °C in Hank's Balanced Salt Solution (HBSS) buffer with 1.0% bovine serum albumin (BSA). The tissue digest was then filtered through a 70 μm cell strainer and washed with HBSS buffer with 1.0% BSA for 5 min at 1300 rpm and at 4 °C. The supernatant was carefully removed and the digested tissue pellet was resuspended to form a single-cell suspension.

2.8. Flow Cytometry Analyses

The retinal cells were stained with fluorescent dyes conjugated to antibodies as follows: CD206-FITC (C068C2; BioLegend, San Diego, CA, USA); 7-AAD (BD Pharmingen, San Jose, CA, USA); CD11b-APC-Cy7 (M1/70; BD Biosciences San Jose, CA, USA); f4/80-PE-Cy7 (BM8; BioLegend). After staining, the cells were washed twice with a FACS buffer (0.5% BSA in PBS buffer) and suspended in the FACS buffer for FACS analyses. Data were collected using a FACS Canto II flow cytometer (BD Biosciences) and analyzed by FCS Express software (De Novo Software, Los Angeles, CA, USA). The 7-AAD− CD11b+ f4/80+ cells were defined as retinal macrophages.

To investigate intracellular cytokine production, retinal cells were treated with a leukocyte activation cocktail with BD GolgiPlug™ (BD Biosciences) for 4.5 h and with a BD Cytofix/Cytoperm Fixation/Permeabilization™ kit (BD Biosciences) and then stained with the following antibodies: TNF-α-FITC (MP6-XT22; eBiosciences, San Diego, CA, USA); IL-10-PE (GK1.5; BioLegend); CD11b-APC-Cy7 (M1/70; BD Biosciences); F4/80-PE-Cy7 (BM8; BioLegend); and 7-AAD

(BD Pharmingen, San Jose, CA, USA). The 7-AAD− CD11b+ f4/80+ cells were defined as retinal macrophages. Data were collected using a FACS Canto II flow cytometer (BD Biosciences) and analyzed by FCS Express software (De Novo Software).

2.9. Analysis of Cytokine Concentrations

The retinal cells were cultured for 24 h in RPMI 1640 medium supplemented with 10% FCS to evaluate the production of inflammatory cytokines. Supernatants were collected and analyzed for cytokine concentrations using a Bio-Plex Pro mouse cytokine assay kit (Bio-Rad).

2.10. Measurements of the Retinal Thickness

One week after the start of light exposure, eye balls were fixed in neutral 10% formalin and decalcified. The tissues were sectioned including the regions from the optic nerve head to the most peripheral, then stained with hematoxylin and eosin. The outer nuclear layer (ONL) thickness in the retinal section was measured in all areas. We randomly selected ten observation points in each image and averaged using WinROOF software (MITANI Corporation).

2.11. Electroretinography (ERG)

After acclimatization, the mice (BALB/c, male) were divided by equal average weights into two groups. The control mice group ($n = 4$) was fed AIN93G diets. The *L. paracasei* KW3110 mice group ($n = 4$) was fed AIN93G containing approximately 1 mg heat-killed *L. paracasei* KW3110/day/mouse. All mice were housed in specific pathogen-free conditions under a 12-h light-dark (about 700 lux) photo cycle and had ad libitum access to water and the diet. Two weeks later, the mice were dark-adapted for 12 h and then placed under dim red illumination before conducting ERGs. The mice were anesthetized with an MMB combination anesthetic containing midazolam (4 mg/kg, SANDOZ, Yamagata, Japan), medetomidine (0.75 mg/kg, Nippon Zenyaku Kogyo Co., Ltd., Fukushima, Japan) and butorphanol tartrate (5 mg/kg, Meiji Seika Pharma, Tokyo, Japan) and placed on a heating pad to maintain their body temperatures at 35–36 °C throughout the experiments. The pupils were dilated with a single drop of a mixed solution of 0.5% tropicamide and 0.5% phenylephrine (Santen Pharmaceutical, Osaka, Japan). The ground and reference electrodes were then placed on the tail and subcutaneously between the eyes, respectively, while the active gold wire electrodes were placed on the cornea. The recordings were performed with a Ganzfeld dome, an acquisition system, and LED stimulators (PuREC, MAYO Corporation, Inazawa, Japan). The amplitude of the a-wave was measured from the baseline to the trough of the a-wave. The amplitude of the b-wave was determined from trough of the a-wave to the peak of the b-wave.

2.12. Statistical Analysis

All values are presented as the mean ± SEM. Statistical differences for the results of Figure 1 were performed using Dunnett's test for post-hoc comparisons. Statistical differences between three groups (control mice group fed a control diet without light exposure, light control mice group fed a control diet with light exposure, and *L. paracasei* KW3110 mice group fed a diet containing *L. paracasei* KW3110 with light exposure) were analyzed by one-way analysis of variance (ANOVA), followed by the Tukey-Kramer test with significance set at $p < 0.05$. Statistical differences between the two groups (light control mice group fed a control diet with light exposure and *L. paracasei* KW3110 mice group fed a diet containing *L. paracasei* KW3110 with light exposure) were determined using an unpaired, two-tailed Student's *t*-test with significance set at $p < 0.05$. All statistical analyses were performed using the Ekuseru-Toukei 2012 software program (Social Survey Research Information, Tokyo, Japan).

3. Results

3.1. L. paracasei KW3110 Activates M2 Macrophages In Vitro and Induces IL-10 Production In Vivo

In order to determine the effects of *L. paracasei* KW3110 on M2 macrophage activation, bone marrow-derived M-CSF-induced M2 macrophages were treated with *L. paracasei* KW3110 and IL-10 levels, as a marker of M2-polarization [27], were measured in culture supernatants. *L. paracasei* KW3110 at 0.1–10 µg/mL induced IL-10 production in a concentration-dependent manner (Figure 1A). In the previous report, our team showed that orally provided *L. paracasei* KW3110 (50 mg/head) interacted with the immune cells in the gut [19]. To examine IL-10 induction of *L. paracasei* KW3110 in vivo, we evaluated IL-10 gene expression in mesenteric lymph nodes (MLNs) at several time points after oral administration of 50 mg/head *L. paracasei* KW3110 in mice. The *IL-10* mRNA level in MLNs significantly increased 6 h after oral administration and decreased to the basal level 24 h after administration (Figure 1B). These results suggest that *L. paracasei* KW3110 activated M2 macrophages inducing the production of IL-10.

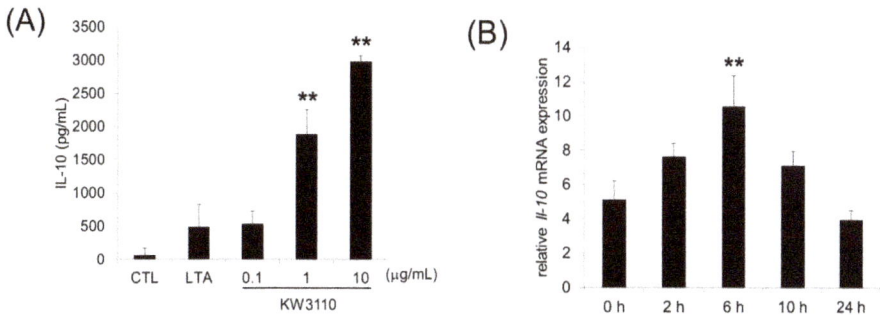

Figure 1. Effects of *L. paracasei* KW3110 on the activation of M2 macrophages and induction of IL-10 production. (**A**) Bone marrow-derived M2 macrophages were stimulated with lipo-teichoic acid (10 ng/mL), and *L. paracasei* KW3110 (0.1, 1, or 10 µg/mL) and the amounts of secreted IL-10 were measured by ELISA. (**B**) Relative *IL-10* mRNA expression was measured using PCR. Values are represented as the mean ± SEM. Significance was assumed if the *p* value was < 0.05. ** *p* < 0.01. LTA, lipo-teichoic acid treated group; CTL, control; KW3110, *L. paracasei* KW3110 treated group.

3.2. L. paracasei KW3110 Induces Retinal M2 Macrophages Following Light Exposure

We next investigated the effects of *L. paracasei* KW3110 on retinal macrophages in a murine light-induced retinopathy model. Flow cytometry analyses revealed that intake of *L. paracasei* KW3110 significantly increased the ratio of f4/80-, CD11b-, and CD206-positive macrophages in the retina to CD11b-positive cells, 3 days after the light exposure compared with the control mice group (Figure 2A,B). We also evaluated the levels of inflammatory cytokines in retinal macrophages. Intake of *L. paracasei* KW3110 significantly decreased the expression of the inflammatory cytokine TNF-α in retinal macrophages compared with that in the mice group fed a control diet (Figure 2C). In addition, the production of IL-1β (Figure 2D left graph) and RANTES (regulated on activation, normal T cell expressed and secreted) (Figure 2D right graph) inflammatory cytokines, were significantly lower in the mice group fed a diet containing *L. paracasei* KW3110 than that in the control group. These data indicate that intake of *L. paracasei* KW3110 induced M2 macrophages and suppressed the production of inflammatory cytokines evoked by blue light exposure.

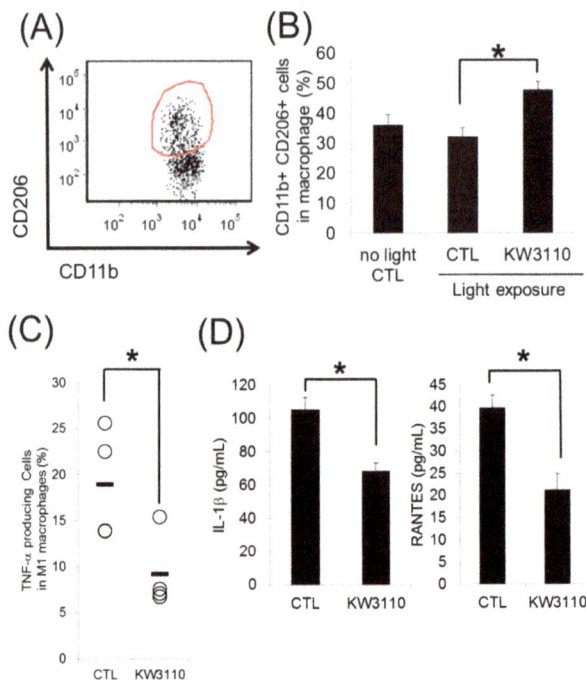

Figure 2. Effect of *L. paracasei* KW3110 on the induction of M2 macrophage in the retina of a light-exposed mouse model. (**A**) Representative flow cytometry data of CD11b-positive and CD206-positive cells from the blue light-exposed mice fed a diet containing *L. paracasei* KW3110 (KW3110) or fed a control diet (CTL). (**B**) The ratio of CD11b-positive, f4/80-positive, and CD206-positive M2 macrophage cells to CD11b-positive and f4/80-positive macrophage cells. (**C**) The ratio of TNF-α-producing cells in CD11b-positive M1 macrophage cells. To detect inflammatory cytokine-producing cells, retinal cells were cultured under stimulation with Leukocyte Activation Cocktail plus BD GolgiPlug™ and analyzed by flow cytometry. (**D**) The production of inflammatory cytokines IL-1β and RANTES in the retinal cell culture medium. CTL, control; KW3110, *L. paracasei* KW3110; TNF-α, tumor necrosis factor-α; IL-1β, interleukin-1β; RANTES, regulated on activation, normal T cell expressed and secreted. Values are presented as the mean ± SEM. Significance was assumed if the *p* value was < 0.05. * *p* < 0.05.

3.3. Intake of L. paracasei KW3110 Suppresses the Photoreceptor Degeneration Induced by Light Exposure

Retinal inflammation was previously suggested to be associated with photoreceptor degeneration [10]. To evaluate the effects of *L. paracasei* KW3110 on light-induced retinal degeneration, we compared the ONL thickness containing photoreceptor cells from the optic nerve head to the periphery in the retina. The ONL thickness in the light-exposure mice group fed a control diet was significantly thinner than that in the non-light exposed mice fed a control diet (Figure 3A,B). In contrast, the ONL thickness in the light-exposure mice fed a diet containing *L. paracasei* KW3110 was maintained at the same thickness as in the non-light exposed mice fed a control diet (Figure 3A,B). The ONL thickness in the light-exposure mice group fed a control diet was significantly thinner than that in the light-exposed mice fed a diet containing *L. paracasei* KW3110 (Figure 3A,B and Figure S1). These results indicate that intake of *L. paracasei* KW3110 attenuated photoreceptor degeneration caused by an excessive blue light exposure.

(A)

Non-light CTL   Light CTL   Light KW3110

(B)

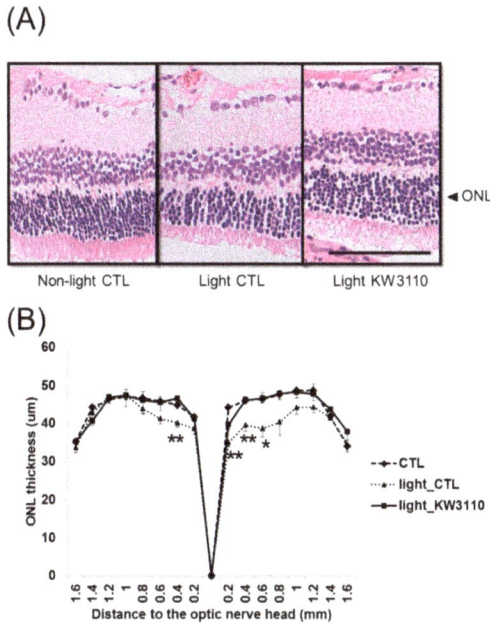

Figure 3. A protective effect of *L. paracasei* KW3110 on light-induced histological retinal changes. (A) Hematoxylin and eosin staining of retinal sections. Arrow heads indicate the outer nuclear layer (ONL). Scale bar represents 100 μm. (B) ONL thickness was lower in mice fed a control diet than in mice fed a diet with *L. paracasei* KW3110. Values are presented as the mean ± SEM. Significance was assumed if the *p* value was < 0.05. *$p < 0.05$; **$p < 0.01$; non-light CTL, no light exposed mice group fed a control diet; light control, light exposed mice group fed a control diet; light KW3110, light exposed mice group fed a diet containing *L. paracasei* KW3110.

3.4. Intake of L. paracasei KW3110 Attenuates the Impairment of Retinal Function

To investigate the effects of intake of *L. paracasei* KW3110 on retinal functions, ERG analyses were performed. In the scotopic ERG, the amplitudes of the a- and b-waves tended to be lower in the mice group fed a control diet than in the mice group fed a diet containing *L. paracasei* KW3110 (Figure 4A,B). In addition, the amplitude of the b-wave in the photopic ERG was significantly lower in the mice group fed a control diet than in the mice group fed a diet containing *L. paracasei* KW3110 (Figure 4C). These results suggest that administration of *L. paracasei* KW3110 has a protective effect in both cone and rod photoreceptor functions.

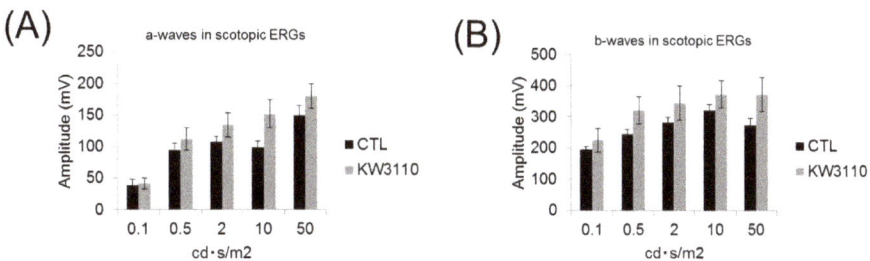

(A)

(B)

Figure 4. *Cont.*

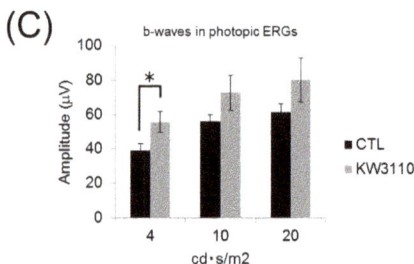

Figure 4. Suppression of visual function impairment by intake of *L. paracasei* KW3110. (**A**) The amplitude of a-waves in the scotopic ERG. (**B**) The amplitude of b-waves in the scotopic ERG. (**C**) The amplitude of b-waves in the photopic ERG. Results are represented as the mean ± SEM. * $p < 0.05$ for the control. CTL, control diet; KW3110, *L. paracasei* KW3110 diet.

4. Discussion

In this study, the in vitro and in vivo experiments suggested that *L. paracasei* KW3110 activated M2 macrophages and induced anti-inflammatory cytokine IL-10 production. We also demonstrated that *L. paracasei* KW3110 had a positive effect on retinal functional restoration in vivo.

In a previous study, ligands for toll-like receptor 2 (TLR2) such as Pam3Cys have been reported to activate mouse dendritic cells and induce IL-10 production through activation of the ERK pathway [28]. In the current study, lipo-teichoic acid (LTA), one of the ligands for TLR2, also slightly activated M2 macrophages and induced IL-10 production. The effect of *L. paracasei* KW3110 on IL-10 production in bone marrow-derived M2 macrophages might have been mediated through the TLR2-dependent ERK pathway since lactic acid bacteria have lipo-teichoic acid. In addition, peptidoglycans in lactic acid bacteria were also reported to increase IL-10 levels via the nucleotide-oligomerization domain receptor 2 (NOD2) pathway [29]. The effects of *L. paracasei* KW3110 on the production of IL-10 might be through the NOD2-dependent pathway at least in part.

In this study, we also showed that *L. paracasei* KW3110 had the potential of activating M2 macrophages and inducing the production of IL-10 in vitro and in vivo (Figure 1A,B). Previously, we reported that orally provided *L. paracasei* KW3110 interacted with the gut immune cells in mice [20]. These results suggested that *L. paracasei* KW3110 could activate the gut immune cells inducing IL-10 production. In a previous report, oral administration of *Pantoea agglomerans*-derived lipopolysaccharide reduced proinflammatory cytokine expression in the blood and reduced the brain Aβ burden and memory impairment [30]. These results suggested that the regulation of cytokine levels, induced by oral administration of food constituents, might have the potential to affect the inflammatory state of the peripheral tissues through systemic blood flow. IL-10 is known as not only one of the M2 macrophage-producing anti-inflammatory cytokines but also one of the factors that induce M2 macrophages [31]. In this study, we demonstrated that intake of *L. paracasei* KW3110 induced CD11b-positive and CD206-positive monocytes which are generally defined as M2 type macrophages in the blue light-exposed retina (Figure 2). Previously, blood-borne macrophages have been reported to integrate into the retina through the optic nerve and the ciliary body in a light-induced retinopathy mouse model [32]. Taken together, *L. paracasei* KW3110 interacts with gut immune cells and might induce M2 macrophages, at least in part, through IL-10 induced from the gut immune cells. Then, those M2 macrophages might be recruited to the retina.

M2 macrophages have been reported to have an anti-inflammatory phenotype when the tissue is damaged [33]. In this study, inflammatory macrophages, i.e., TNF-α-producing macrophages were decreased in the retina of the mice group fed a diet containing *L. paracasei* KW3110 compared with that of mice fed a control diet under the same blue light exposure conditions (Figure 2C). IL-1β and RANTES, which were known as inflammatory phenotype markers in the stressed retina [34–36], were also significantly lower in the mice group fed a diet containing *L. paracasei* KW3110 (Figure 2D). In addition, we recently reported that *L. paracasei* KW3110 activated human peripheral blood mononuclear cell- (human-PBMCs) derived M2 macrophages and mitigated VDT load-induced ocular disorders, including eye fatigue, in humans [24]. These results suggested that *L. paracasei* KW3110 induced anti-inflammatory M2 macrophages in the stress conditioned retina.

Intake of *L. paracasei* KW3110 also suppressed light-induced ONL thinning (Figure 3). The ONL is composed of photoreceptor cell bodies and the ONL thickness has been reported to decrease in response to light-induced photoreceptor loss [37]. Although further studies, including analyses of apoptotic cell death of photoreceptors, are needed, intake of *L. paracasei* KW3110 might attenuate photoreceptor loss. We also showed that intake of *L. paracasei* KW3110 could mitigate the impairments of the retinal function evaluated by ERG (Figure 4). The a-wave responses as shown by scotopic ERG indicate rod photoreceptor function and the b-wave responses as shown by scotopic ERG indicate the subsequent responses of photoreceptor function. The b-wave responses as shown by photopic ERG indicate the subsequent response evoked from the cone photoreceptor function [38]. Taken together, it is suggested that intake of *L. paracasei* KW3110 has a protective effect in both cone and rod photoreceptor functions.

Retinal phototoxicity models in small rodents, including a mouse model of light-induced retinopathy, have been widely used in the majority of studies. However, previous studies have demonstrated that light-induced damaged retina showed various morphological patterns in different animal models [39–42]. In rats and mice, the light-induced damages in the rod photoreceptors have been reported to be more sensitive than in cone photoreceptors [43] while in chickens and pigeons cone photoreceptors have been reported to be damaged first [44]. Therefore, further studies using the larger animals are needed to confirm the preventive effects of *L. paracasei* KW3110 on light-induced inflammation and degeneration in the retina.

5. Conclusions

In summary, *L. paracasei* KW3110 induced retinal M2 macrophages in a murine model of light-induced retinopathy. In addition, oral intake of *L. paracasei* KW3110 had a positive effect on retinal morphology and function. These findings suggested that *L. paracasei* KW3110 might have potential as a dietary food supplement to prevent retinal degeneration through regulating inflammation in response to blue-light damage.

Supplementary Materials: The following are available online at http://www.mdpi.com/2072-6643/10/12/1991/s1. Figure S1: Representative images of H&E staining for retinal sections from the optic nerve head to the most peripheral area.

Author Contributions: Y.M., K.J. and D.F. designed this study and conducted the in vitro experiments and in vivo experiments using a murine model of light-induced retinopathy; Y.M. and T.K. designed a murine model of light-induced retinopathy and conducted the electroretinography experiments and wrote the manuscript; Y.M. and O.K. wrote the manuscript.

Funding: This work was supported by Kirin Company, Limited.

Acknowledgments: We thank Konomi Ohshio and Masato Otake of Kirin Company for their excellent technical support, and valuable discussions.

Conflicts of Interest: Morita, Y., Fujiwara, D. and Kanauchi, O. are employed by Kirin Company Ltd. Jounai, K. is employed by Koiwai Dairy Products Co Ltd. The preventive effects of *L. paracasei* KW3110 have been applied for a patent (Application no. 2017-172620).

References

1. Kishi, S.; Li, D.; Takahashi, M.; Hashimoto, H. Photoreceptor damage after prolonged gazing at a computer game display. *Jpn. J. Ophthalmol.* **2010**, *54*, 514–516. [CrossRef] [PubMed]
2. Suzuki, M.; Tsujikawa, M.; Itabe, H.; Du, Z.J.; Xie, P.; Matsumura, N.; Fu, X.; Zhang, R.; Sonoda, K.H.; Egashira, K.; et al. Chronic photo-oxidative stress and subsequent MCP-1 activation as causative factors for age-related macular degeneration. *J. Cell Sci.* **2012**, *125*, 2407–2415. [CrossRef] [PubMed]

3. Glazer-Hockstein, C.; Dunaief, J.L. Could blue light-blocking lenses decrease the risk of age-related macular degeneration? *Retina* **2006**, *26*, 1–4. [CrossRef]

4. Reme, C.E.; Grimm, C.; Hafezi, F.; Marti, A.; Wenzel, A. Apoptotic cell death in retinal degenerations. *Prog. Retin. Eye Res.* **1998**, *17*, 443–464. [CrossRef]

5. Wong, W.L.; Su, X.; Li, X.; Cheung, C.M.; Klein, R.; Cheng, C.Y.; Wong, T.Y. Global prevalence of age-related macular degeneration and disease burden projection for 2020 and 2040: A systematic review and meta-analysis. *Lancet Glob. Health* **2014**, *2*, 106–116. [CrossRef]

6. Choi, W.; Kim, J.C.; Kim, W.S.; Oh, H.J.; Yang, J.M.; Lee, J.B.; Yoon, K.C. Clinical Effect of Antioxidant Glasses Containing Extracts of Medicinal Plants in Patients with Dry Eye Disease: A Multi-Center, Prospective, Randomized, Double-Blind, Placebo-Controlled Trial. *PLoS ONE* **2015**, *10*, e0139761. [CrossRef]

7. Uchino, Y.; Uchino, M.; Dogru, M.; Fukagawa, K.; Tsubota, K. Improvement of accommodation with anti-oxidant supplementation in visual display terminal users. *J. Nutr. Health Aging* **2012**, *16*, 478–481. [CrossRef]

8. Platania, C.B.M.; Fidilio, A.; Lazzara, F.; Piazza, C.; Geraci, F.; Giurdanella, G.; Leggio, G.M.; Salomone, S.; Drago, F.; Bucolo, C. Retinal Protection and Distribution of Curcumin in Vitro and in Vivo. *Front. Pharmacol.* **2018**, *9*, 670. [CrossRef]

9. Kurihara, T.; Westenskow, P.D.; Gantner, M.L.; Usui, Y.; Schultz, A.; Bravo, S.; Aguilar, E.; Wittgrove, C.; Friedlander, M.; Paris, L.P.; et al. Hypoxia-induced metabolic stress in retinal pigment epithelial cells is sufficient to induce photoreceptor degeneration. *Elife* **2016**, *5*, e14319. [CrossRef]

10. Jiao, H.; Natoli, R.; Valter, K.; Provis, J.M.; Rutar, M. Spatiotemporal Cadence of Macrophage Polarisation in a Model of Light-Induced Retinal Degeneration. *PLoS ONE* **2015**, *10*, e0143952. [CrossRef]

11. Rutar, M.; Natoli, R.; Provis, J.M. Small interfering RNA-mediated suppression of Ccl2 in Muller cells attenuates microglial recruitment and photoreceptor death following retinal degeneration. *J. Neuroinflamm.* **2012**, *9*, 221. [CrossRef] [PubMed]

12. Noailles, A.; Fernandez-Sanchez, L.; Lax, P.; Cuenca, N. Microglia activation in a model of retinal degeneration and TUDCA neuroprotective effects. *J. Neuroinflammation* **2014**, *11*, 186. [CrossRef] [PubMed]

13. Martinez, F.O.; Helming, L.; Gordon, S. Alternative activation of macrophages: An immunologic functional perspective. *Annu. Rev. Immunol.* **2009**, *27*, 451–483. [CrossRef] [PubMed]

14. Biswas, S.K.; Mantovani, A. Macrophage plasticity and interaction with lymphocyte subsets: Cancer as a paradigm. *Nat. Immunol.* **2010**, *11*, 889–896. [CrossRef] [PubMed]

15. Sica, A.; Mantovani, A. Macrophage plasticity and polarization: In vivo veritas. *J. Clin. Investig.* **2012**, *122*, 787–795. [CrossRef]

16. Satoh, T.; Kidoya, H.; Naito, H.; Yamamoto, M.; Takemura, N.; Nakagawa, K.; Yoshioka, Y.; Morii, E.; Takakura, N.; Takeuchi, O.; et al. Critical role of Trib1 in differentiation of tissue-resident M2-like macrophages. *Nature* **2013**, *495*, 524–528. [CrossRef]

17. London, A.; Itskovich, E.; Benhar, I.; Kalchenko, V.; Mack, M.; Jung, S.; Schwartz, M. Neuroprotection and progenitor cell renewal in the injured adult murine retina requires healing monocyte-derived macrophages. *J. Exp. Med.* **2011**, *208*, 23–39. [CrossRef]

18. Kubota, S.; Kurihara, T.; Ebinuma, M.; Kubota, M.; Yuki, K.; Sasaki, M.; Noda, K.; Ozawa, Y.; Oike, Y.; Ishida, S.; et al. Resveratrol prevents light-induced retinal degeneration via suppressing activator protein-1 activation. *Am. J. Pathol.* **2010**, *177*, 1725–1731. [CrossRef]

19. Sasaki, M.; Yuki, K.; Kurihara, T.; Miyake, S.; Noda, K.; Kobayashi, S.; Ishida, S.; Tsubota, K.; Ozawa, Y. Biological role of lutein in the light-induced retinal degeneration. *J. Nutr. Biochem.* **2012**, *23*, 423–429. [CrossRef]

20. Ichikawa, S.; Miyake, M.; Fujii, R.; Konishi, Y. Orally administered Lactobacillus paracasei KW3110 induces in vivo IL-12 production. *Biosci. Biotechnol. Biochem.* **2009**, *73*, 1561–1565. [CrossRef]

21. Wakabayashi, H.; Nariai, C.; Takemura, F.; Nakao, W.; Fujiwara, D. Dietary supplementation with lactic acid bacteria attenuates the development of atopic-dermatitis-like skin lesions in NC/Nga mice in a strain-dependent manner. *Int. Arch. Allergy Immunol.* **2008**, *145*, 141–151. [CrossRef] [PubMed]

22. Fujiwara, D.; Inoue, S.; Wakabayashi, H.; Fujii, T. The anti-allergic effects of lactic acid bacteria are strain dependent and mediated by effects on both Th1/Th2 cytokine expression and balance. *Int. Arch. Allergy Immunol.* **2004**, *135*, 205–215. [CrossRef] [PubMed]

23. Fujiwara, D.; Wakabayashi, H.; Watanabe, H.; Nishida, S.; Iino, H. A Double-blind Trial of *Lactobacillus paracasei* Strain KW3110 Administration for Immunomodulation in Patients with Pollen Allergy. *Allergol. Int.* **2005**, *54*, 143–149. [CrossRef]

24. Morita, Y.; Jounai, K.; Miyake, M.; Inaba, M.; Kanauchi, O. Effect of Heat-Killed Lactobacillus paracasei KW3110 Ingestion on Ocular Disorders Caused by Visual Display Terminal (VDT) Loads: A Randomized, Double-Blind, Placebo-Controlled Parallel-Group Study. *Nutrients* **2018**, *10*, 1058. [CrossRef] [PubMed]

25. Fleetwood, A.J.; Lawrence, T.; Hamilton, J.A.; Cook, A.D. Granulocyte-macrophage colony-stimulating factor (CSF) and macrophage CSF-dependent macrophage phenotypes display differences in cytokine profiles and transcription factor activities: Implications for CSF blockade in inflammation. *J. Immunol.* **2007**, *178*, 5245–5252. [CrossRef] [PubMed]

26. Fleetwood, A.J.; Dinh, H.; Cook, A.D.; Hertzog, P.J.; Hamilton, J.A. GM-CSF- and M-CSF-dependent macrophage phenotypes display differential dependence on type I interferon signaling. *J. Leukoc. Biol.* **2009**, *86*, 411–421. [CrossRef] [PubMed]

27. Mantovani, A.; Sozzani, S.; Locati, M.; Allavena, P.; Sica, A. Macrophage polarization: Tumor-associated macrophages as a paradigm for polarized M2 mononuclear phagocytes. *Trends Immunol.* **2002**, *23*, 549–555. [CrossRef]

28. Kaji, R.; Kiyoshima-Shibata, J.; Nagaoka, M.; Nanno, M.; Shida, K. Bacterial teichoic acids reverse predominant IL-12 production induced by certain lactobacillus strains into predominant IL-10 production via TLR2-dependent ERK activation in macrophages. *J. Immunol.* **2010**, *184*, 3505–3513. [CrossRef]

29. Macho Fernandez, E.; Valenti, V.; Rockel, C.; Hermann, C.; Pot, B.; Boneca, I.G.; Grangette, C. Anti-inflammatory capacity of selected lactobacilli in experimental colitis is driven by NOD2-mediated recognition of a specific peptidoglycan-derived muropeptide. *Gut* **2011**, *60*, 1050–1059. [CrossRef]

30. Kobayashi, Y.; Inagawa, H.; Kohchi, C.; Kazumura, K.; Tsuchiya, H.; Miwa, T.; Okazaki, K.; Soma, G.I. Oral administration of Pantoea agglomerans-derived lipopolysaccharide prevents metabolic dysfunction and Alzheimer's disease-related memory loss in senescence-accelerated prone 8 (SAMP8) mice fed a high-fat diet. *PLoS ONE* **2018**, *13*, e0198493. [CrossRef]

31. Mia, S.; Warnecke, A.; Zhang, X.M.; Malmstrom, V.; Harris, R.A. An optimized protocol for human M2 macrophages using M-CSF and IL-4/IL-10/TGF-beta yields a dominant immunosuppressive phenotype. *Scand. J. Immunol.* **2014**, *79*, 305–314. [CrossRef] [PubMed]

32. Joly, S.; Francke, M.; Ulbricht, E.; Beck, S.; Seeliger, M.; Hirrlinger, P.; Hirrlinger, J.; Lang, K.S.; Zinkernagel, M.; Odermatt, B.; et al. Cooperative phagocytes: Resident microglia and bone marrow immigrants remove dead photoreceptors in retinal lesions. *Am. J. Pathol.* **2009**, *174*, 2310–2323. [CrossRef] [PubMed]

33. Miron, V.E.; Boyd, A.; Zhao, J.W.; Yuen, T.J.; Ruckh, J.M.; Shadrach, J.L.; van Wijngaarden, P.; Wagers, A.J.; Williams, A.; Franklin, R.J.M.; et al. M2 microglia and macrophages drive oligodendrocyte differentiation during CNS remyelination. *Nat. Neurosci.* **2013**, *16*, 1211–1218. [CrossRef] [PubMed]

34. Kuse, Y.; Tsuruma, K.; Kanno, Y.; Shimazawa, M.; Hara, H. CCR3 Is Associated with the Death of a Photoreceptor Cell-line Induced by Light Exposure. *Front. Pharmacol.* **2017**, *8*, 207. [CrossRef] [PubMed]

35. Duncan, D.S.; McLaughlin, W.M.; Vasilakes, N.; Echevarria, F.D.; Formichella, C.R.; Sappington, R.M. Constitutive and Stress-induced Expression of CCL5 Machinery in Rodent Retina. *J. Clin. Cell Immunol.* **2017**, *8*. [CrossRef] [PubMed]

36. Narimatsu, T.; Negishi, K.; Miyake, S.; Hirasawa, M.; Osada, H.; Kurihara, T.; Tsubota, K.; Ozawa, Y. Blue light-induced inflammatory marker expression in the retinal pigment epithelium-choroid of mice and the protective effect of a yellow intraocular lens material in vivo. *Exp. Eye Res.* **2015**, *132*, 48–51. [CrossRef] [PubMed]

37. Hafezi, F.; Steinbach, J.P.; Marti, A.; Munz, K.; Wang, Z.Q.; Wagner, E.F.; Aguzzi, A.; Reme, C.E. The absence of c-fos prevents light-induced apoptotic cell death of photoreceptors in retinal degeneration in vivo. *Nat. Med.* **1997**, *3*, 346–349. [CrossRef]

38. Lei, B.; Yao, G.; Zhang, K.; Hofeldt, K.J.; Chang, B. Study of rod- and cone-driven oscillatory potentials in mice. *Investig. Ophthalmol. Vis. Sci.* **2006**, *47*, 2732–2738. [CrossRef]

39. Narimatsu, T.; Ozawa, Y.; Miyake, S.; Kubota, S.; Yuki, K.; Nagai, N.; Tsubota, K. Biological effects of blocking blue and other visible light on the mouse retina. *Clin. Exp. Ophthalmol.* **2014**, *42*, 555–563. [CrossRef]

40. Tanito, M.; Kaidzu, S.; Anderson, R.E. Protective effects of soft acrylic yellow filter against blue light-induced retinal damage in rats. *Exp. Eye Res.* **2006**, *83*, 1493–1504. [CrossRef]

41. Tanito, M.; Kaidzu, S.; Anderson, R.E. Delayed loss of cone and remaining rod photoreceptor cells due to impairment of choroidal circulation after acute light exposure in rats. *Investig. Ophthalmol. Vis. Sci.* **2007**, *48*, 1864–1872. [CrossRef] [PubMed]

42. Saenz-de-Viteri, M.; Heras-Mulero, H.; Fernandez-Robredo, P.; Recalde, S.; Hernandez, M.; Reiter, N.; Moreno-Orduna, M.; Garcia-Layana, A. Oxidative stress and histological changes in a model of retinal phototoxicity in rabbits. *Oxid. Med. Cell Longev.* **2014**, *2014*, 637137. [CrossRef]

43. Sparrow, J.R.; Parish, C.A.; Hashimoto, M.; Nakanishi, K. A2E, a lipofuscin fluorophore, in human retinal pigmented epithelial cells in culture. *Investig. Ophthalmol. Vis. Sci.* **1999**, *40*, 2988–2995. [PubMed]

44. Marshall, J.; Mellerio, J.; Palmer, D.A. Damage to pigeon retinae by moderate illumination from fluorescent lamps. *Exp. Eye Res.* **1972**, *14*, 164–169. [CrossRef]

nutrients

MDPI

Article

Chrysin Ameliorates Malfunction of Retinoid Visual Cycle through Blocking Activation of AGE-RAGE-ER Stress in Glucose-Stimulated Retinal Pigment Epithelial Cells and Diabetic Eyes

Min-Kyung Kang, Eun-Jung Lee, Yun-Ho Kim, Dong Yeon Kim, Hyeongjoo Oh, Soo-Il Kim and Young-Hee Kang *

Department of Food Science and Nutrition, Hallym University, Chuncheon 24252, Korea;
mitoly@hallym.ac.kr (M.-K.K.); reydmswjd@naver.com (E.-J.L.); royalskim@hallym.ac.kr (Y.-H.K.);
ehddus3290@naver.com (D.Y.K.); ohhyeongju@gmail.com (H.O.); ky4850@naver.com (S.-I.K.)
* Correspondence: yhkang@hallym.ac.kr; Tel.: +82-33-248-2132

Received: 22 May 2018; Accepted: 6 August 2018; Published: 8 August 2018

Abstract: Diabetes-associated visual cycle impairment has been implicated in diabetic retinopathy, and chronic hyperglycemia causes detrimental effects on visual function. Chrysin, a naturally occurring flavonoid found in various herbs, has anti-inflammatory, antioxidant, and neuroprotective properties. The goal of the current study was to identify the retinoprotective role of chrysin in maintaining robust retinoid visual cycle-related components. The in vitro study employed human retinal pigment epithelial (RPE) cells exposed to 33 mM of glucose or advanced glycation end products (AGEs) in the presence of 1–20 µM chrysin for three days. In the in vivo study, 10 mg/kg of chrysin was orally administrated to db/db mice. Treating chrysin reversed the glucose-induced production of vascular endothelial growth factor, insulin-like growth factor-1, and pigment epithelium-derived factor (PEDF) in RPE cells. The outer nuclear layer thickness of chrysin-exposed retina was enhanced. The oral gavage of chrysin augmented the levels of the visual cycle enzymes of RPE65, lecithin retinol acyltransferase (LRAT), retinol dehydrogenase 5 (RDH5), and rhodopsin diminished in db/db mouse retina. The diabetic tissue levels of the retinoid binding proteins and the receptor of the cellular retinol-binding protein, cellular retinaldehyde-binding protein-1, interphotoreceptor retinoid-binding protein and stimulated by retinoic acid 6 were restored to those of normal mouse retina. The presence of chrysin demoted AGE secretion and AGE receptor (RAGE) induction in glucose-exposed RPE cells and diabetic eyes. Chrysin inhibited the reduction of PEDF, RPE 65, LRAT, and RDH5 in 100 µg/mL of AGE-bovine serum albumin-exposed RPE cells. The treatment of RPE cells with chrysin reduced the activation of endoplasmic reticulum (ER) stress. Chrysin inhibited the impairment of the retinoid visual cycle through blocking ER stress via the AGE-RAGE activation in glucose-stimulated RPE cells and diabetic eyes. This is the first study demonstrating the protective effects of chrysin on the diabetes-associated malfunctioned visual cycle.

Keywords: advanced glycation end products; chyrsin; diabetic retinopathy; endoplasmic reticulum; retinal pigment epithelium; visual cycle

---

1. Introduction

Diabetic retinopathy (DR) is a microvascular eye disease involving retinal neurodegeneration [1]. DR is associated strongly with a prolonged duration of hyperglycemia and hypertension, in which serious damage occurs to the retina, consequently causing vision loss and blindness [2,3]. In the early stage of DR, hyperglycemia can cause blood vessels in the retina to leak fluid, making retina and macula swell [1,4]. In the advanced stage of DR, abnormal new blood vessels can outgrow on the surface of the retina, and scar tissues and tiny exudate particles can form in the retina [1,3,5]. Here, new vessels are fragile, and occasionally bleed into the vitreous, which is called vitreous hemorrhage [4]. These abnormal alterations in the macula and retina can steal central and peripheral vision [2–4]. Pathophysiological factors germane to the development of DR include genetic and epigenetic factors, free radicals, advanced glycosylation end products (AGEs), and inflammatory factors [1]. Increasing evidence highlights inflammation in regard to the induction of diabetes-mediated biochemical and molecular alterations in the retina, ultimately contributing to retinal complications and vision loss [1,4]. However, the molecular mechanisms underlying inflammatory pathways are not concretely defined in DR. On the other hand, the aberrant functions in the mitochondria and endoplasmic reticulum (ER) are recognized as key players promoting the apoptotic demise of retinal vascular and neuronal cells in diabetic eyes [6,7]. The apoptotic death of retinal cells such as photoreceptors, neurons, and vascular cells directly affects visual function in DR and retinitis pigmentosa [7,8].

Several studies have addressed the putative role of ER stress in the visual system [9–11]. The retinal pigment epithelial (RPE) cells support the light-sensitive photoreceptor cells in the retina, which are crucial for visual function [12]. The ER stress in these cells promotes RPE injury in DR [13,14]. Inhibition of aldose reductase alleviates hyperglycemia-induced RPE cell death and ER stress, and prevents retinal degeneration in the diabetic eye [15]. In addition, methylglyoxal, the AGE adduct in DR, reduces RPE cell viability via the ER stress-dependent intracellular reactive oxygen species (ROS) production, mitochondrial membrane potential loss, and intracellular calcium increase [16]. Thus, reducing or blocking ER stress may be a therapeutic option for preventing DR. The retinoid visual cycle spanning photoreceptor cells and the underlying RPE is the cyclical processing of retinol by which 11-cis-retinal is regenerated from all-trans-retinal, following a photoisomerization event entailing enzymes broadly classified as acyltransferases, short-chain dehydrogenases/reductases, and carotenoid/retinoid isomerases/oxygenases [12,17,18]. Several animal models determine the mechanisms that underlie RPE65-associated retinal dystrophies [17]. Recessive blinding diseases are attributed to mutations or deficiency in RPE65, a key isomerase converting light-insensitive all-trans-retinyl ester to light-sensitive 11-cis-retinol for continued visual function [17–19].

Numerous pharmacologic agents such as antivascular endothelial growth factor (VEGF), aldose reductase inhibitors, and protein kinase C inhibitors have been suggested as therapeutic potential ones for ocular diseases [20–22]. However, these inhibitors have failed to demonstrate significant efficacy in the treatment of DR in clinical trials. Natural products have been developed as agents with minimal side effects for intraocular proliferation and angiogenesis [23,24]. Evidence is emerging that lutein and zeaxanthin protect against visual disorders, including age-related macular degeneration, retinitis pigmentosa, and DR [25,26]. The mechanism(s) underscoring the prevention of eye diseases by hydroxycarotenoids may come from their antioxidant, neuroprotective, and anti-inflammatory functions in the retina [26]. Chrysin (Figure 1A) is a flavone-type flavonoid that is present in honey, propolis, honeycomb, and passion flowers, and exhibits multiple biological effects, including anti-inflammation and neuroprotection [27,28]. One safety study reports that the recommended daily concentrations of chrysin are 0.5 g to 3 g [29]. Following intake by humans, chrysin has low oral bioavailability and rapid fecal elimination, in which its major form in plasma is chrysin sulphate [28,30]. Our previous study showed that the multifunctional chrysin exerted the retinoprotection through inhibiting diabetes-associated retinal neovascularization and blood–retinal barrier breakdown [31]. However, whether chrysin is capable of preventing vision loss in diabetic ocular diseases remains unclear. The current study attempted to investigate that chrysin ameliorated glucose-induced visual damage in RPE cells and in mouse diabetic models. This study examined the mechanisms germane to visual cycle dysfunction in RPE, and the contribution of AGE-RAGE system and ER stress to the visual impairment in DR.

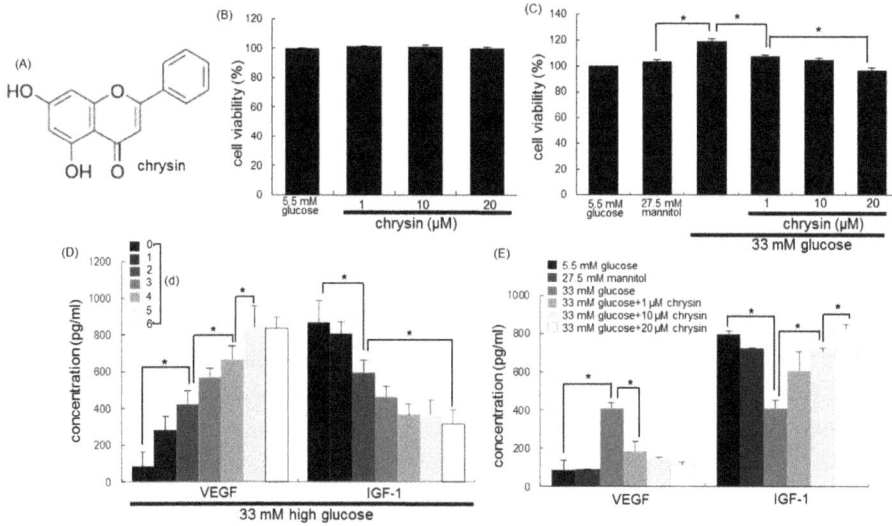

Figure 1. Chemical structure (**A**) and retinal pigment epithelial cytotoxicity (**B**) of chrysin, and the inhibitory effects of chrysin on cell proliferation by glucose (**C**), and temporal responses (**D**) and the inhibition of antivascular endothelial growth factor (VEGF) and insulin-like growth factor-1 (IGF-1) secretion by chrysin (**E**) in human retinal pigment epithelial (RPE) cells. Cells were incubated with 33 mM of glucose in the absence and presence of between 1–20 μM chrysin for up to six days. Cells were also incubated with 5.5 mM of glucose and 27.5 mM of mannitol as osmotic controls. After RPE cells were cultured in high glucose media, cell viability was measured by using MTT (3-(4,5-dimethylthiazol-2-yl)-2,5-diphenyltertrazolium bromide) assay (**B** and **C**, 100% viability with 5.5 mM of glucose). The secretion of VEGF and IGF-1 was measured with commercial ELISA kits (**D** and **E**). * Values in bar graphs (mean ± standard error of the mean (SEM), n = nine independent experiments) indicate significant different at $p < 0.05$.

2. Materials and Methods

2.1. Materials

Fetal bovine serum (FBS), trypsin-ethylenediaminetetraacetic acid (EDTA), and penicillin–streptomycin were supplied by Lonza (Walkersvillle, MD, USA). Dulbecco's modified eagle medium (DMEM, low glucose) media, mannitol, D-glucose, tunicamycin, and thapsigargin were obtained from Sigma-Aldrich Chemical (St. Louis, MO, USA), as were all other reagents, unless specifically stated elsewhere. Mouse monoclonal antibodies of pigment epithelium-derived factor (PEDF, Lot# E2813), RPE65 (Lot# I2816), interphotoreceptor retinoid-binding protein (IRBP, Lot# A1315), and receptor for advanced glycation end product (RAGE, Lot# J2616) were provided from Santa Cruz Biotechnology (Santa Cruz, CA, USA). Mouse monoclonal antibodies of rhodopsin and cellular retinaldehyde-binding protein-1 (CRALBP, Lot# ab15015), rabbit monoclonal retinol dehydrogenase 5 (RDH5, Lot# ab200197) antibody, and rabbit polyclonal antibodies of activating transcription factor 6 (ATF6, Lot# ab37149) and 78 kDa glucose-regulated protein/binding immunoglobulin protein (GRP78/Bip, Lot# ab53068) were obtained by Abcam Biochemicals (Cambridge, UK). Rabbit polyclonal antibodies of advanced glycation end product (AGEs, Lot# AG01111024) and stimulated by retinoic acid 6 (STRA6, Lot# BS12351R) were purchased from Bioss (Boston, MA, USA). Rabbit polyclonal cellular retinol-binding protein (CRBP, Lot # NBP2-20132) antibody was supplied from Novus Biologicals (Littleton, CO, USA). Rabbit polyclonal inositol-requiring enzyme 1α (IRE1α) antibody (Lot# 3294S) was provided from cell signaling (Danvers, MA, USA), and rabbit polyclonal lecithin retinol acyltransferase (LRAT,

Lot# MBS8508176) antibody was purchased from Mybiosource (San Diego, CA, USA). All of the antibodies for Western blot analysis were diluted at 1:1000 ratio, according to the manufacture's instruction. Horseradish peroxidase (HRP)-conjugated goat anti-rabbit immunoglobulin (Ig)G and goat anti-mouse were purchased from Jackson ImmumnoReserch Laboratories (West Grove, PA, USA). Advanced glycation end product-bovine serum albumin (AGE-BSA) was provided by Merck Millipore (Billerica, MA, USA).

Chrysin (Sigma-Aldrich Chemical) was dissolved in dimethyl sulfoxide (DMSO) for live culture with cells; a final culture concentration of DMSO was <0.5%.

2.2. RPE Cell Culture

Primary human RPE cells were obtained from Lonza (Walkersvillle, MD, USA). Cells were grown in DMEM containing 2% FBS, 100 U/mL of penicillin, 100 µg/mL of streptomycin, 2 mM of glutamine, and 1 µg/mL of human fibroblast growth factor basic at 37 °C humidified atmosphere of 5% CO_2 in air. RPE cells were subcultured at 90% confluence and used for further experiments within 10 passages. To induce hyperglycemia, RPE cells were incubated in media containing normal glucose (5.5 mM of D-glucose), 27.5 mM of mannitol (+5.5 mM D-glucose) as an osmotic control, or high glucose (33 mM of D-glucose) in the absence and presence of 1–20 µM chrysin for up to 72 h.

The RPE cell viability was determined by assaying with MTT (3-(4,5-dimethylthiazol-2-yl)-2,5-diphenyltertrazolium bromide). RPE cells seeded at a density of 1×10^4 cells/mL on a 24-well plate were treated with 1–20 µM chrysin in different glucose media. Cells were incubated with 1 mg/mL of MTT solution at 37 °C for 3 h, forming an insoluble purple formazan product that was dissolved in 250 µL of isopropanol. Optical density was measured using a microplate reader at the wavelength of 570 nm. Chrysin at the doses of 1–20 µM had no cytotoxicity (Figure 1B). Thus, the current experiments employed chrysin in the range of 1–20 µM.

2.3. In Vivo Animal Experiments

Adult male db/db mice (C57BLKS/+Leprdb Iar; Jackson Laboratory, Sacramento, CA, USA) and their age-matched non-diabetic db/m littermates (C57BLKS/J; Jackson Laboratory) were employed in the present study. Mice were supplied by the animal facility of Hallym University, kept on a 12 h light/12 h dark cycle at 23 ± 1 °C with 50 ± 10% relative humidity under specific pathogen-free conditions, and fed a standard pellet laboratory chow diet (Cargill Agri Purina, Biopia, Korea). This study was performed with seven-week-old db/db mice, because they begin to develop diabetes (hyperglycemia) at the age of seven to eight weeks. The animals were allowed to acclimatize for a week before commencing the feeding experiments. Mice were divided into three subgroups (*n* = 9 for each subgroup). Mice of the first group were non-diabetic db/m control mice, and db/db mice were divided into two subgroups. One group of db/db mice was orally administrated 10 mg/kg chrysin daily for 10 weeks via gavage. The other group of db/db mice was administrated 0.02% DMSO as the chrysin vehicle. All of the mice were sacrificed after anesthesia with ketamine/Rompun cocktail (40 mg of ketamine and 10 mg of Rompun/kg BW) at 17–18 weeks of age. No mice were dead, and no apparent signs of exhaustion were observed during the experimental period. Chrysin treatment lowered blood levels of glycated hemoglobin HbA1C (~11.5%) and fasting glucose (~580–700 mg/dL) markedly elevated in db/db mice, indicating that chrysin had a glucose-lowering effect [32]. In addition, the 24-h urine volume of diabetic mice were higher (~20-fold) than non-diabetic controls, while chrysin administration reduced the volume by ~50% [32].

All of the animal experiments were approved by the Committee on Animal Experimentation of Hallym University, and performed in compliance with the University's Guidelines for the Care and Use of Laboratory Animals (hallym 2013–125).

2.4. Western Blot Analysis

Western blot analysis was conducted using whole cell lysates prepared from RPE cells at a density of 3.5×10^5 cells. Mouse eye extracts were also prepared from mice that were supplemented with 10 mg/kg chrysin for 10 weeks. Whole cell lysates and mouse eye extracts were prepared in a lysis buffer containing 1 M of β-glycerophosphate, 1% of β-mercaptoethanol, 0.5 M of NaF, 0.1 M of Na_3VO_4, and protease inhibitor cocktail. Cell lysates and eye tissue extracts containing equal amounts of proteins and equal volumes of culture medium supernatants were electrophoresed on 6–15% SDS-PAGE and transferred onto a nitrocellulose membrane. Non-specific binding was blocked with 5% of skim milk for 3 h. The membrane was incubated overnight at 4 °C with each primary antibody of target proteins and washed in a Tris buffered saline-Tween 20 (TBS-T) for 10 min. The membrane was then incubated for 1 h with a secondary antibody of goat anti-rabbit IgG or goat anti-mouse IgG conjugated to HRP. Each target protein level was determined by using immobilon western chemiluminescent HRP substrate (Millipore Corporation, Billerica, MA, USA) and Agfa X-ray film (Agfa-Gevaert, Mortsel, Belgium). Incubation with mouse monoclonal β-actin antibody (Sigma-Aldrich Chemical) was also performed for comparative controls.

2.5. Immunohistochemical Staining

For the immunohistochemical analysis, paraffin-embedded mouse eye tissue sections (5-μm thick) were employed. The sections were placed on glass slides, deparaffinated, and hydrated with xylene and graded alcohol. The sections were pre-incubated in a boiled sodium citrate buffer (10 mM of sodium citrate, 0.05% Tween 20, pH 6.0) for antigen retrieval. A specific primary antibody against RPE65 (mouse monoclonal, Santa Cruz Biotechnology, Santa Cruz, CA, USA; 1:200 dilution) and rhodopsin (mouse monoclonal, Lot# ab5417, Abcam Biochemicals, Cambridge, UK; 1:200 dilution) was incubated overnight with the tissue sections. For the measurement of RPE65 and rhodopsin expression, the tissue section was double-stained with fluorescein isothiocyanate (FITC)-conjugated anti-mouse IgG and with Cy3-conjugated anti-mouse IgG. Nuclear staining was performed with 4′,6-diamidino-2-phenylindole (DAPI, Santa Cruz Biotechnology, Santa Cruz, CA, USA). The stained tissue sections were examined using an optical Axiomager microscope system (Zeiss, Göttingen, Germany), and five images (×200) were taken for each section.

2.6. Enzyme-Linked Immunosorbent Assay (ELISA)

Following culture protocols, the secretion of PEDF, VEGF, and insulin-like growth factor-1 (IGF-1) in RPE cells was determined in collected culture medium supernatants by using ELISA kits (R&D Systems, Minneapolis, MN, USA), according to the manufacturer's instructions.

2.7. Data Analysis

The results are presented as mean ± SEM. Statistical analyses were carried out by using the Statistical Analysis statistical software package version 6.12 (SAS Institute, Cary, NC, USA). Significance was determined by one-way analysis of variance, followed by Duncan's multiple range test for multiple comparisons. Differences were considered significant at $p < 0.05$.

3. Results

3.1. Modulation of Production of VEGF and IGF-1 by Chrysin

When RPE cells were incubated in media containing 33 mM of glucose for three days, the proliferation of RPE cells was observed (Figure 1C). However, the presence of nontoxic chrysin at 1–20 μM dose-dependently attenuated their proliferation. RPE cells secrete a variety of cytokines, growth factors, and extracellular matrix components, all of which contribute to retinal and choroidal neovascularization [33,34]. This study investigated that chrysin influenced the production of VEGF and IGF-1 in glucose-exposed RPE cells. The temporal production of VEGF and IGF was examined in glucose-exposed RPE cells for six days. The VEGF secretion was steadily up-regulated for six days (Figure 1D). On the contrary, the secretion of IGF-1 was continuously diminished by glucose stimulation (Figure 1D). The addition of 1–20 μM of chrysin to 33 mM of glucose-exposed RPE cells curtailed the VEGF secretion in a dose-dependent manner, while the IGF-1 production was nearly completely restored by treating ≥1 μM of chrysin to the RPE cells (Figure 1E).

3.2. Restoration of PEDF Production by Chrysin

Outer nuclear layer (ONL) thickness measured in μm was reduced in the diabetic retina (Figure 2A). This could be due to the marked loss of photoreceptors in the ONL in diabetic retina. However, the ONL thickness of the chrysin-administrated retina was enhanced. PEDF plays a clinical role in choroidal neovascularization by suppressing retinal neovascularization and endothelial cell proliferation [35]. The exposure of RPE cells to 33 mM of glucose for six days temporally reduced the PEDF secretion, as evidenced by Western blot analysis and ELISA (Figure 2B,C). When 1–20 μM chrysin was added to glucose-stimulated cells, its secretion was enhanced in a dose-dependent manner (Figure 2D,E). In addition, the eye tissue level of PEDF was dampened in db/db mice (Figure 2F). In contrast, oral administration of 10 mg/kg of chrysin restored PEDF to the level of db/m control mice.

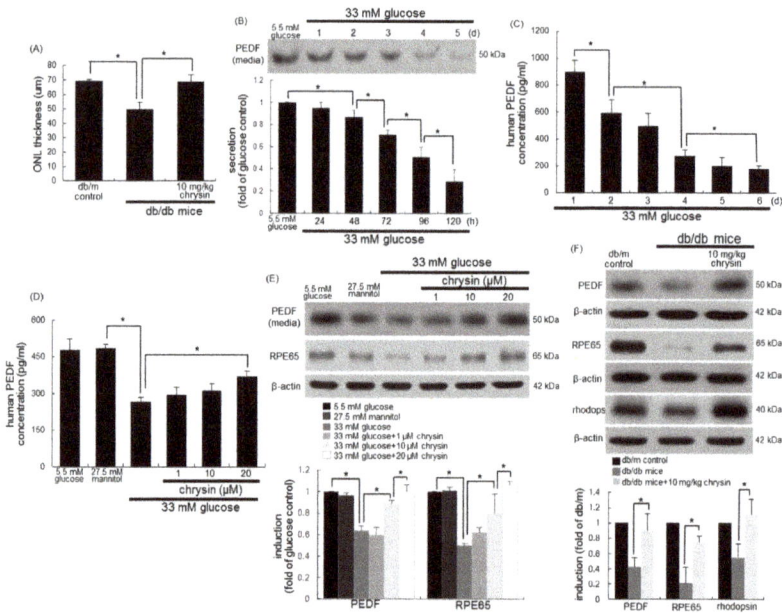

Figure 2. Restoration of outer nuclear layer (ONL) thickness (**A**), time course responses of production of pigment epithelium-derived factor (PEDF) by glucose (**B,C**) and the induction of PEDF, RPE65, and rhodopsin by chrysin (**D–F**). Human retinal pigment epithelial (RPE) cells were incubated with 33 of mM glucose in the absence and presence of 1–20 μM chrysin for up to 6 days. Cells were also incubated with 5.5 mM of glucose and 27.5 mM of mannitol as osmotic controls. The db/db mice were orally supplemented with 10 mg/kg of chrysin daily for 10 weeks. The db/m mice were introduced as control animals. The secretion of PEDF was measured with ELISA kits (**C,D**). Culture media, cell lysates, and retinal tissue extracts were subject to Western blot analysis with a primary antibody against PEDF, RPE65, and rhodopsin (**B,E,F**). β-actin protein was used as a cellular internal control for RPE cells. Bar graphs (mean ± SEM, n = nine independent experiments) in the bottom panels represent the densitometric results of upper blot bands. * values in bar graphs indicate significant different at $p < 0.05$.

3.3. Protective Effects of Chrysin on RPE65 Induction

The visual cycle occurring between the photoreceptors and RPE regenerates 11-cis-retinal through a series of steps involving specialized enzymes and retinoid binding proteins [18,36]. RPE65 is a critical enzyme responsible for the conversion of all-trans-retinyl esters to 11-cis-retinal during phototransduction [17]. When RPE cells were stimulated with 33 mM of glucose, the cellular expression of RPE65 markedly decreased (Figure 2E). However, treatment with ≥1 µM of chrysin elevated the RPE65 expression to the glucose control level. This study attempted to show that chrysin ameliorated the RPE65 induction in diabetic animal retina. As expected, the eye tissue levels of RPE65 and rhodopsin dropped in db/db mice (Figure 2F). When db/db mice were orally administrated with 10 mg/kg of chrysin for 10 weeks, the tissue levels of these proteins were boosted. Also, the induction of RPE65 and rhodopsin was examined in mouse retina by using a double immunohistochemical staining of FITC and Cy3. In db/m control retina, the RPE65 in the photoreceptor inner/outer segment (IS/OS) layer was strongly green-stained, and the retinal rhodopsin was red-stained in retina (Figure 3). However, there was a weak staining of RPE and rhodopsin in the IS/OS layer of diabetic mouse retina, as compared to that of the db/m control. Oral supplementation of chrysin to db/db mice improved the induction of RPE65 and rhodopsin, in which the double staining of FITC and Cy3 was indistinguishable from that of the db/m control (Figure 3).

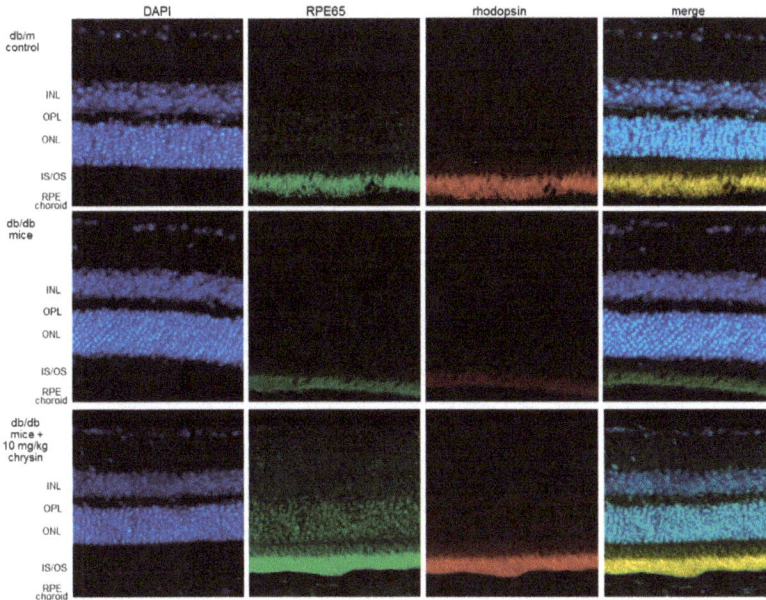

Figure 3. Induction of RPE65 and rhodopsin by chrysin. The db/db mice were orally supplemented with 10 mg/kg of chrysin daily for 10 weeks. The db/m mice were introduced as control animals. Histological sections of mouse retina were immunohistochemically double-stained using a primary antibody of RPE5 and rhodopsin. The RPE65 was identified as fluorescein isothiocyanate (FITC) green staining, and the rhodopsin localization was detected with Cy3 red staining. The sections were counterstained with 4′,6-diamidino-2-phenylindole (DAPI, blue) for the nuclear staining. The triple staining of DAPI-FITC-Cy3 was merged. Magnification: 200×. Retinal layers are labeled as follows: inner nuclear layer (INL), outer plexiform layer (OPL), outer nuclear layer (ONL), photoreceptor inner segment/outer segment (IS/OS), and retinal pigment epithelium (RPE).

3.4. Protective Effects of Chrysin on Induction of Visual Cycle-Related Proteins

This study examined whether the hyperglycemic insult influenced the induction of other visual cycle-related enzymes of LRAT and RDH5, and whether chrysin improved the dysfunction of retinoid visual cycle by glucose. The eye tissue levels of LRAT and RDH5 declined in diabetic mice (Figure 4A), while chrysin increased their levels. In addition, the tissue levels of the retinoid binding proteins of CRBP, CRALBP, and IRBP were diminished in eyes of db/db mice (Figure 4B). In contrast, the levels of these binding proteins were nearly completely restored to those of normal mouse retina. This study further examined the level of the vitamin A receptor of STRA6 in diabetic retina. There was a marked loss of STRA6 observed in diabetic eyes (Figure 4C). In chrysin-treated mouse eyes, the STRA6 level was significantly elevated. Accordingly, chrysin ameliorated the diabetes-associated malfunction of the retinoid visual cycle.

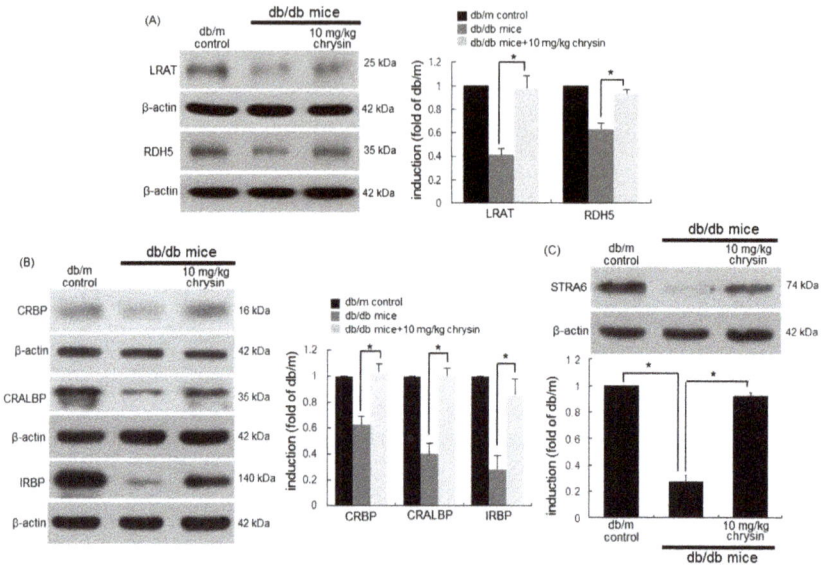

Figure 4. Elevation of retinal tissue induction of visual cycle enzymes (**A**), and retinoid binding proteins (**B**), and stimulated by retinoic acid 6 (STRA6, **C**) in chrysin-treated mice. The db/db mice were orally supplemented with 10 mg/kg of chrysin daily for 10 weeks. The db/m mice were introduced as control animals. Mouse retinal tissue extracts were subject to Western blot analysis with a primary antibody against each target protein of lecithin-retinol acyltransferase (LRAT), retinol dehydrogenase 5 (RDH5), cellular retinol binding protein (CRBP), cellular retinaldehyde-binding protein (CRALBP), interphotoreceptor retinoid-binding protein (IRBP), or STRA6. β-actin protein was used as an internal control. Bar graphs (mean ± SEM, n = nine independent experiments) in the bottom or right panels represent densitometric results of upper blot bands. * Values in bar graphs indicate a significant difference at $p < 0.05$.

3.5. Blockade of AGE-Mediated Malfunction of Visual Cycle by Chrysin

High glucose prompted the AGE production and RAGE induction in RPE cells in a time course-dependent manner for six days (Figure 5A). The cellular induction of AGEs and RAGE was highly enhanced on two days post-stimulation with 33 mM of glucose. However, RPE cells exposed to glucose in the presence of 1–20 μM chrysin for three days demoted the AGE secretion and RAGE induction (Figure 5B). Consistently, the tissue levels of AGEs and RAGE were enhanced in diabetic eyes

(Figure 5C). In contrast, their levels were diminished in diabetic mice orally treated with 10 mg/kg of chrysin for 10 weeks (Figure 5C).

To investigate the involvement of AGEs in the glucose-induced malfunction of the visual cycle, 100 μg/mL of AGE-BSA were treated to RPEC for three days in the absence and presence of 1–20 μM chrysin. The PEDF secretion declined in AGE-BSA-treated RPE cells (Figure 5D). In addition, the induction of visual cycle-related proteins of RPE 65, LRAT, and RDH5 decreased in AGE-exposed RPE cells. In contrast, the presence of 20 μM of chrysin elevated the induction of these proteins dampened by 100 μg/mL of AGE-BSA (Figure 5D). Therefore, AGEs produced from RPE cells may be responsible for the glucose-induced loss of visual cycle-related proteins.

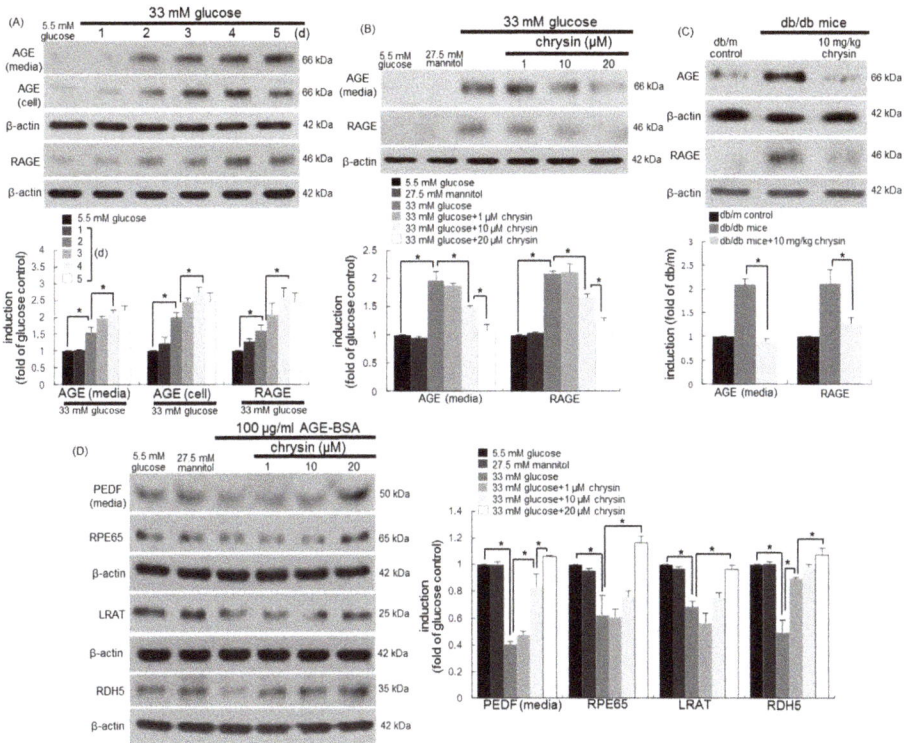

Figure 5. Temporal induction of advanced glycation end products (AGE) and receptor for advanced glycation end product (RAGE) by glucose (**A**), the inhibitory effects of chrysin on their induction (**B**,**C**), and restoration by chrysin of PEDF and retinal visual cycle enzymes in AGE-BSA-exposed human retinal pigment epithelial (RPE) cells (**D**). RPE cells were incubated with 33 mM of glucose in the absence and presence of 1–20 μM chrysin for up to five days. Cells were also incubated with 5.5 mM of glucose and 27.5 mM of mannitol as osmotic controls. The db/db mice were orally supplemented with 10 mg/kg of chrysin daily for 10 weeks. The db/m mice were introduced as control animals. Culture media, cell lysates, and mouse retinal tissue extracts were subject to Western blot analysis with a primary antibody against AGEs, RAGE, PEDF, RPE65, lecithin retinol acyltransferase (LRAT), and retinol dehydrogenase 5 (RDH5). β-actin protein was used as an internal control. Bar graphs (mean ± SEM, *n* = nine independent experiments) in the bottom or right panel represent densitometric results of upper or left blot bands. * Values in bar graphs indicate a significant difference at *p* < 0.05.

3.6. Involvement of ER Stress in Loss of Visual Cycle Proteins

Aberrant functions take place in ER by glucose insult, which promotes the apoptotic demise of retinal cells in diabetic eyes and impairs visual function [6,7]. This study examined whether ER stress influenced the induction of the retinoid visual cycle enzymes of RPE65, LRAT, and RDH5 located on the smooth ER of RPE cells. As expected, the ER stress inducer tunicamycin resulted in marked ATF6 activation on the day two in RPE cells cultured for up to five days (Figure 6A). During ER stress, the gradual loss of RPE65, LRAT, and RDH5 temporally occurred in tunicamycin-exposed RPEC (Figure 6A). Marked reduction of these proteins was observed at three days-post stimulation. In addition, another ER stress inducer thapsigargin, which is a specific inhibitor of the sarcoplasmic/endoplasmic reticulum Ca^{2+}-ATPase, caused the gradual reduction of the visual cycle enzymes of RPE65, LRAT, and RDH5 in RPE cells for five days (Figure 6B). These proteins were near-completely lost at three days-post stimulation with 10 nM of thapsigargin. Furthermore, the PEDF secretion was diminished in RPE cells stimulated with both tunicamycin and thapsigargin for up to five days (Figure 6C).

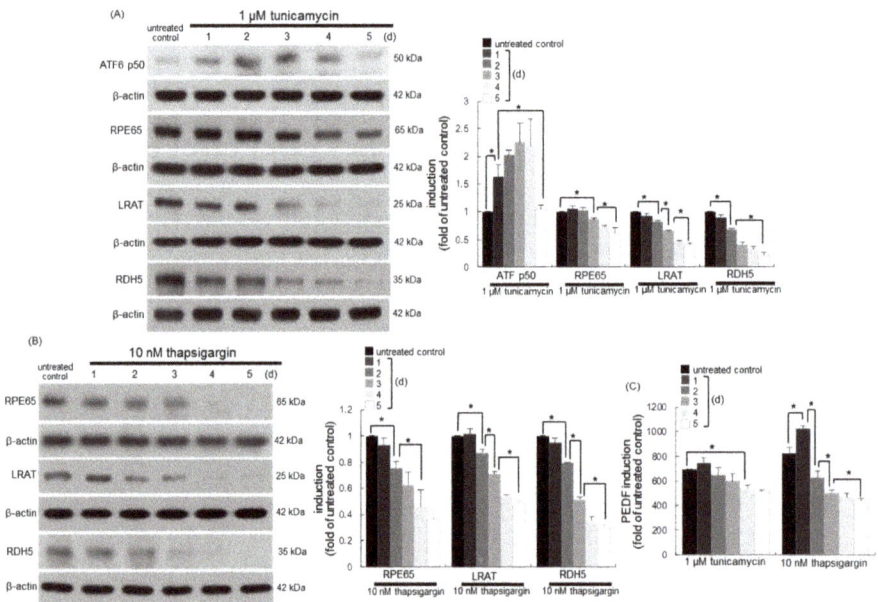

Figure 6. Temporal inhibition of retinal visual cycle enzymes (**A,B**) and PEDF (**C**) in endoplasmic reticulum (ER) stress-faced human retinal pigment epithelial (RPE) cells. RPE cells were incubated in the absence and presence of 1 µM of tunicamycin and 10 nM of thapsigargin for up to five days. Cell lysates were subject to Western blot analysis with a primary antibody against activating transcription factor 6 (ATF6) p50, RPE65, LRAT, and RDH5 (**A,B**). β-actin protein was used as an internal control. Bar graphs (mean ± SEM, *n* = nine independent experiments) in the right represent the densitometric results of left blot bands (**A,B**). The secretion of PEDF in RPE cell culture media was measured with an ELISA kit (**C**). * Values in bar graphs indicate a significant difference at *p* < 0.05.

3.7. Inhibition of ER Stress-Mediated Loss of Visual Cycle Proteins by Chrysin

This study investigated whether glucose and AGEs induced ER stress just as tunicamycin and thapsigargin did. High glucose activated ATF6 in RPE cells from one day-post stimulation (Figure 7A). Another ER stress sensor protein of IRE1α was induced in 33 mM of glucose-exposed RPE cells (Figure 7B). When 1–20 µM chrysin was treated to glucose-stimulated RPE cells, the ATF6 activation and IRE1α induction were dose-dependently attenuated (Figure 7B). Accordingly, chrysin may improve visual function through suppressing the ER stress of RPE cells in diabetic eyes. In fact, the ER stress-related sensors and chaperone of GRP78/Bip, ATF6, and IRE1α were clearly detected in diabetic eyes (Figure 7C). The eye tissue levels of these proteins were dampened in db/db mice orally administrated with 10 mg/kg of chrysin for 10 weeks. Finally, 100 μg/mL of AGE-BSA per se prompted both ATF6 activation and IRE1α induction in RPE cells, evoking ER stress (Figure 7D). However, the chrysin treatment of glucose-stimulated RPE cells reduced the induction and activation of ER stress-related sensors. Collectively, chrysin may ameliorate the malfunction of the retinoid visual cycle in diabetic retina through combating AGE-induced ER stress.

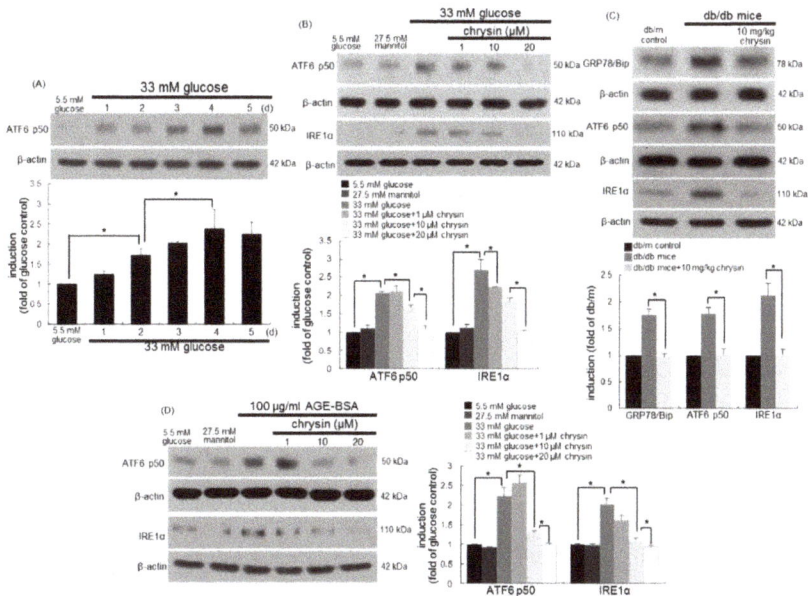

Figure 7. Induction of ER stress by glucose and AGE-BSA and its inhibition by chrysin. Human retinal pigment epithelial (RPE) cells were incubated with 33 mM of glucose (**A,B**) or 100 μg/mL of AGE-BSA (**D**) in the absence and presence of between 1–20 μM of chrysin for up to five days. Cells were also incubated with 5.5 mM of glucose and 27.5 mM of mannitol as osmotic controls. The db/db mice were orally supplemented with 10 mg/kg of chrysin daily for 10 weeks (**C**). The db/m mice were introduced as control animals. Cell lysates and tissue extracts were subject to Western blot analysis with a primary antibody against ATF6 p50, inositol-requiring enzyme 1α (IRE1α), and 78 kDa glucose-regulated protein/binding immunoglobulin protein (GRP78/Bip). β-actin protein was used as an internal control. Bar graphs (mean ± SEM, *n* = nine independent experiments) in the bottom represent the densitometric results of upper blot or left bands (**A,B**). * Values in bar graphs indicate a significant difference at *p* < 0.05.

4. Discussion

Ten major findings were extracted from this study. (1) Glucose influenced the production of VEGF, IGF-1, and PEDF in RPE cells, which was reversed by treating chrysin. (2) The ONL thickness of chrysin-administrated retina was enhanced. (3) The RPE65 reduction in glucose-exposed RPE cells was enhanced by chrysin. (4) Oral administration of chrysin for 10 weeks augmented the protein levels of RPE65, and diminished rhodopsin in db/db mouse retina. (5) The reduced eye tissue levels of LRAT and RDH5 increased in chrysin-administrated diabetic mice. (6) The diabetic tissue levels of CRBP, CRALBP, IRBP, and STRA6 were restored to those of normal mouse retina. (7) The presence of chrysin demoted the AGE secretion and RAGE induction in RPE cells exposed to glucose. (8) Supplementing chrysin to diabetic mice diminished the eye tissue levels of AGEs and RAGE. (9) The reduced induction of PEDF, RPE65, LRAT, and RDH5 was elevated by chrysin in AGE-BSA-exposed RPE cells. (10) The treatment of RPE cells with chrysin reduced the activation of ER stress. These results indicated that chrysin may ameliorate the malfunction of retinoid visual cycle in diabetic retina through combating AGE-induced ER stress (Figure 8).

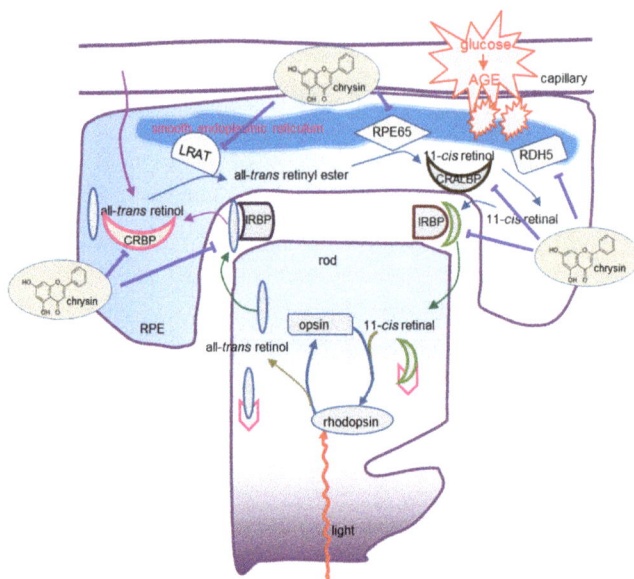

Figure 8. Schematic diagram showing the inhibitory effects of chrysin on the malfunction of retinoid visual cycle and its mechanistic actions in glucose/AGE-exposed human retinal pigment epithelial (RPE) cells. Chrysin inhibited the ER stress activated by glucose/AGEs. All of the arrows indicate increase, activation or induction; ⊥ indicates inhibition or blockade.

There is increasing evidence that angiogenic growth factors and cytokines, including VEGF and IGF-1, play crucial roles in proliferative DR development [37]. The intraocular neovascularization is counteracted by the formation of anti-angiogenic factors such as PEDF and transforming growth factor-β (TGF-β). These factors could be used as markers for disease prognosis and therapy. However, several inhibitors of angiogenic factors, including VEGF and placental growth factor, have failed to demonstrate significant efficacy in the treatment of DR in clinical trials [38]. Natural compounds with minimal side effects for intraocular angiogenesis may serve as specific and efficacious agents for a potential DR therapy [39]. This study showed that chrysin inhibited the induction of angiogenic VEGF in glucose-stimulated RPE cells, while the induction of antiangiogenic PEDF was up-regulated. Accordingly, the molecules that shift the balance toward PEDF and away from VEGF may prove useful tools in retinal neovascularization. Interestingly, this study showed that glucose temporally attenuated the induction of angiogenic IGF-1 of RPE cells, which was reversed by treating chrysin. A recent study shows that IGF-1 protects RPE cells from amiodarone-mediated injury via activation of PI3K/Akt signaling [40]. In this study, IGF-1 had potential as a protective agent for deterring the AGE-mediated toxicity of glucose.

This study identified protective mechanisms of chrysin that are germane to visual cycle malfunction in RPE. Pharmacotherapy with visual cycle modulators, including oral retinoids, may improve visual acuity and visual fields in blinding diseases that lack effective treatment options [41]. These modulators may show the side effects and lack of proof of efficacy in humans.

Chrysin improved the dysfunction of visual cycle by glucose through enhancing the reduced expression of the rod visual cycle enzymes of RPE65, LRAT, and RDH5 localized on the smooth ER of RPE cells. This study further found that chrysin boosted the induction of all of the retinoid binding proteins of CRALBP and CRBP in RPE cells and IRBP in the subretinal space that diminished in diabetic retina. Unfortunately, this study did not examine the accumulation of all-trans-retinol and all-trans-retinyl esters in RPE cells due to a lack of LRAT, RPE65, and RDH5 enzymes, and did not measure the tissue level of the visual chromophore of 11-cis-retinal, which is the light-sensitive component of visual pigments. Nevertheless, chrysin could facilitate the efficient retinoid cycling between the rod outer segment and the RPE through maintaining optimal levels of visual cycle enzymes in RPE cells, as well as retinoid binding proteins. Several studies have demonstrated that dietary flavonoids attenuate retinal degeneration through deterring the apoptosis of RPE cells and photoreceptors via the inhibition of retinal oxidative stress and inflammation [42–44]. However, little is known about the protective roles of dietary compounds in the alteration of rod visual cycle components. This is the first study demonstrating the protective effects of chrysin on the malfunction of visual cycle components that are present in diabetic RPE and photoreceptors.

The molecular mechanisms underlying the inflammatory pathways associated with DR are not concretely defined. In addition, a mutual connection between oxidative stress and major metabolic abnormalities has been implicated in the development of DR [45,46]. Curcumin inhibits oxidative stress and protects Müller cells in diabetic retina [39,47]. Recently, there has been much advance study in the possible molecular mechanisms leading to autophagy that are involved in the pathophysiology of DR [48]. On the other hand, the AGE-RAGE system plays a crucial role in eliciting oxidative stress and inflammatory reactions, and is involved in diabetic damage [49]. This study revealed that the AGE-RAGE system was activated in glucose-stimulated RPE cells and diabetic eyes, ultimately inducing visual cycle dysfunction in retina. Chrysin diminished the AGE accumulation and RAGE induction in glucose-exposed RPE cells and diabetic eyes. Various inhibitors of the AGE-RAGE system have been evaluated as their therapeutic utility for DR [49]. The antioxidant sulforaphane that is present in edible cruciferous vegetables inhibits AGE-induced pericyte injury through blocking the AGE-RAGE axis in pericytes, which is a novel therapeutic target for the treatment of DR [50].

Numerous studies have addressed that ER stress plays a putative role in the visual function [6,9,51]. Oxidative stress and ER stress contribute to the progression of age-related macular degeneration, which is characterized by retinal degeneration resulting in the loss of central vision [52]. In this study, AGEs promoted the induction of ER stress sensor proteins of ATF6 and IRE1α in RPE cells. In addition, the ER stress inducers of tunicamycin and thapsigargin dampened the expression of visual cycle components in retinal pigment epithelium. These results identified ER stress as a negative regulator of the RPE visual cycle. Since the dysfunction of any retinoid cycle enzymes in the RPE can cause ocular diseases, the blockade of glucose-triggered ER stress by chrysin may be a manipulative strategy to protect retinal degeneration and the vision loss of DR. Moreover, ER stress is a critical adverse component of RPE cells and photoreceptors [53,54]. One investigation shows that the tea polyphenol (−)epigallocatechin gallate inhibited ER stress-mediated apoptotic cell death via the proper calcium homeostasis and decreased ROS production in age-related macular degeneration [55]. Bilberry extract attenuates photoinduced apoptosis and visual dysfunction via ER stress attenuation in the retina [54]. The accumulation of misfolded rhodopsin within the ER is a prominent cause of retinitis pigmentosa [56]. This study found that the tissue level of rhodopsin declined in diabetic eyes, which was ameliorated by implementing the chrysin strategy. The chrysin supplementation could enhance the clearance of misfolded rhodopsin and maintain proper unfolded protein response signaling in the ER.

In summary, this study investigated the capability of chrysin in counteracting the diabetes-mediated malfunction of the visual cycle in glucose-exposed RPE cells and diabetic mice. Chrysin reciprocally influenced the production of VEGF, IGF-1, and PEDF in glucose-stimulated RPE cells and diabetic eyes. This compound neutralized the hyperglycemia-elicited reduction of the retinoid visual cycle components of RPE65, LRAT, RDH5, CRBP, CRALBP, IRBP, and STRA6 in retina. In addition, chrysin encumbered the formation of AGEs and RAGE in diabetic eyes due to glucose insults, leading to the loss of RPE enzymes for the visual cycle. Furthermore, chrysin blocked the ER stress of RPE cells evoked by glucose and AGEs in diabetic mice. Therefore, chrysin was a therapeutic drug antagonizing the malfunction of the retinal visual cycle, leading to the loss of retinal vision in cellular or animal models of diabetic complications. It can be assumed that the protective effects of chrysin on the malfunctioned visual cycle in diabetic eyes may be, at least partly, due to its glucose-lowering effects.

Author Contributions: M.-K.K. and Y.-H.K. (Young-Hee Kang) designed research; M.-K.K., E.-J.L., Y.-H.K. (Yun-Ho Kim), D.Y.K., H.O., and S.-I.K. conducted research; M.-K.K., and D.Y.K. analyzed data; M.-K.K., and Y.-H.K. (Young-Hee Kang) wrote the paper. Y.-H.K. (Young-Hee Kang) had primary responsibility for final content. All authors read and approved the final manuscript.

Funding: This work was supported by the National Research Foundation of Korea (NRF) grants funded by the Korea government (MEST) (2015R1A2A2A01006666, and NRF-2017R1A6A3A04011473).

Conflicts of Interest: The authors declare that they have no conflict of interest.

References

1. Roy, S.; Kern, T.S.; Song, B.; Stuebe, C. Mechanistic insights into pathological changes in the diabetic retina: Implications for targeting diabetic retinopathy. *Am. J. Pathol.* **2017**, *187*, 9–19. [CrossRef] [PubMed]
2. Shah, A.R.; Gardner, T.W. Diabetic retinopathy: Research to clinical practice. *Clin. Diabetes Endocrinol.* **2017**, *3*, 9. [CrossRef] [PubMed]
3. Wong, T.Y.; Cheung, C.M.; Larsen, M.; Sharma, S.; Simó, R. Diabetic retinopathy. *Nat. Rev. Dis. Prim.* **2016**, *2*, 16012. [CrossRef] [PubMed]
4. Nentwich, M.M.; Ulbig, M.W. Diabetic retinopathy-ocular complications of diabetes mellitus. *World J. Diabetes* **2015**, *6*, 489–499. [CrossRef] [PubMed]
5. Crawford, T.N.; Alfaro, D.V., III; Kerrison, J.B.; Jablon, E.P. Diabetic retinopathy and angiogenesis. *Curr. Diabetes Rev.* **2009**, *5*, 8–13. [CrossRef] [PubMed]
6. Oshitari, T.; Hata, N.; Yamamoto, S. Endoplasmic reticulum stress and diabetic retinopathy. *Vasc. Health Risk Manag.* **2008**, *4*, 115–122. [CrossRef] [PubMed]
7. Roy, S.; Trudeau, K.; Roy, S.; Tien, T.; Barrette, K.F. Mitochondrial dysfunction and endoplasmic reticulum stress in diabetic retinopathy: Mechanistic insights into high glucose-induced retinal cell death. *Curr. Clin. Pharmacol.* **2013**, *8*, 278–284. [CrossRef] [PubMed]
8. Wert, K.J.; Lin, J.H.; Tsang, S.H. General pathophysiology in retinal degeneration. *Dev. Ophthalmol.* **2014**, *53*, 33–43. [PubMed]
9. Periyasamy, P.; Shinohara, T. Age-related cataracts: Role of unfolded protein response, Ca^{2+} mobilization, epigenetic DNA modifications, and loss of Nrf2/Keap1 dependent cytoprotection. *Prog. Retin. Eye. Res.* **2017**, *60*, 1–19. [CrossRef] [PubMed]
10. Boya, P.; Esteban-Martínez, L.; Serrano-Puebla, A.; Gómez-Sintes, R.; Villarejo-Zori, B. Autophagy in the eye: Development, degeneration, and aging. *Prog. Retin. Eye Res.* **2016**, *55*, 206–245. [CrossRef] [PubMed]
11. Cohen, S.R.; Gardner, T.W. Diabetic retinopathy and diabetic macular edema. *Dev. Ophthalmol.* **2016**, *55*, 137–146. [PubMed]
12. Wright, C.B.; Redmond, T.M.; Nickerson, J.M. A history of the classical visual cycle. *Prog. Mol. Biol. Transl. Sci.* **2015**, *134*, 433–448. [PubMed]
13. Du, M.; Wu, M.; Fu, D.; Yang, S.; Chen, J.; Wilson, K.; Lyons, T.J. Effects of modified LDL and HDL on retinal pigment epithelial cells: A role in diabetic retinopathy? *Diabetologia* **2013**, *56*, 2318–2328. [CrossRef] [PubMed]
14. Kim, D.I.; Park, M.J.; Choi, J.H.; Lim, S.K.; Choi, H.J.; Park, S.H. Hyperglycemia-induced GLP-1R downregulation causes RPE cell apoptosis. *Int. J. Biochem. Cell Biol.* **2015**, *59*, 41–51. [CrossRef] [PubMed]
15. Chang, K.C.; Snow, A.; LaBarbera, D.V.; Petrash, J.M. Aldose reductase inhibition alleviates hyperglycemic effects on human retinal pigment epithelial cells. *Chem. Biol. Interact.* **2015**, *234*, 254–260. [CrossRef] [PubMed]

16. Chan, C.M.; Huang, D.Y.; Huang, Y.P.; Hsu, S.H.; Kang, L.Y.; Shen, C.M.; Lin, W.W. Methylglyoxal induces cell death through endoplasmic reticulum stress-associated ROS production and mitochondrial dysfunction. *J. Cell. Mol. Med.* **2016**, *20*, 1749–1760. [CrossRef] [PubMed]

17. Cai, X.; Conley, S.M.; Naash, M.I. RPE65: Role in the visual cycle, human retinal disease, and gene therapy. *Ophthalmic. Genet.* **2009**, *30*, 57–62. [CrossRef] [PubMed]

18. Kiser, P.D.; Golczak, M.; Maeda, A.; Palczewski, K. Key enzymes of the retinoid (visual) cycle in vertebrate retina. *Biochim. Biophys. Acta.* **2012**, *1821*, 137–151. [CrossRef] [PubMed]

19. Travis, G.H.; Golczak, M.; Moise, A.R.; Palczewski, K. Diseases caused by defects in the visual cycle: Retinoids as potential therapeutic agents. *Ann. Rev. Pharmacol. Toxicol.* **2007**, *47*, 469. [CrossRef] [PubMed]

20. Boyer, D.S.; Hopkins, J.J.; Sorof, J.; Ehrlich, J.S. Anti-vascular endothelial growth factor therapy for diabetic macular edema. *Ther. Adv. Endocrinol. Metab.* **2013**, *4*, 151–169. [CrossRef] [PubMed]

21. Agarwal, A.; Parriott, J.; Demirel, S.; Argo, C.; Sepah, Y.J.; Do, D.V.; Nguyen, Q.D. Nonbiological pharmacotherapies for the treatment of diabetic macular edema. *Expert. Opin. Pharmacother.* **2015**, *16*, 2625–2635. [CrossRef] [PubMed]

22. Obrosova, I.G.; Kador, P.F. Aldose reductase/polyol inhibitors for diabetic retinopathy. *Curr. Pharm. Biotechnol.* **2011**, *12*, 373–385. [CrossRef] [PubMed]

23. Falkenstein, I.A.; Cheng, L.; Wong-Staal, F.; Tammewar, A.M.; Barron, E.C.; Silva, G.A.; Li, Q.X.; Yu, D.; Hysell, M.; Liu, G.; et al. Toxicity and intraocular properties of a novel long-acting anti-proliferative and anti-angiogenic compound IMS2186. *Curr. Eye Res.* **2008**, *33*, 599–609. [CrossRef] [PubMed]

24. Jo, H.; Jung, S.H.; Yim, H.B.; Lee, S.J.; Kang, K.D. The effect of baicalin in a mouse model of retinopathy of prematurity. *BMB Rep.* **2015**, *48*, 271–276. [CrossRef] [PubMed]

25. Jia, Y.P.; Sun, L.; Yu, H.S.; Liang, L.P.; Li, W.; Ding, H.; Song, X.B.; Zhang, L.J. The Pharmacological effects of lutein and zeaxanthin on visual disorders and cognition diseases. *Molecules* **2017**, *22*, 610. [CrossRef] [PubMed]

26. Neelam, K.; Goenadi, C.J.; Lun, K.; Yip, C.C.; Au Eong, K.G. Putative protective role of lutein and zeaxanthin in diabetic retinopathy. *Br. J. Ophthalmol.* **2017**, *101*, 551–558. [CrossRef] [PubMed]

27. Feng, X.; Qin, H.; Shi, Q.; Zhang, Y.; Zhou, F.; Wu, H.; Ding, S.; Niu, Z.; Lu, Y.; Shen, P. Chrysin attenuates inflammation by regulating M1/M2 status via activating PPARγ. *Biochem. Pharmacol.* **2014**, *89*, 503–514. [CrossRef] [PubMed]

28. Nabavi, S.F.; Braidy, N.; Habtemariam, S.; Orhan, I.E.; Daglia, M.; Manayi, A.; Gortzi, O.; Nabavi, S.M. Neuroprotective effects of chrysin: From chemistry to medicine. *Neurochem. Int.* **2015**, *90*, 224–231. [CrossRef] [PubMed]

29. Samarghandian, S.; Farkhondeh, T.; Azimi-Nezhad, M. Protective effects of chrysin against drugs and toxic agents. *Dose Response* **2017**, *15*. [CrossRef] [PubMed]

30. Walle, T.; Otake, Y.; Brubaker, J.A.; Walle, U.K.; Halushka, P.V. Disposition and metabolism of the flavonoid chrysin in normal volunteers. *Br. J. Clin. Pharmacol.* **2001**, *51*, 143–146. [CrossRef] [PubMed]

31. Kang, M.K.; Park, S.H.; Kim, Y.H.; Lee, E.J.; Antika, L.D.; Kim, D.Y.; Choi, Y.J.; Kang, Y.H. Dietary compound chrysin inhibits retinal neovascularization with abnormal capillaries in db/db mice. *Nutrients* **2016**, *8*, 782. [CrossRef] [PubMed]

32. Kang, M.K.; Park, S.H.; Choi, Y.J.; Shin, D.; Kang, Y.H. Chrysin inhibits diabetic renal tubulointerstitial fibrosis through blocking epithelial to mesenchymal transition. *J. Mol. Med.* **2015**, *93*, 759–772. [CrossRef] [PubMed]

33. Wilkinson-Berka, J.L.; Wraight, C.; Werther, G. The role of growth hormone, insulin-like growth factor and somatostatin in diabetic retinopathy. *Curr. Med. Chem.* **2006**, *13*, 3307–3317. [CrossRef] [PubMed]

34. Sall, J.W.; Klisovic, D.D.; O'Dorisio, M.S.; Katz, S.E. Somatostatin inhibits IGF-1 mediated induction of VEGF in human retinal pigment epithelial cells. *Exp. Eye Res.* **2004**, *79*, 465–476. [CrossRef] [PubMed]

35. Mori, K.; Duh, E.; Gehlbach, P.; Ando, A.; Takahashi, K.; Pearlman, J.; Mori, K.; Yang, H.S.; Zack, D.J.; Ettyreddy, D.; et al. Pigment epithelium-derived factor inhibits retinal and choroidal neovascularization. *J. Cell. Physiol.* **2001**, *188*, 253–263. [CrossRef] [PubMed]

36. Parker, R.O.; Crouch, R.K. Retinol dehydrogenases (RDHs) in the visual cycle. *Exp. Eye Res.* **2010**, *91*, 788–792. [CrossRef] [PubMed]

37. Simó, R.; Carrasco, E.; García-Ramírez, M.; Hernández, C. Angiogenic and antiangiogenic factors in proliferative diabetic retinopathy. *Curr. Diabetes Rev.* **2006**, *2*, 71–98. [CrossRef] [PubMed]

38. Nguyen, Q.D.; De Falco, S.; Behar-Cohen, F.; Lam, W.C.; Li, X.; Reichhart, N.; Ricci, F.; Pluim, J.; Li, W.W. Placental growth factor and its potential role in diabetic retinopathy and other ocular neovascular diseases. *Acta Ophthalmol.* **2018**, *96*, e1–e9. [CrossRef] [PubMed]

39. Peddada, K.V.; Brown, A.; Verma, V.; Nebbioso, M. Therapeutic potential of curcumin in major retinal pathologies. *Int. Ophthalmol.* **2018**. [CrossRef] [PubMed]

40. Liao, R.; Yan, F.; Zeng, Z.; Wang, H.; Qiu, K.; Xu, J.; Zheng, W. Insulin-like growth factor-1 activates PI3K/Akt signalling to protect human retinal pigment epithelial cells from amiodarone-induced oxidative injury. *Br. J. Pharmacol.* **2018**, *175*, 125–139. [CrossRef] [PubMed]

41. Hussain, R.M.; Gregori, N.Z.; Ciulla, T.A.; Lam, B.L. Pharmacotherapy of retinal disease with visual cycle modulators. *Expert. Opin. Pharmacother.* **2018**, *19*, 471–481. [CrossRef] [PubMed]

42. Ha, J.H.; Shil, P.K.; Zhu, P.; Gu, L.; Li, Q.; Chung, S. Ocular inflammation and endoplasmic reticulum stress are attenuated by supplementation with grape polyphenols in human retinal pigmented epithelium cells and in C57BL/6 mice. *J. Nutr.* **2014**, *144*, 799–806. [CrossRef] [PubMed]

43. Bian, M.; Zhang, Y.; Du, X.; Xu, J.; Cui, J.; Gu, J.; Zhu, W.; Zhang, T.; Chen, Y. Apigenin-7-diglucuronide protects retinas against bright light-induced photoreceptor degeneration through the inhibition of retinal oxidative stress and inflammation. *Brain Res.* **2017**, *1663*, 141–150. [CrossRef] [PubMed]

44. Hytti, M.; Szabó, D.; Piippo, N.; Korhonen, E.; Honkakoski, P.; Kaarniranta, K.; Petrovski, G.; Kauppinen, A. Two dietary polyphenols, fisetin and luteolin, reduce inflammation but augment DNA damage-induced toxicity in human RPE cells. *J. Nutr. Biochem.* **2017**, *42*, 37–42. [CrossRef] [PubMed]

45. Kowluru, R.A.; Mishra, M. Oxidative stress, mitochondrial damage and diabetic retinopathy. *Biochim. Biophys. Acta.* **2015**, *1852*, 2474–2483. [CrossRef] [PubMed]

46. Chen, M.; Curtis, T.M.; Stitt, A.W. Advanced glycation end products and diabetic retinopathy. *Curr. Med. Chem.* **2013**, *20*, 3234–3240. [CrossRef] [PubMed]

47. Zuo, Z.F.; Zhang, Q.; Liu, X.Z. Protective effects of curcumin on retinal Müller cell in early diabetic rats. *Int. J. Ophthalmol.* **2013**, *6*, 422–424. [PubMed]

48. Rosa, M.D.; Distefano, G.; Gagliano, C.; Rusciano, D.; Malaguarnera, L. Autophagy in diabetic retinopathy. *Curr. Neuropharmacol.* **2016**, *14*, 810–825. [CrossRef] [PubMed]

49. Wu, Y.; Tang, L.; Chen, B. Oxidative stress: Implications for the development of diabetic retinopathy and antioxidant therapeutic perspectives. *Oxid. Med. Cell. Longev.* **2014**, *2014*, 752387. [CrossRef] [PubMed]

50. Maeda, S.; Matsui, T.; Ojima, A.; Takeuchi, M.; Yamagishi, S. Sulforaphane inhibits advanced glycation end product-induced pericyte damage by reducing expression of receptor for advanced glycation end products. *Nutr. Res.* **2014**, *34*, 807–813. [CrossRef] [PubMed]

51. Elmasry, K.; Ibrahim, A.S.; Saleh, H.; Elsherbiny, N.; Elshafey, S.; Hussein, K.A.; Al-Shabrawey, M. Role of endoplasmic reticulum stress in 12/15-lipoxygenase-induced retinal microvascular dysfunction in a mouse model of diabetic retinopathy. *Diabetologia* **2018**, *61*, 1220–1232. [CrossRef] [PubMed]

52. Minasyan, L.; Sreekumar, P.G.; Hinton, D.R.; Kannan, R. Protective mechanisms of the mitochondrial-derived peptide humanin in oxidative and endoplasmic reticulum stress in RPE cells. *Oxid. Med. Cell. Longev.* **2017**, *2017*, 1675230. [CrossRef] [PubMed]

53. Li, J.; Cai, X.; Xia, Q.; Yao, K.; Chen, J.; Zhang, Y.; Naranmandura, H.; Liu, X.; Wu, Y. Involvement of endoplasmic reticulum stress in all-trans-retinal-induced retinal pigment epithelium degeneration. *Toxicol. Sci.* **2015**, *143*, 196–208. [CrossRef] [PubMed]

54. Osada, H.; Okamoto, T.; Kawashima, H.; Toda, E.; Miyake, S.; Nagai, N.; Kobayashi, S.; Tsubota, K.; Ozawa, Y. Neuroprotective effect of bilberry extract in a murine model of photo-stressed retina. *PLoS ONE* **2017**, *12*, e0178627. [CrossRef] [PubMed]

55. Karthikeyan, B.; Harini, L.; Krishnakumar, V.; Kannan, V.R.; Sundar, K.; Kathiresan, T. Insights on the involvement of (-)-epigallocatechin gallate in ER stress-mediated apoptosis in age-related macular degeneration. *Apoptosis* **2017**, *22*, 72–85. [CrossRef] [PubMed]

56. Griciuc, A.; Aron, L.; Ueffing, M. ER stress in retinal degeneration: A target for rational therapy? *Trends Mol. Med.* **2011**, *17*, 442–451. [CrossRef] [PubMed]

nutrients

MDPI

Article

Lutein and Zeaxanthin Isomers Protect against Light-Induced Retinopathy via Decreasing Oxidative and Endoplasmic Reticulum Stress in BALB/cJ Mice

Minzhong Yu [1,2,*], Weiming Yan [1,3] and Craig Beight [1,4]

[1] Department of Ophthalmic Research, Cole Eye Institute, Cleveland Clinic Foundation, Cleveland, OH 44195, USA; ywming@fmmu.edu.cn (W.Y.); Beightc2@ccf.org (C.B.)
[2] Department of Ophthalmology, Cleveland Clinic Lerner College of Medicine of Case Western Reserve University, Cleveland, OH 44195, USA
[3] Department of Clinical Medicine, Faculty of Aerospace Medicine, Key Laboratory of Aerospace Medicine of the National Education Ministry, Fourth Military University, Xi'an 710032, China
[4] Louis Stokes Cleveland Veterans Affairs Medical Center, Cleveland, OH 44195, USA
* Correspondence: YUM@ccf.org; Tel.: +1-216-444-3071

Received: 14 May 2018; Accepted: 22 June 2018; Published: 28 June 2018

Abstract: Oxidative stress (OS) and endoplasmic reticulum stress (ERS) are the major factors underlying photoreceptor degeneration. Lutein, RR-zeaxanthin (3R,3'R-zeaxanthin) and RS (meso)-zeaxanthin (3R,3'S-RS- zeaxanthin) (L/Zi) could protect against cell damage by ameliorating OS in retina. In this study, we examined the effect of L/Zi supplementation in a mouse model of photoreceptor degeneration and investigated whether the treatment of L/Zi ameliorated OS and ERS. BALB/cJ mice after light exposure were used as the animal model. The protective effects of L/Zi were observed by electroretinography (ERG) and terminal deoxyuridine triphosphate nick-end labeling (TUNEL) analysis. The underlying mechanisms related to OS and ERS were explored by Western blotting. After L/Zi treatment, the ERG amplitudes were significantly higher, and the number of TUNEL-positive cells was significantly reduced compared to that of the vehicle group. Western blotting results revealed that OS was ameliorated according to the significant downregulation of phosphorylated c-Jun N-terminal kinase (p-JNK), and significant upregulation of nuclear factor erythroid 2-related factor 2 (Nrf2). In addition, ERS was reduced according to the significant downregulation of 78 kDa glucose-regulated protein (GRP78), phosphorylated protein kinase RNA-like endoplasmic reticulum kinase (p-PERK), activating transcription factor 4 (ATF4) and activating transcription factor (ATF6). Our data shows that L/Zi provided functional and morphological preservation of photoreceptors against light damage, which is probably related to its mitigation of oxidative and endoplasmic reticulum stress.

Keywords: lutein; RR-zeaxanthin; mesozeaxanthin (RS zeaxanthin); light damage; photoreceptor degeneration; oxidative stress; endoplasmic reticulum stress; electroretinography

1. Introduction

Excessive light exposure could often result in photoreceptor degeneration [1]. Accumulated light-induced damage is related to the development of age-related macular degeneration (AMD) [2], a globally prevalent retinal disease that causes blindness [3]. Although the details of the pathogenesis of light-induced retinal damage remain unclear, light-induced oxidative stress is known to cause photoreceptor loss [4,5]. During light exposure, the production of reactive oxygen species (ROS) is increased which causes retinal degeneration [6]. Quite a few antioxidants have been confirmed to be effective in reducing photoreceptor degeneration in animal models of light-induced damage [7,8].

In addition to the role of oxidative stress (OS) involved in the pathogenesis of photoreceptor degeneration, endoplasmic reticulum stress (ERS) also contributed to retinal degeneration in a variety of conditions such as age-related macular degeneration [9,10], retinitis pigmentosa (RP), diabetic retinopathy [11,12], and glaucoma [13,14]. Light exposure was also reported to induce ERS as well as abnormal ER membranes and endomembranes in 661W cells [15]. In addition, the levels of ERS protein markers are up-regulated in mouse retinas following exposure to light [16]. Consistent with this hypothesis, a number of studies have demonstrated that early administration of the agents that inhibit ERS could significantly decrease the rate of photoreceptor cell death in animal models of light-induced retinopathy [15,17].

Some carotenoids are highly concentrated in the light-exposed structures in plants and in the human retina [18]. Lutein, zeaxanthin and mesozeaxanthin (L/Zi) belong to the class of xanthophyll carotenoids, which are found in a number of fruits and vegetables [19] and are the major carotenoids in the human retina [20,21]. Higher dietary lutein and zeaxanthin intake can reduce the incidence of AMD, while low level of L/Zi has been associated with AMD [22]. Furthermore, administration of L/Zi has been showed to protect against retinal cell damage in diabetic retinopathy [23]. Barker et'al. [24] reported that lutein or zeaxanthin supplementation protected the fovea against acute blue light-induced retinal damage in rhesus monkeys, which is mainly attributed to the anti-oxidative properties of L/Zi [18,25]. However, the mechanisms of ERS underlying the effect of L/Zi in the amelioration of light-induced retinopathy have not been fully elucidated [21]. The BALB/cJ mouse exposed to light is a commonly used model characterized by photoreceptor degeneration [26]. The purpose of this study is to investigate the protective effects of L/Zi supplementation on this mouse model of early retinal cell degeneration and the underlying mechanisms of OS and ERS.

2. Materials and Methods

2.1. Animals and Experiment Design

Male BALB/cJ mice (9–13 weeks old) (Stock Number: 000651) were purchased from the Jackson Laboratory (Bar Harbor, ME, USA) and were housed under the same conditions in a low-illuminance (extracage/intracage: 13 lx/1 lx) vivarium under cyclic light (14 h light and 10 h dark) [27]. All animal experiments were performed in accordance with the Association for Research in Vision and Ophthalmology (ARVO) Statement for the Use of Animals in Ophthalmic and Vision Research, and were approved by the Institutional Animal Care and Use Committee of the Cleveland Clinic Foundation (IACUC protocol #2013-0933).

L/Zi, (10 mg/kg of body weight, OmniActive Health Technologies Ltd., Maharashtra, India) dissolved in sunflower oil (1 mg/mL, SFO, Sigma, St. Louis, MO, USA) or equal volume of SFO as vehicle was administered by daily oral gavage to BALB/cJ mice in treatment group (*n* = 7) and vehicle group (*n* = 7), respectively for a 5-day period from Day 1 to Day 5. The dose of L/Zi was chosen after a preliminary experiment (Data not shown), which is the maximum dose without the adverse effect on body weight. The number of the animals in our study was conservative enough to achieve the statistical power of at least 0.8.

2.2. Light Exposure

Mice were dark-adapted for 12 h, and then the pupils were dilated with 1% Tropicamide Ophthalmic Solution (Bausch & Lomb, Rochester, NY, USA). Mice were exposed to blue light (5000 lx) for 1 h. The blue light was obtained by filtering white fluorescent light by a filter which transmits light between 380 and 570 nm (Midnight Blue 5940, Solar Graphics, Clearwater, FL, USA). The mice were free to move in the light chamber during the light exposure. After the light exposure, the mice were returned to the low-illuminance animal room where they were housed.

2.3. Electroretinography (ERG)

After overnight dark adaptation, mice were anesthetized with a mixture of ketamine (80 mg/kg) and xylazine (16 mg/kg) diluted in saline. The pupils were dilated with 1% mydriacyl (tropicamide), 1% cyclopentolate HCl, 2.5% phenylephrine HCl) and the corneal surface was anesthetized with 0.5% proparacaine HCl eye drops. Stimulus response functions were obtained under dark- and then light-adapted conditions, and ERG a-wave and b-wave amplitudes in multiple flash luminances were measured and analyzed using published procedures [28].

2.4. TUNEL Analysis and Measurement of the Thickness of Outer Nuclear Layer (ONL)

Apoptosis was detected using the In Situ Cell Death Detection Kit (Roche Applied Science, Indianapolis, IN, USA). After the ERG recording, the eyes of the mice were immediately enucleated under euthanized, and then fixed in 4% paraformaldehyde for 2 h. The eyes were dehydrated in graded sucrose solutions (10–30%) and embedded in Optimal Cutting Temperature (OCT) compound. Retinal sections of 10 μm in thickness were cut near the optic nerve and incubated with freshly prepared 0.1% Triton X-100/0.1% sodium citrate permeabilization solution for 2 min on ice. After rinsing with phosphate buffered saline (PBS) for 3 times, sections were incubated with the TUNEL reaction mixture for 60 min at 37 °C in the dark and then rinsed with PBS 3 times. Sections were mounted with VECTASHIELD mounting medium with 4′,6-diamidino-2-phenylindole (DAPI) (Burlingame, CA, USA), and visualized at 400× magnification with the fluorescence microscope (BX60, Olympus, Tokyo, Japan). The pictures were taken in the area next to the optic nerve head. The number of apoptotic cells was counted in three sections of each eye and averaged.

For the measurement of the thickness of ONL, the DAPI-stained pictures taken in the TUNEL procedure were used. The thickness of the ONL was measured at 200 μm from the edge of the optic disc on either side of the optic nerve head using ImageJ 1.48v software (National Institutes of Health, Bethesda, MD, USA). Four sections from each retina were measured to calculate the mean of ONL thickness of that retina.

2.5. Western Blotting

The retinas were collected and homogenized on ice with the protein extraction reagent. Lysates were then centrifuged at 16,000 rpm at 4 °C for 25 min to obtain the supernatant. The protein content of the retinal extracts was measured by Pierce 660 nm Protein Assay Reagent (Thermo Scientific, Rockford, IL, USA). Equal amounts of protein (15 μg) of each extract in Laemmli Sample buffer was heated on a boiling water bath for 7 min and thereafter electrophoresed on 8–16% gradient SDS-polyacrylamide gel. After electrophoresis, proteins were transferred to a polyvinylidenedifluoride (PVDF) membrane. The membrane was blocked with Tris-buffered saline containing 0.1% Tween-20 (TBST) and 5% dried non-fat milk at room temperature for 2 h. Blocked membrane was incubated with indicated primary antibodies (anti-p-JNK1+p-JNK2+p-JNK3, 1:2000, #ab124956; anti-GRP78/BiP, 1:2000, #ab21685; anti-ATF6, 1:1500, #ab37149; Abcam, Cambridge, MA, USA. anti-Nrf2, 1:1000, #sc-722; anti-p-PERK, 1:500, #sc-32577; anti-ATF4, 1:1000, #sc-200; Santa Cruz Biotechnology, Dallas, TX, USA) at 4 °C overnight, followed by incubating with the goat-anti-rabbit horseradish peroxidase-conjugated secondary antibody for 16 h at 4 °C. Membrane was washed 3 times with TBST and then incubated in TBST containing 1:7000 diluted Goat-anti-rabbit IgG - Horseradish Peroxidase (HRP) (sc-2004, Santa Cruz Biotechnology, Dallas, TX, USA) for 1 h at 25 °C. Membrane was washed again with TBS-TW20 (3 times) and antigen-antibody complexes were visualized by the enhanced chemiluminescence-2 (ECL-2, Thermo Scientific, Rockford, CA, USA). Beta-actin (#4970, Cell Signaling Technology, Danvers, MA, USA) was used as internal control.

2.6. Statistical Analysis

For the analysis of ERG data, two-way repeated measures Analysis of Variance (ANOVA) was used. The power analysis was conducted by the F-test of one-way ANOVA, where we considered numbers as outcome and groups as the factor. All other comparisons were made by one-way ANOVA. A *p* value of less than 0.05 was considered statistically significant.

3. Results

3.1. Effect of L/Zi on Retinal Function

ERG was used to compare outer retinal function of mice. Before the light exposure, the a-wave and b-wave amplitudes of both dark-adapted and light-adapted ERG of the vehicle group and the treatment group were almost the same, with no statistical difference ($p > 0.05$). Intense light exposure led to significant reductions in a- and b-wave amplitudes in vehicle group under both dark-adapted and light-adapted conditions ($p < 0.05$), while the amplitudes were significantly higher in L/Zi treated group under all luminances (all $p < 0.01$). In addition, our data shows that L/Zi treatment ameliorated the decrease of light-adapted ERG b-wave amplitude more markedly than that of the dark-adapted ERG b-wave amplitude, indicating that the protective effect of L/Zi to the cone system was better than to the rod system (Figure 1).

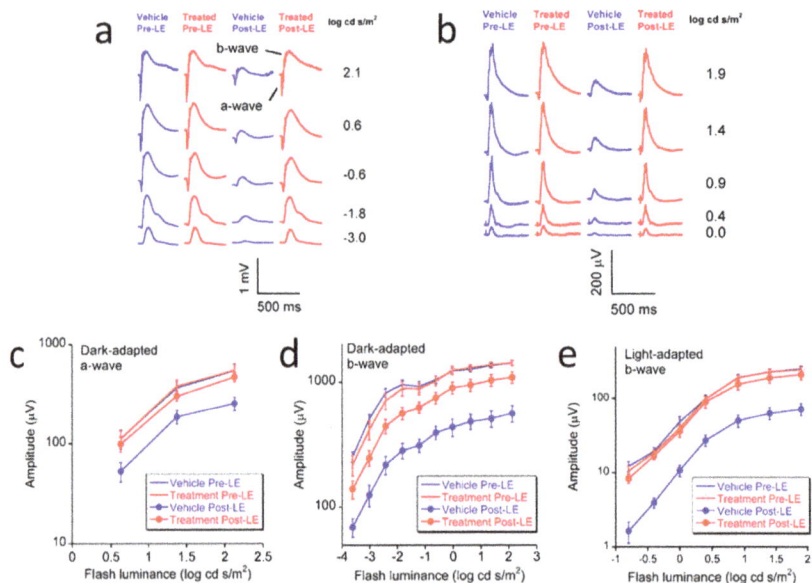

Figure 1. Lutein, RR-zeaxanthin (3R,3′R-zeaxanthin) and RS (meso)-zeaxanthin (3R,3′S-RS-zeaxanthin) (L/Zi) treatment rescued retinal function in light-damaged retinas. Typical waveforms of electroretinogram (ERG) and luminance-response curves of BALB/cJ mice before (9–11 weeks old) and after (11–13 weeks old) blue light exposure in vehicle (sunflower oil) and L/Zi treatment groups ($n = 7$ in each group). (**a**) Typical dark-adapted ERGs waveforms; (**b**) Typical light-adapted ERGs waveforms; (**c**) Luminance-response curves of a-wave amplitudes in dark-adapted conditions which are associated with the responses from rod and cone photoreceptors; (**d**) Luminance-response curves of b-wave amplitudes in dark-adapted condition which are associated with the responses from bipolar cells in both rod and cone pathways; (**e**) Luminance-response curves of b-wave amplitudes in light-adapted condition which are associated with the responses from bipolar cells in the cone pathway. Error bars indicate standard errors. Pre-LE: Before light exposure. Post-LE: After light exposure.

3.2. Effect of L/Zi on Cellular Apoptosis

TUNEL assay was performed on retinal sections to examine the TUNEL-positive cells, which include apoptotic cells [29] and different types of dying cells [30]. In the light-damaged retinas, significantly more TUNEL-positive cells were found in the retinal sections of vehicle group, predominantly in the outer nuclear layers (ONL). In comparison, TUNEL-positive cells were scarce in the retina of mice in L/Zi treatment group. Quantitative analysis revealed a significant difference of TUNEL positive cells in retinal section between vehicle and treatment group after light damage (vehicle group versus treatment group, $p < 0.01$) (Figure 2a,b).

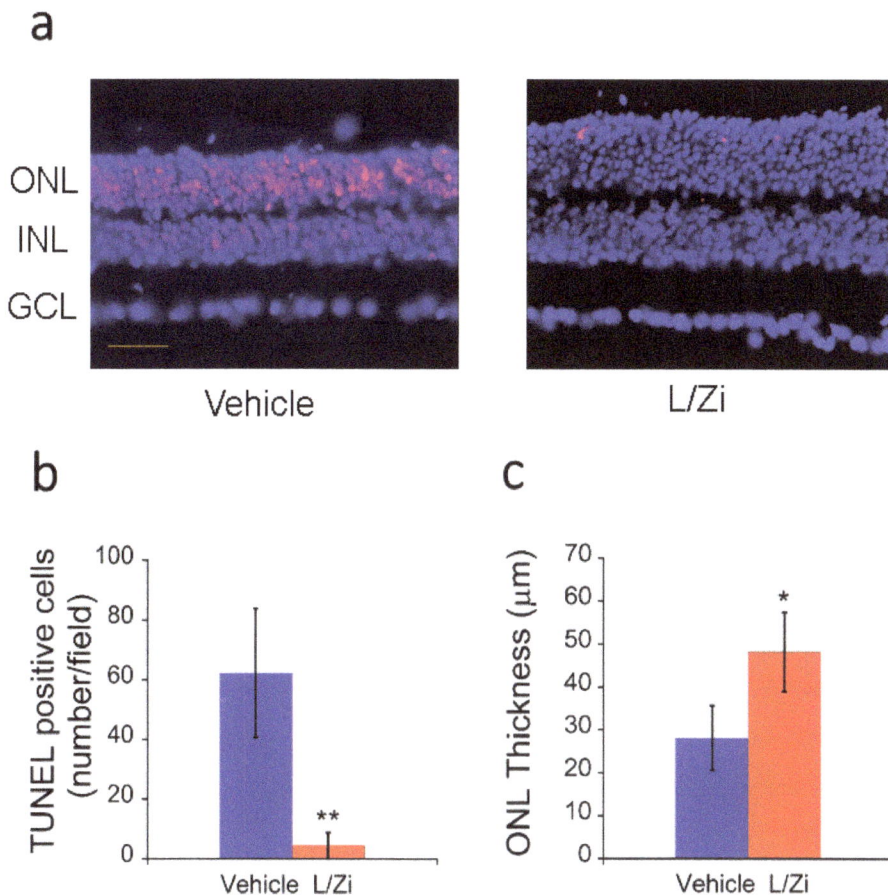

Figure 2. L/Zi treatment reduced cellular death in light-damaged retinas. (a) Representative images (20×) of the staining with TUNEL assay. The dead cells are shown as red spots. Scale bar: 30 µm; (b) Means of the number of TUNEL-positive cells in retinal sections of BALB/cJ mice treated with vehicle or L/Zi ($n = 3$ in each group); (c) Means of the ONL thickness in retinal sections of BALB/cJ mice treated with vehicle or L/Zi ($n = 3$ in each group). ONL: outer nuclear layer; INL: inner nuclear layer; GCL: ganglion cell layer. Error bars: Standard deviations. * $p < 0.05$ vs. vehicle group. ** $p < 0.01$ vs. vehicle group.

3.3. Effect of L/Zi on Outer Nuclear Layer Thickness

The ONL thicknesses were measured in the DAPI-stained images. In the light-damaged retinas, the ONL thickness was less in the vehicle group than that in L/Zi treatment group. Quantitative analysis revealed a significant difference of ONL thickness in retinal section between vehicle and treatment group after light damage (vehicle group versus treatment group, $p < 0.05$) (Figure 2c).

3.4. Effect of L/Zi on OS in Light-Damaged Retinas

C-Jun N-terminal kinase (JNK) and Nuclear factor (erythroid-derived 2)-like 2 (Nrf2) were chosen as protein markers of OS in our study. Western blotting results showed that L/Zi treatment significantly downregulated p-JNK and reduced the ratio of the protein densities of p-JNK to t-JNK (vehicle group versus treatment group, all $p < 0.01$), while the t-JNK were not changed significantly between the two groups ($p > 0.05$). In addition, L/Zi significantly upregulated Nrf2 in light-damaged retinas (vehicle group versus treatment group, $p < 0.05$) (Figure 3).

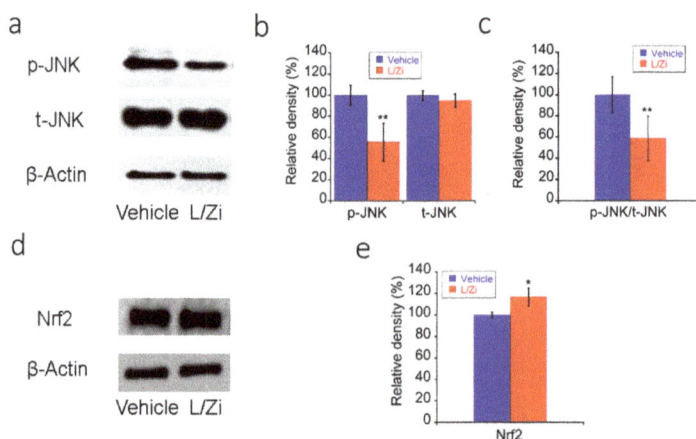

Figure 3. L/Zi treatment down-regulated the markers of oxidative stress in light-damaged retinas. (**a**) Representative images of Western blotting of phosphorylated c-Jun N-terminal kinase (p-JNK); (**b**) Relative protein expression of p-JNK; (**c**) Ratio of protein densities of p-JNK to total (t)-JNK in light-damaged mice ($n = 3$ in each group); (**d**) Representative images of Western blotting of Nuclear factor (erythroid-derived 2)-like 2 (Nrf2); (**e**) Relative protein density of Nrf2 in light-damaged mice ($n = 3$ in each group). * $p < 0.05$ vs. vehicle group; ** $p < 0.01$ vs. vehicle group.

3.5. The Effect of L/Zi on Light-Induced ERS

In our study, the expression levels of four ERS protein markers, including 78 kDa glucose-regulated protein (GRP78), activating transcription factor 6 (ATF6), phosphorylated protein kinase RNA-like endoplasmic reticulum kinase (p-PERK) and activating transcription factor 4 (ATF4), were determined by Western blotting to investigate whether L/Zi treatment reduced ERS in light-damaged retinas. Western blotting results revealed the expression of these ERS markers were significantly reduced by L/Zi treatment (vehicle group versus treatment group: GPR78, $p < 0.01$; ATF6, $p < 0.05$; p-PERK, $p < 0.01$; ATF4, $p < 0.01$) (Figure 4).

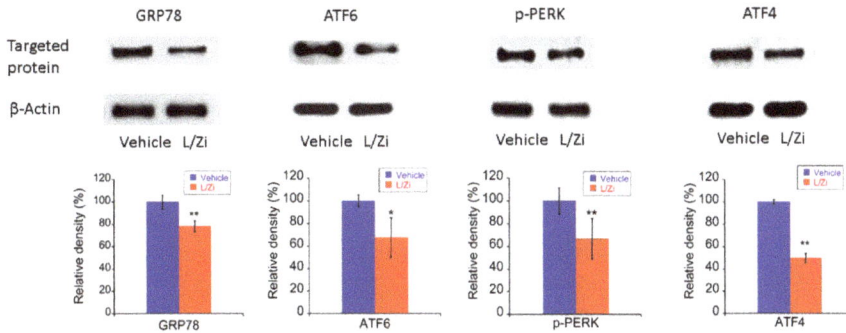

Figure 4. L/Zi treatment down-regulated the markers of ERS in light-damaged retinas. Representative images of Western blotting and relative protein expression of 78 kDa glucose-regulated protein (GRP78), activating transcription factor 6 (ATF6), phosphorylated protein kinase RNA-like endoplasmic reticulum kinase (p-PERK) and activating transcription factor 4 (ATF4) in light-damaged retinas, respectively (*n* = 3 in each group). Error bars indicate standard deviations. * *p* < 0.05 vs. vehicle group; ** *p* < 0.01 vs. vehicle group.

4. Discussion

Animal models of light-induced retinopathy have been used to study phototoxic retinal damage and associated mechanisms, including oxidative stress responses and metabolic abnormalities [31–34]. It was observed that many genes were upregulated in light-damaged retinas of Balb/c mice.

The upregulated genes are in the categories of anti-oxidants, anti-apoptosis, chloride channels, transcription factors, secreted signaling molecules, and inflammation. The upregulated genes may affect the fate of photoreceptors by photo-oxidative stress [35]. A number of studies have shown that OS is one of the main factors which causes light-induced disruption of photoreceptor outer segment and retinal degeneration [36–39]. In addition, the aggregation of S-opsin and ERS was observed in light-damaged models in vitro and in vivo [40–42]. More and more evidence shows that the activation of unfolded protein response on exposure to oxidative stress is an adaptive mechanism to preserve cell function and survival [43]. Reactive oxygen species (ROS), produced in the ER or other organelles, can target the ER calcium channels which causes the release of calcium from ER. The calcium is then obtained by mitochondria, which increases the mitochondrial metabolism and production of ROS. ROS can trigger ER stress [44]. Both of OS and ERS could finally induce cellular apoptosis. Antioxidants with the abilities of ameliorating the OS or ERS have been reported to reduce the light-induced damage on retinas [7,15]. The xanthophyll carotenoids—lutein, zeaxanthin and mesozeaxanthin (L/Zi) have higher concentration in visual system (eye and brain) compared to other carotenoids in the blood [21]. Specifically, L/Zi comprise 80% to 90% of xanthophyll carotenoids in human eyes [20], and decreased level of these antioxidants has been associated with AMD [25]. In addition, administration of L/Zi has been showed to protect against retinal cell damage in many eye pathologies, such as diabetic retinopathy [23].

In our study, the protective effect of L/Zi on retina was observed in an animal model of retinal degeneration induced by extensive blue light exposure, which was confirmed by the significant improvement of dark-adapted and light-adapted ERG a-wave and b-wave amplitudes, and reduction of the number of TUNEL-positive cells in the retinas. We further explored the effect of L/Zi on protein markers of OS to understand the exact underlying mechanism of cellular protection provided by L/Zi. c-Jun N-terminal kinase (JNK) is one the three well-defined subgroups of mitogen-activated protein kinase (MAPK) [45]. The activation of JNK induces a cascade of phosphorylation events, leading to the formation of phosphorylated (p)-JNK, which has been associated to the OS [46]. Nuclear factor (erythroid-derived 2)-like 2 (Nrf2) is a transcription factor recognized as a pivotal element of cellular defense responses to an increased level of ROS [47]. In an inactive status, Nrf2 protein binds to Keap1 [48]. When Nrf2 is activated, Nrf2 separates from Keap1 and moves into the cell nucleus by phosphorylation of Nrf2. The released Nrf2 then binds with antioxidant response element to promote the expression of phase II enzymes and endogenous antioxidants to restore the homeostasis of reactive oxygen species [49,50]. Our Western blotting data of the light-damaged retinas treated with L/Zi showed that p-JNK expression was significantly downregulated. This is consistent with the finding that lutein and zeaxanthin downregulated the lipopolysaccharide-induced increase of p-JNK levels in cultured human uveal melanocytes [51]. Furthermore, our Western blotting data also revealed that the Nrf2 expression was significantly upregulated in mice with L/Zi treatment, which is consistent with the study in mouse microglial (BV-2) cells [52]. These data implied that the MAPK and Nrf2 pathways, which are involved in OS response, may be positively and negatively related to the photoreceptor degeneration in light-damaged retina, and the treatment effect of L/Zi may be related to the regulation of these cellular pathways.

ERS results from protein misfolding in the ER, which has interaction with OS [44,53] and causes cell death [54]. Its role in the light-induced retinal damage has also been illustrated [40]. GRP78, p-PERK and ATF6 are part of protein markers of ER stress. Specifically, GRP78 is a master regulator of the unfolded protein response in ER [55]. PKR-like ER kinase (PERK) is one of the major signaling pathways of ERS, which regulates cellular protein synthesis related to the influx of proteins into the lumen of the stressed ER [56,57]. ATF4 is a downstream transcription factor in PERK signaling pathway. Activated ATF4 increases CCAAT-enhancer-binding protein homologous protein (CHOP) expression, which in turn regulates the expression of a number of stress-induced target genes and amplify the signal initiated by the original stress [58]. ERS can also activate the Activating Transcription Factor 6 (ATF6) which optimizes protein folding and degradation [59–61]. The Western blotting results in our study demonstrated that L/Zi reduce ERS in light-damaged retinas by downregulating key protein markers of ERS, including GRP78, p-PERK and ATF6 pathway. A similar reduction of these ERS biomarkers was observed in diabetic mice treated by wolfberry with high content of lutein and zeaxanthin [62]. Our data indicated that L/Zi might exert its protective effects at least partially through reducing ERS.

5. Conclusions

Treatment with L/Zi could protect photoreceptors against degeneration induced by high intensity of blue light. The treatment effect of this xanthophyll carotenoid is probably related to the decrease of OS and ERS pathways. Future studies are needed to explore this study of L/Zi in other animal models of inherited retinal degeneration, which could form the basis of clinical trials of different types of retinal degenerations that affect humans.

Author Contributions: M.Y. was responsible for conception and design of the experiments, obtained the main research grant, performed some of the experiments, analyzed and interpreted all of the data, and revised the manuscript. W.Y. interpreted all of the data, wrote and revised the manuscript. C.B. conducted some of the experiments.

Funding: This research was funded by OmniActive Health Technologies Ltd., India Research Grant, NIH P30EY025585 and An Unrestricted Grant to the Department of Ophthalmology-CCLCM-CWRU from Research to Prevent Blindness.

Conflicts of Interest: The authors declare no conflict of interest.

References

1. Aziz, M.K.; Ni, A.; Esserman, D.A.; Chavala, S.H. Evidence of early ultrastructural photoreceptor abnormalities in light-induced retinal degeneration using spectral domain optical coherence tomography. *Br. J. Ophthalmol.* **2014**, *98*, 984–989. [CrossRef] [PubMed]
2. Dong, Z.; Li, J.; Leng, Y.; Sun, X.; Hu, H.; He, Y.; Tan, Z.; Ge, J. Cyclic intensive light exposure induces retinal lesions similar to age-related macular degeneration in APPswe/PS1 bigenic mice. *BMC Neurosci.* **2012**, *13*, 34. [CrossRef] [PubMed]

3. Cipriani, V.; Hogg, R.E.; Sofat, R.; Moore, A.T.; Webster, A.R.; Yates, J.R.W.; Fletcher, A.E. Association of C-Reactive Protein Genetic Polymorphisms with Late Age-Related Macular Degeneration. *JAMA Ophthalmol.* **2017**, *135*, 909–916. [CrossRef] [PubMed]

4. Nagar, S.; Noveral, S.M.; Trudler, D.; Lopez, K.M.; McKercher, S.R.; Han, X.; Yates, J.R.R.; Pina-Crespo, J.C.; Nakanishi, N.; Satoh, T.; et al. MEF2D haploinsufficiency downregulates the NRF2 pathway and renders photoreceptors susceptible to light-induced oxidative stress. *Proc. Natl. Acad. Sci. USA* **2017**, *114*, E4048–E4056. [CrossRef] [PubMed]

5. Bian, M.; Du, X.; Cui, J.; Wang, P.; Wang, W.; Zhu, W.; Zhang, T.; Chen, Y. Celastrol protects mouse retinas from bright light-induced degeneration through inhibition of oxidative stress and inflammation. *J. Neuroinflamm.* **2016**, *13*. [CrossRef] [PubMed]

6. Kuse, Y.; Ogawa, K.; Tsuruma, K.; Shimazawa, M.; Hara, H. Damage of photoreceptor-derived cells in culture induced by light emitting diode-derived blue light. *Sci. Rep.* **2014**, *4*, 5223. [CrossRef] [PubMed]

7. Osada, H.; Okamoto, T.; Kawashima, H.; Toda, E.; Miyake, S.; Nagai, N.; Kobayashi, S.; Tsubota, K.; Ozawa, Y. Neuroprotective effect of bilberry extract in a murine model of photo-stressed retina. *PLoS ONE* **2017**, *12*, e0178627. [CrossRef] [PubMed]

8. Qi, S.; Wang, C.; Song, D.; Song, Y.; Dunaief, J.L. Intraperitoneal injection of (-)-Epigallocatechin-3-gallate protects against light-induced photoreceptor degeneration in the mouse retina. *Mol. Vis.* **2017**, *23*, 171–178. [PubMed]

9. Libby, R.T.; Gould, D.B. Endoplasmic reticulum stress as a primary pathogenic mechanism leading to age-related macular degeneration. *Adv. Exp. Med. Biol.* **2010**, *664*, 403–409. [CrossRef] [PubMed]

10. Salminen, A.; Kauppinen, A.; Hyttinen, J.M.; Toropainen, E.; Kaarniranta, K. Endoplasmic reticulum stress in age-related macular degeneration: Trigger for neovascularization. *Mol. Med.* **2010**, *16*, 535–542. [CrossRef] [PubMed]

11. Sanchez-Chavez, G.; Hernandez-Ramirez, E.; Osorio-Paz, I.; Hernandez-Espinosa, C.; Salceda, R. Potential Role of Endoplasmic Reticulum Stress in Pathogenesis of Diabetic Retinopathy. *Neurochem. Res.* **2016**, *41*, 1098–1106. [CrossRef] [PubMed]

12. Yang, L.; Wu, L.; Wang, D.; Li, Y.; Dou, H.; Tso, M.O.; Ma, Z. Role of endoplasmic reticulum stress in the loss of retinal ganglion cells in diabetic retinopathy. *Neural Regen. Res.* **2013**, *8*, 3148–3158. [CrossRef] [PubMed]

13. Ha, Y.; Liu, H.; Xu, Z.; Yokota, H.; Narayanan, S.P.; Lemtalsi, T.; Smith, S.B.; Caldwell, R.W.; Caldwell, R.B.; Zhang, W. Endoplasmic reticulum stress-regulated CXCR3 pathway mediates inflammation and neuronal injury in acute glaucoma. *Cell Death Dis.* **2015**, *6*, e1900. [CrossRef] [PubMed]

14. Anholt, R.R.; Carbone, M.A. A molecular mechanism for glaucoma: Endoplasmic reticulum stress and the unfolded protein response. *Trends Mol. Med.* **2013**, *19*, 586–593. [CrossRef] [PubMed]

15. Li, G.Y.; Fan, B.; Jiao, Y.Y. Rapamycin attenuates visible light-induced injury in retinal photoreceptor cells via inhibiting endoplasmic reticulum stress. *Brain Res.* **2014**, *1563*, 1–12. [CrossRef] [PubMed]

16. Yang, L.P.; Wu, L.M.; Guo, X.J.; Li, Y.; Tso, M.O. Endoplasmic reticulum stress is activated in light-induced retinal degeneration. *J. Neurosci. Res.* **2008**, *86*, 910–919. [CrossRef] [PubMed]

17. Shimazawa, M.; Sugitani, S.; Inoue, Y.; Tsuruma, K.; Hara, H. Effect of a sigma-1 receptor agonist, cutamesine dihydrochloride (SA4503), on photoreceptor cell death against light-induced damage. *Exp. Eye Res.* **2015**, *132*, 64–72. [CrossRef] [PubMed]

18. Bernstein, P.S.; Li, B.; Vachali, P.P.; Gorusupudi, A.; Shyam, R.; Henriksen, B.S.; Nolan, J.M. Lutein, zeaxanthin, and meso-zeaxanthin: The basic and clinical science underlying carotenoid-based nutritional interventions against ocular disease. *Prog. Retin. Eye Res.* **2016**, *50*, 34–66. [CrossRef] [PubMed]

19. Eisenhauer, B.; Natoli, S.; Liew, G.; Flood, V.M. Lutein and Zeaxanthin-Food Sources, Bioavailability and Dietary Variety in Age-Related Macular Degeneration Protection. *Nutrients* **2017**, *9*. [CrossRef] [PubMed]

20. Vishwanathan, R.; Schalch, W.; Johnson, E.J. Macular pigment carotenoids in the retina and occipital cortex are related in humans. *Nutr. Neurosci.* **2016**, *19*, 95–101. [CrossRef] [PubMed]

21. Mares, J. Lutein and Zeaxanthin Isomers in Eye Health and Disease. *Annu. Rev. Nutr.* **2016**, *36*, 571–602. [CrossRef] [PubMed]

22. Tan, J.S.; Wang, J.J.; Flood, V.; Rochtchina, E.; Smith, W.; Mitchell, P. Dietary antioxidants and the long-term incidence of age-related macular degeneration: The Blue Mountains Eye Study. *Ophthalmology* **2008**, *115*, 334–341. [CrossRef] [PubMed]

23. Neelam, K.; Goenadi, C.J.; Lun, K.; Yip, C.C.; Au Eong, K.G. Putative protective role of lutein and zeaxanthin in diabetic retinopathy. *Br. J. Ophthalmol.* **2017**, *101*, 551–558. [CrossRef] [PubMed]

24. Barker, F.M., 2nd; Snodderly, D.M.; Johnson, E.J.; Schalch, W.; Koepcke, W.; Gerss, J.; Neuringer, M. Nutritional manipulation of primate retinas, V: Effects of lutein, zeaxanthin, and n-3 fatty acids on retinal sensitivity to blue-light-induced damage. *Investig. Ophthalmol. Vis. Sci.* **2011**, *52*, 3934–3942. [CrossRef] [PubMed]

25. Chew, E.Y.; Clemons, T.E.; Agron, E.; Launer, L.J.; Grodstein, F.; Bernstein, P.S.; Age-Related Eye Disease Study 2 Research Group. Effect of Omega-3 Fatty Acids, Lutein/Zeaxanthin, or Other Nutrient Supplementation on Cognitive Function: The AREDS2 Randomized Clinical Trial. *JAMA* **2015**, *314*, 791–801. [CrossRef] [PubMed]

26. Lee, B.L.; Kang, J.H.; Kim, H.M.; Jeong, S.H.; Jang, D.S.; Jang, Y.P.; Choung, S.Y. Polyphenol-enriched *Vaccinium uliginosum* L. fractions reduce retinal damage induced by blue light in A2E-laden ARPE19 cell cultures and mice. *Nutr. Res.* **2016**, *36*, 1402–1414. [CrossRef] [PubMed]

27. Bell, B.A.; Kaul, C.; Bonilha, V.L.; Rayborn, M.E.; Shadrach, K.; Hollyfield, J.G. The BALB/c mouse: Effect of standard vivarium lighting on retinal pathology during aging. *Exp. Eye Res.* **2015**, *135*, 192–205. [CrossRef] [PubMed]

28. Kang, K.; Tarchick, M.J.; Yu, X.; Beight, C.; Bu, P.; Yu, M. Carnosic acid slows photoreceptor degeneration in the Pde6b(rd10) mouse model of retinitis pigmentosa. *Sci. Rep.* **2016**, *6*, 22632. [CrossRef] [PubMed]

29. Gavrieli, Y.; Sherman, Y.; Ben-Sasson, S.A. Identification of programmed cell death in situ via specific labeling of nuclear DNA fragmentation. *J. Cell Biol.* **1992**, *119*, 493–501. [CrossRef] [PubMed]

30. Grasl-Kraupp, B.; Ruttkay-Nedecky, B.; Koudelka, H.; Bukowska, K.; Bursch, W.; Schulte-Hermann, R. In situ detection of fragmented DNA (TUNEL assay) fails to discriminate among apoptosis, necrosis, and autolytic cell death: A cautionary note. *Hepatology* **1995**, *21*, 1465–1468. [PubMed]

31. Gu, R.; Tang, W.; Lei, B.; Ding, X.; Jiang, C.; Xu, G. Glucocorticoid-Induced Leucine Zipper Protects the Retina from Light-Induced Retinal Degeneration by Inducing Bcl-xL in Rats. *Investig. Ophthalmol. Vis. Sci.* **2017**, *58*, 3656–3668. [CrossRef] [PubMed]

32. Noell, W.K.; Walker, V.S.; Kang, B.S.; Berman, S. Retinal damage by light in rats. *Investig. Ophthalmol.* **1966**, *5*, 450–473.

33. Brandli, A.; Johnstone, D.M.; Stone, J. Remote Ischemic Preconditioning Protects Retinal Photoreceptors: Evidence from a Rat Model of Light-Induced Photoreceptor Degeneration. *Investig. Ophthalmol. Vis. Sci.* **2016**, *57*, 5302–5313. [CrossRef] [PubMed]

34. Organisciak, D.T.; Vaughan, D.K. Retinal light damage: Mechanisms and protection. *Prog. Retin. Eye Res.* **2010**, *29*, 113–134. [CrossRef] [PubMed]

35. Chen, L.; Wu, W.; Dentchev, T.; Zeng, Y.; Wang, J.; Tsui, I.; Tobias, J.W.; Bennett, J.; Baldwin, D.; Dunaief, J.L. Light damage induced changes in mouse retinal gene expression. *Exp. Eye Res.* **2004**, *79*, 239–247. [CrossRef] [PubMed]

36. Roehlecke, C.; Schumann, U.; Ader, M.; Knels, L.; Funk, R.H. Influence of blue light on photoreceptors in a live retinal explant system. *Mol. Vis.* **2011**, *17*, 876–884. [PubMed]

37. Ding, Y.; Aredo, B.; Zhong, X.; Zhao, C.X.; Ufret-Vincenty, R.L. Increased susceptibility to fundus camera-delivered light-induced retinal degeneration in mice deficient in oxidative stress response proteins. *Exp. Eye Res.* **2017**, *159*, 58–68. [CrossRef] [PubMed]

38. Demontis, G.C.; Longoni, B.; Marchiafava, P.L. Molecular steps involved in light-induced oxidative damage to retinal rods. *Investig. Ophthalmol. Vis. Sci.* **2002**, *43*, 2421–2427.

39. Saenz-de-Viteri, M.; Heras-Mulero, H.; Fernandez-Robredo, P.; Recalde, S.; Hernandez, M.; Reiter, N.; Moreno-Orduna, M.; Garcia-Layana, A. Oxidative stress and histological changes in a model of retinal phototoxicity in rabbits. *Oxid. Med. Cell. Longev.* **2014**. [CrossRef] [PubMed]

40. Nakanishi, T.; Shimazawa, M.; Sugitani, S.; Kudo, T.; Imai, S.; Inokuchi, Y.; Tsuruma, K.; Hara, H. Role of endoplasmic reticulum stress in light-induced photoreceptor degeneration in mice. *J. Neurochem.* **2013**, *125*, 111–124. [CrossRef] [PubMed]

41. Marsili, S.; Genini, S.; Sudharsan, R.; Gingrich, J.; Aguirre, G.D.; Beltran, W.A. Exclusion of the unfolded protein response in light-induced retinal degeneration in the canine T4R RHO model of autosomal dominant retinitis pigmentosa. *PLoS ONE* **2015**, *10*, e0115723. [CrossRef] [PubMed]

42. Ooe, E.; Tsuruma, K.; Kuse, Y.; Kobayashi, S.; Shimazawa, M.; Hara, H. The involvement of ATF4 and S-opsin in retinal photoreceptor cell damage induced by blue LED light. *Mol. Vis.* **2017**, *23*, 52–59. [PubMed]

43. Malhotra, J.D.; Kaufman, R.J. Endoplasmic reticulum stress and oxidative stress: A vicious cycle or a double-edged sword? *Antioxid. Redox Signal.* **2007**, *9*, 2277–2293. [CrossRef] [PubMed]

44. Dandekar, A.; Mendez, R.; Zhang, K. Cross talk between ER stress, oxidative stress, and inflammation in health and disease. *Methods Mol. Biol.* **2015**, *1292*, 205–214. [CrossRef] [PubMed]

45. Kucinski, I.; Dinan, M.; Kolahgar, G.; Piddini, E. Chronic activation of JNK JAK/STAT and oxidative stress signalling causes the loser cell status. *Nat. Commun.* **2017**, *8*. [CrossRef] [PubMed]

46. Jeong, C.B.; Won, E.J.; Kang, H.M.; Lee, M.C.; Hwang, D.S.; Hwang, U.K.; Zhou, B.; Souissi, S.; Lee, S.J.; Lee, J.S. Microplastic Size-Dependent Toxicity, Oxidative Stress Induction, and p-JNK and p-p38 Activation in the Monogonont Rotifer (*Brachionus koreanus*). *Environ. Sci. Technol.* **2016**, *50*, 8849–8857. [CrossRef] [PubMed]

47. Duan, F.F.; Guo, Y.; Li, J.W.; Yuan, K. Antifatigue Effect of Luteolin-6-C-Neohesperidoside on Oxidative Stress Injury Induced by Forced Swimming of Rats through Modulation of Nrf2/ARE Signaling Pathways. *Oxid. Med. Cell. Longev.* **2017**. [CrossRef] [PubMed]

48. Itoh, K.; Wakabayashi, N.; Katoh, Y.; Ishii, T.; Igarashi, K.; Engel, J.D.; Yamamoto, M. Keap1 represses nuclear activation of antioxidant responsive elements by Nrf2 through binding to the amino-terminal Neh2 domain. *Genes Dev.* **1999**, *13*, 76–86. [CrossRef] [PubMed]

49. Kansanen, E.; Kuosmanen, S.M.; Leinonen, H.; Levonen, A.L. The Keap1-Nrf2 pathway: Mechanisms of activation and dysregulation in cancer. *Redox Biol.* **2013**, *1*, 45–49. [CrossRef] [PubMed]

50. Zhang, M.; An, C.; Gao, Y.; Leak, R.K.; Chen, J.; Zhang, F. Emerging roles of Nrf2 and phase II antioxidant enzymes in neuroprotection. *Prog. Neurobiol.* **2013**, *100*, 30–47. [CrossRef] [PubMed]

51. Chao, S.C.; Vagaggini, T.; Nien, C.W.; Huang, S.C.; Lin, H.Y. Effects of Lutein and Zeaxanthin on LPS-Induced Secretion of IL-8 by Uveal Melanocytes and Relevant Signal Pathways. *J. Ophthalmol.* **2015**. [CrossRef] [PubMed]

52. Wu, W.; Li, Y.; Wu, Y.; Zhang, Y.; Wang, Z.; Liu, X. Lutein suppresses inflammatory responses through Nrf2 activation and NF-kappaB inactivation in lipopolysaccharide-stimulated BV-2 microglia. *Mol. Nutr. Food Res.* **2015**, *59*, 1663–1673. [CrossRef] [PubMed]

53. Cao, S.S.; Kaufman, R.J. Endoplasmic reticulum stress and oxidative stress in cell fate decision and human disease. *Antioxid. Redox Signal.* **2014**, *21*, 396–413. [CrossRef] [PubMed]

54. Sano, R.; Reed, J.C. ER stress-induced cell death mechanisms. *Biochim. Biophys. Acta* **2013**, *1833*, 3460–3470. [CrossRef] [PubMed]

55. Gorbatyuk, M.S.; Gorbatyuk, O.S. The Molecular Chaperone GRP78/BiP as a Therapeutic Target for Neurodegenerative Disorders: A Mini Review. *J. Genet. Syndr. Gene Ther.* **2013**, *4*. [CrossRef] [PubMed]

56. Hamanaka, R.B.; Bobrovnikova-Marjon, E.; Ji, X.; Liebhaber, S.A.; Diehl, J.A. PERK-dependent regulation of IAP translation during ER stress. *Oncogene* **2009**, *28*, 910–920. [CrossRef] [PubMed]

57. Liu, Z.; Lv, Y.; Zhao, N.; Guan, G.; Wang, J. Protein kinase R-like ER kinase and its role in endoplasmic reticulum stress-decided cell fate. *Cell Death Dis.* **2015**, *6*. [CrossRef] [PubMed]

58. B'Chir, W.; Maurin, A.C.; Carraro, V.; Averous, J.; Jousse, C.; Muranishi, Y.; Parry, L.; Stepien, G.; Fafournoux, P.; Bruhat, A. The eIF2alpha/ATF4 pathway is essential for stress-induced autophagy gene expression. *Nucleic Acids Res.* **2013**, *41*, 7683–7699. [CrossRef] [PubMed]

59. Wu, J.; Rutkowski, D.T.; Dubois, M.; Swathirajan, J.; Saunders, T.; Wang, J.; Song, B.; Yau, G.D.; Kaufman, R.J. ATF6alpha optimizes long-term endoplasmic reticulum function to protect cells from chronic stress. *Dev. Cell* **2007**, *13*, 351–364. [CrossRef] [PubMed]

60. Glembotski, C.C. Roles for ATF6 and the sarco/endoplasmic reticulum protein quality control system in the heart. *J. Mol. Cell. Cardiol.* **2014**, *71*, 11–15. [CrossRef] [PubMed]

61. Shen, J.; Snapp, E.L.; Lippincott-Schwartz, J.; Prywes, R. Stable binding of ATF6 to BiP in the endoplasmic reticulum stress response. *Mol. Cell. Biol.* **2005**, *25*, 921–932. [CrossRef] [PubMed]

62. Tang, L.; Zhang, Y.; Jiang, Y.; Willard, L.; Ortiz, E.; Wark, L.; Medeiros, D.; Lin, D. Dietary wolfberry ameliorates retinal structure abnormalities in db/db mice at the early stage of diabetes. *Exp. Biol. Med.* **2011**, *236*, 1051–1063. [CrossRef] [PubMed]

nutrients

MDPI

Review

Nutritional Strategies to Prevent Lens Cataract: Current Status and Future Strategies

Andrea J Braakhuis [1], Caitlin I Donaldson [1], Julie C Lim [2] and Paul J Donaldson [2,*]

[1] Discipline of Nutrition, Faculty of Medical and Health Sciences, the University of Auckland, Auckland 1142, New Zealand; a.braakhuis@auckland.ac.nz (A.J.B.); c-donaldson@hotmail.com (C.I.D.)

[2] Department of Physiology, Faculty of Medical and Health Sciences, New Zealand National Eye Centre, the University of Auckland, Auckland 1142, New Zealand; j.lim@auckland.ac.nz

* Correspondence: p.donaldson@auckland.ac.nz; Tel.: +64-9-923-4625

Received: 26 April 2019; Accepted: 23 May 2019; Published: 27 May 2019

Abstract: Oxidative stress and the subsequent oxidative damage to lens proteins is a known causative factor in the initiation and progression of cataract formation, the leading cause of blindness in the world today. Due to the role of oxidative damage in the etiology of cataract, antioxidants have been prompted as therapeutic options to delay and/or prevent disease progression. However, many exogenous antioxidant interventions have to date produced mixed results as anti-cataract therapies. The aim of this review is to critically evaluate the efficacy of a sample of dietary and topical antioxidant interventions in the light of our current understanding of lens structure and function. Situated in the eye behind the blood-eye barrier, the lens receives it nutrients and antioxidants from the aqueous and vitreous humors. Furthermore, being a relatively large avascular tissue the lens cannot rely of passive diffusion alone to deliver nutrients and antioxidants to the distinctly different metabolic regions of the lens. We instead propose that the lens utilizes a unique internal microcirculation system to actively deliver antioxidants to these different regions, and that selecting antioxidants that can utilize this system is the key to developing novel nutritional therapies to delay the onset and progression of lens cataract.

Keywords: dietary antioxidants; antioxidant supplements; lens; cataract

1. Introduction

Lens cataract is the leading cause of visual impairment and blindness worldwide [1,2]. It has been estimated that over 68% of people over 79 years of age have some form of reduced lens opacity or cataract [3], with disease incidence increasing with age. Clinically four main forms of lens cataract are recognised: sub-capsular, cortical, nuclear and mixed (nuclear and cortical). Of these classes, diabetic cortical cataract and age related nuclear (ARN) cataract are the most common. Population growth, sedentary lifestyles, unhealthy diets and an increasing prevalence of obesity are increasing the number of people with diabetes mellitus. Worldwide more than 285 million people are affected by diabetes mellitus with the number expected to increase to 439 million by 2030 [4]. A frequent complication of both type 1 and type 2 diabetes is diabetic cortical cataracts which occur 2–5 times more frequently in patients with diabetes, and occur at an earlier age [5]. In parallel, our aging population is growing at a remarkable rate with the population aged 60 years or over predicted to double to 2.1 billion by 2050 [6]. Age is major risk factor for cataracts with the current estimate of cataracts to afflict more than 20 million people worldwide. Given our globally aging population, the social and economic costs of lens cataract are quite staggering and the demand for cataract surgery far exceeds limited public health resources. In the USA alone, cataract surgery is the most commonly performed surgical procedure and costs around $3.5 billion per year. In Australia, where it has been estimated that the population will increase by 22% between 1996 and 2021, the incidence of age-related cataract will disproportionately increase by 76% during the same period [7]. While in New Zealand, it is predicted that by 2020 almost 22,800 New Zealanders will have vision loss and about 2,000 will be blind, due to cataract [8]. Annually some 16,000 cataract surgeries are performed in New Zealand, and in 2009–2010 more than 30 million dollars was spent on public inpatient and day stay services for cataract operations [8].

Thus, it is expected that the burden of cataract-impaired vision will increasingly outstrip the resources available for its surgical treatment. The alternative approach is to delay the onset of cataract. It has been predicted that delaying the onset of cataract by 10 years will halve the incidence of ARN cataract, greatly reducing the need for, and expense associated with, surgical intervention [9]. Since cataract is a progressive disease of old age, any medical therapy devised would need to be taken by a large cohort of the population over many years to be effective. Hence, it is likely that the most effective treatment will be in the form of nutritional supplements, either delivered orally or via eye drops. However, while there is evidence that diet can influence the onset and incidence of cataract (see Table 1), the results from trials into the use of nutritional supplements to prevent cataract have had mixed results (see Table 2). For example, the protective effects of antioxidant Vitamin supplements have been advocated in a number of studies [10–12], but a randomized, placebo-controlled trial of Vitamin E and cataract [13], and the Age-Related Eye Disease Study involving use of a high-dose formulation of Vitamin C, Vitamin E, and beta carotene [14], revealed no significant benefits of antioxidants. Furthermore, in large, prospective, population-based cohorts it was found that high-dose Vitamin C or Vitamin E supplements had a statistically significant increased risk of age-related cataract [15], most likely due to the pro-oxidative properties of Vitamins C and E at high doses reported in *in vitro* studies [16,17]. While in a large-scale randomized trial of women at high risk of cardiovascular disease, daily supplementation with a combination of folic acid, vitamin B_6, and vitamin B_{12} had no significant effect on cataract, but may have increased the risk of cataract extraction [18]. However, high intake of dietary vitamin K_1 was associated with a reduced risk of cataracts in an elderly Mediterranean population even after adjusting by other potential confounders [19]. These inconsistencies are in part firstly because lens cataract manifests as multiple disease phenotypes and different nutritional supplements may be more effective on a specific type of cataract, and secondly from a general lack of knowledge on how nutrients are delivered to, and metabolized in the distinctly different regions of the lens. Confounding this variability are differences in the methodology of the studies used to assess the efficacy of a dietary intervention on reducing the incidence of cataract.

In this review, we first characterize the etiology of the main types of lens cataract, and then provide a summary of the evidence for and against the efficacy of nutritional status to delay or prevent the onset of these different sub-types of lens cataract. Then we provide an overview of lens structure and function to highlight how an increased understanding of these properties will be required to design more targeted nutritional strategies to reduce the incidence of cataract.

2. The Etiology of Lens Cataract

Loss of lens transparency manifests itself clinically as cataract, the leading cause of blindness worldwide [2,20]. The two most common types of age-related cataract, cortical and nuclear cataract, have different damage phenotypes and are associated with a loss of transparency in the lens cortex and nucleus, respectively [21].

2.1. Cortical Cataract

Clinically, cortical cataract presents as wedge shaped or radial spoke opacifications in the lens cortex (Figure 1A), and is particularly prevalent in the elderly or diabetic patients [22]. Cortical cataracts tend to be associated with significant astigmatic shifts [23], caused by asymmetrical refractive index changes within the lens cortex [24]. This change in refractive index is most likely due to accumulation of fluid in the lens cortex since, at the cellular level, this light scattering is due to a discrete localized zone of tissue liquefaction surrounded by cells that have a normal morphological structure [25].

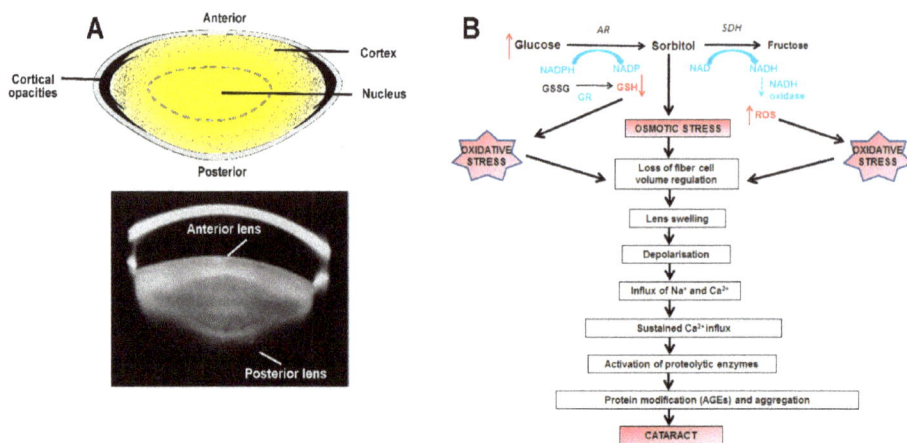

Figure 1. Cortical cataracts. (**A**) Location of the cortical cataract subtype. *Top panel:* diagram showing the opacities that form in the lens cortex. *Lower panel:* Scheimpflug slit-lamp photographic image revealing a cortical cataract. (**B**) Molecular mechanisms involved in the pathogenesis of diabetic cortical cataract. An increase in glucose leads to a decrease in GSH and an increase in reactive oxygen species (ROS) as indicated by the red arrows. The induced osmotic and oxidative stress work synergistically to inhibit the ability of fibre cells to regulate their volume. This leads to cell swelling, depolarization and an influx of sodium and calcium ions. The accumulation of calcium ions results in the activation of calcium-dependent proteases, which target cytoskeletal and crystallin proteins. Furthermore, proteins are modified by the formation of advanced glycation end (AGEs) products, which are known to alter the structure and function of crystallins, resulting in an increase in insoluble proteins, the formation of high molecular weight aggregates, and cataract.

Animal models such as the streptozotocin (STZ) rat, a model used to chemically induce Type 1 diabetes by destroying pancreatic betacells, and galactose fed animals, a model for obese Type 2 diabetes, have been used to understand the mechanisms of diabetic cataract formation. These models have been commonly used because diabetes or galactosemia can be induced rapidly and effectively resulting in formation of "fast" sugar lens cataract. From these animal studies, the consensus view was that high levels of the impermeable osmolyte, sorbitol, produced from excess glucose by the enzyme aldose reductase (AR), initiates osmotic stress, resulting in the attraction of fluid, lens fiber cell swelling and tissue liquefaction [26–28]. Based on this view, considerable attention was focused on the development and testing of AR inhibitors, which were proven to be very successful in ameliorating diabetic cataract in rats and dogs [29], but ineffective in humans. The failure of aldose reductase inhibitors to slow the progression of cataract in humans lies with the differences in AR activity and polyol accumulation between rats and humans, with humans exhibiting low levels of AR activity relative to rat. In addition, the acute animal models replicate the fast development of cataract that occurs in diabetic patients with uncontrolled hyperglycemia. However, most diabetic patients are able to control their blood glucose reasonably well and so such acute cataract development is rarely seen. Instead, the majority of adult diabetic patients typically develop cataract after having suffered from diabetes for several years. Therefore, while the initiating mechanism in the development of diabetic cataract is osmotic stress, it is now believed that oxidative stress generated by polyol pathway activity [30], impairs the ability of the lens over time to regulate its volume, resulting in slow developing cataract formation [31]. This suggests that osmotic and oxidative stress work synergistically to cause a loss of cell volume and tissue liquefaction in the lens cortex (Figure 1B).

2.2. Age-Related Nuclear (ARN) Cataract

Age-related nuclear cataract is initiated in the central core of the lens, which contains primary fiber cells that were initially laid down during embryonic development (Figure 2A). Clinically, age-related nuclear cataract appears as a browning or brunescence of the lens nucleus [32]. In contrast to cortical cataract, the morphology of the cells in the lens nucleus from nuclear cataract patients reveal no major structural distortions [33,34]. Instead, nuclear cataract is associated with the extensive loss of protein sulfhydryl groups, with over 90% of cysteine residues and ~50% of methionine residues being found oxidized in nuclear proteins in lenses obtained from patients with ARN cataract [35–38]. Accompanying the loss of protein sulfhydryl groups is an increase in protein-thiol mixed disulfides [39,40], and an increase in the water insoluble fraction [32,38], which culminate in the formation of protein-protein disulfides (PSSP), and other cross-linkages that lead to protein aggregation and light scattering. This series of biochemical changes has been extensively reviewed [41–43] and there is general agreement that oxidative stress is the major contributing factor to age-related nuclear cataract formation (Figure 2B). The unresolved question in nuclear cataract research now centers on how the normally robust oxygen radical scavenger systems present in the lens fail with advancing age, and initiate the observed protein aggregation specifically in the nucleus of the lens.

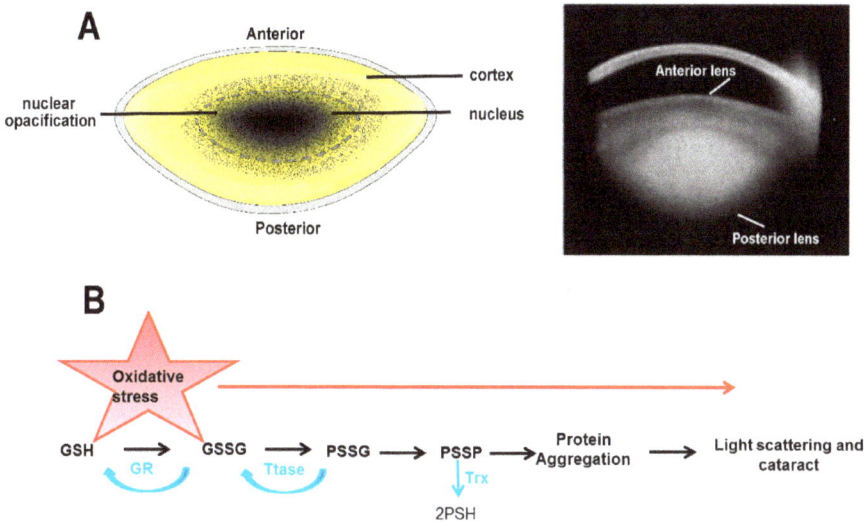

Figure 2. Nuclear cataracts. (**A**) Location of the nuclear cataract subtype. *Left panel:* diagram showing the opacities that form in the lens nucleus. *Right panel:* Scheimpflug slit-lamp photographic image revealing a nuclear cataract. (**B**) Molecular mechanisms involved in the pathogenesis of age-related nuclear cataract. GSH levels are maintained at high levels within the lens by a combination of pathways including regeneration of oxidised GSH (GSSG) back to GSH via the enzyme glutathione reductase (GR) as well as repair enzymes thioltransferase (Ttase), which dethiolate protein mixed disulfides, such as protein bound GSH (PSSG), and thioredoxin (TrX), that dethiolates protein-protein disulphides (PSSP). In age related nuclear cataract, depletion of GSH levels in the nucleus, but not the lens cortex, results in significant oxidation of nuclear proteins, an increase in protein mixed disulphides, formation of protein-protein disulfide bonds, protein aggregation, loss of protein solubility, increased yellowing of the lens nucleus, and eventual nuclear cataract formation.

At younger ages, the human lens is normally protected against oxidative damage by a robust oxygen radical scavenger system, which utilizes GSH as its principal antioxidant to detoxify reactive oxygen species (ROS) [44]. A GSH concentration gradient exists in the lens, with GSH levels highest in the cortex and lowest in the nucleus. The high levels of GSH in the outer cortex are maintained by a combination of synthesis of GSH from its precursor amino acids; cysteine, glutamate and glycine [45]; direct uptake of GSH from the aqueous [46,47]; and the recycling of GSSG to GSH by glutathione reductase (GR), which utilizes NADPH, generated by the metabolism of glucose, as a reducing equivalent [44,48]. In contrast, cells in the lens nucleus have lost their capacity to synthesise GSH [49], and rely on anaerobic metabolism to produce their energy requirements [50]. Therefore, mature fiber cells can only maintain their GSH levels by delivery of GSH to the nucleus and/or local regeneration of GSH by GR. In ARN cataract, cortical fiber cells maintain their GSH levels, but GSH levels in the nucleus are significantly depleted (< 2mM) [51,52]. These observations have led to the hypothesis that ARN cataract is due to an inability to maintain GSH levels in the lens nucleus due to a failure to deliver GSH to the lens centre [53]. As a result, dietary supplementation with antioxidants has been extensively studied as a potential way to delay or prevent cataracts by countering the negative effects of oxidative stress. In this regard, GSH supplementation would appear to be the most obvious antioxidant to administer to restore the depleted levels of GSH observed in the nucleus of the old lens. However, GSH is poorly absorbed by an oral route mainly due to the action of an intestinal enzyme, the γ-glutamyl transpeptidase (GGT) which degrades GSH [54]. As a result, it's low bioavailability would mean that very little GSH would reach the lens, let alone the lens nucleus, and so other strategies for enhancing antioxidant levels in the lens have been investigated.

3. Evidence for and Against Nutritional Strategies to Prevent Lens Cataract

Although the pathogenesis of cortical and age related nuclear cataract are different, oxidative damage has been implicated as an underlying cause of the distinctly different damage phenotypes [30,31,42,43]. Because of this link between oxidative stress and cataract formation, it has been proposed that topical application and/or dietary interventions with antioxidants can be used as therapies to delay or prevent cataract progression. Hence, multiple studies have been conducted to assess the efficacy of such nutritional interventions. Unfortunately, a simplistic view of supplementing the lens with antioxidants has proven to be ineffective, and in some cases detrimental, in delaying cataract progression. This in part may be due to the emerging recognition that reactive oxygen species act as important modulators of redox signalling critical in maintaining normal metabolism and cellular processes. As such, antioxidant supplementation may in fact be counterproductive, eliminating physiological reactive oxygen species required for normal redox signaling and cellular function [55–58]. Therefore, in developing an antioxidant based cataract therapy, it is important to identify not only an appropriate antioxidant to minimise oxidative stress specifically in the lens, but also to consider how this antioxidant will be delivered at levels that will effectively restore antioxidant balance in the different lens regions to delay the formation of a specific type of cataract.

Effectiveness of Antioxidant Therapies on Cataract Progression—A Survey of the Literature

To gain a snap shot of the current effectiveness of antioxidant therapies to reduce the incidence of cataract formation, a survey of the literature was performed. This survey of the literature utilized Google Scholar to extract key references from texts and publications published in English from 2000 onwards, and utilized the search terms cataracts / vision AND antioxidants / supplements / nutrition. Research involving case-control, observational and randomised control design were included, but only if they investigated dietary antioxidants (including lutein, xanthine, Vitamin C, E, A, selenium, polyphenols) and/or fruit and vegetable intake on cataract incidence or progression. The initial search generated 15,900 publications, which once customized to publication date, was reduced to 14,500 articles. Of this search, the first 2900 articles were reviewed by title, representing up to page 30 of the database search. 2845 publications were excluded on the basis of the predefined criteria (predominantly a lack of relevancy, i.e. non-cataract), leaving 55 publications that were subjected to independent review in full by two authors (CD, AB). Disagreement between authors on their inclusion/exclusion was discussed, and a further 29 publications were excluded following full text review. In addition to the primary research articles, there were 14 reviews on the topic that proved useful for background evidence [1,3,59–70]. For ease of presentation, the results of this search are separated into the effects of whole dietary (Table 1), or topical supplement (Table 2) interventions on the incidence or progression of age-related cataracts.

Table 1. Studies investigating antioxidant-rich food on disease progression.

Vitamin C

Author	Any CAT	NUC	CX	PSC
Christen, 2008 [71]	RR 1.00; 95% CI 0.86–1.16; $P = 0.61$	-	-	-
Dherani, 2008 [72]	OR 0.64; 95% CI 0.48–0.85; $P < 0.01$	OR 0.62; 95% CI 0.4–0.96; $P = 0.06$	OR 0.62; 95% CI 0.40–0.97; $P = 0.10$	OR 0.59; 95% CI 0.35–0.99; $P = 0.10$
Jaques, 2001 [73]	-	OR 0.31; 95% CI 0.16–0.58; $P = 0.003$	-	-
Pastor-Valero, 2013 [74]	OR 0.46; 95% CI 0.24–0.88; $P = 0.047$	-	-	-
Ravindran, 2011 [75]	Plasma levels: OR 0.61; 95% CI 0.57–0.82; $P < 0.0001$ Dietary OR 0.78; 95% CI 0.62–0.98; $P = 0.006$	OR 0.66; 95% CI 0.54–0.80; $P < 0.0001$	OR 0.70; 95% CI 0.54–0.90; $P < 0.002$	OR 0.58; 95% CI 0.45–0.74; $P < 0.00003$
Tan, 2008 [76]	-	OR 0.55; 95% CI 0.36–0.86; $P = 0.045$	OR 0.94; 95% CI 0.63–1.40; $P = 0.698$	OR 1.15; 95% CI 0.06–2.23; $P = 0.954$
Theodoropoulou, 2014 [77]	OR 0.50; 95% CI 0.39–0.64; $P < 0.001$	OR 0.55; 95% CI 0.41–0.72; $P < 0.001$	OR 0.62; 95% CI 0.37–1.04; $P = 0.071$	OR 0.30; 95% CI 0.19–0.49; P<0.001
Valero, 2002 [12]	OR 0.70; 95% CI 0.44–1.13; $P = 0.04$	OR 0.56; 95% CI 0.38–0.82	OR 0.92; 95% CI 0.60–1.40	OR 0.75; 95% CI 0.51–1.05
Yoshida, 2007 [78]	Men: OR 0.65; 95% CI 0.42–0.97; $P = 0.094$ Women: OR 0.59; 95% CI 0.43–0.89; $P = 0.047$	-	-	-

Vitamin E

Author	Any CAT	NUC	CX	PSC
Christen, 2008 [71]	RR 0.92; 95% CI 0.80–1.06; $P = 0.39$	-	-	-
Jaques, 2001 [73]	-	OR 0.45; 95% CI 0.23–0.86; $P = 0.06$	-	-
Pastor-Valero, 2013 [74]	OR 0.46; 95% CI 0.24–0.88; $P = 0.944$	-	-	-
Tan, 2008 [76]	-	OR 0.73; 95% CI 0.47–1.13; $P = 0.155$	OR 0.91; 95% CI 0.62–1.33; $P = 0.944$	OR 0.95; 95% CI 0.50–1.83; P=0.597
Theodoropoulou, 2014 [77]	OR 0.50; 95% CI 0.38–0.66; $P < 0.001$	OR 0.50; 95% CI 0.36–0.69; $P < 0.001$	OR 0.71; 95% CI 0.41–1.25; $P = 0.238$	OR 0.42; 95% CI 0.26–0.68; $P < 0.001$
Valero, 2002 [12]	OR 0.77; 95% CI 0.48–1.24; $P = 0.60$	OR 0.81; 95% CI 0.50–1.28	OR 1.00; 95% CI 0.59–1.72	OR 1.16; 95% CI 0.71–1.90

Table 1. *Cont.*

Vitamin A (Retinol)

Author	Any CAT	NUC	CX	PSC
Dherani, 2008 [72]	OR 0.58; 95% CI 0.37–0.91; $P < 0.02$	OR 0.56; 95% CI 0.33–0.46; $P = 0.04$	OR 0.69; 95% CI 0.38–1.26; $P = 0.20$	OR 0.69; 95% CI 0.39–1.23; $P = 0.20$
Tan, 2008 [76]	-	OR 066; 95% CI 0.42–10.3; $P = 0.056$	OR 0.84; 95% CI 0.56–1.25; $P = 0.305$	OR 1.04; 95% CI 0.54–2.02; $P = 0.604$
Theodoropoulou, 2014 [77]	OR 1.47; 95% CI 1.150–1.88; $P = 0.002$	OR 1.46; 95% CI 1.11–1.92; $P = 0.007$	OR 1.02; 95% CI 0.51–2.02; $P = 0.962$	OR 1.88; 95% CI 1.35–2.63; $P < 0.001$
Valero, 2002 [12]	OR 0.82; 95% CI 0.50–1.03; $P = 0.21$	Plasma levels: OR 1.67; 95% CI 1.02–2.72	Plasma levels: OR 1.82; 95% CI 1.09–3.08	Plasma levels: OR 1.22; 95% CI 0.73–2.03

Selenium

Author	Any CAT	NUC	CX	PSC
Valero, 2002 [12]	OR 0.97; 95% CI 0.60–1.58; $P = 0.34$	OR 0.71; 95% CI 0.48–1.04	OR 0.88; 95% CI 0.58–2.46	OR 1.03; 95% CI 0.70–1.51

Carotenoids

Author	Any CAT	NUC	CX	PSC
Christen, 2008 [71]	Lutein/zeaxanthin: RR 0.82; 95% CI 0.71–0.95; $P = 0.045$ Alpha-carotene: RR 0.96; 95% CI 0.84–1.11; $P = 0.77$ Beta-carotene: RR 0.89; 95% CI 0.77–1.02; $P = 0.27$ Beta-cryptoxanthin: RR 0.92; 95% CI 0.80–1.06; $P = 0.19$ Lycopene: RR 0.96; 95% CI 0.52–30.7; $P = 0.6$	-	-	-
Delcourt, 2006 [79]	Plasma levels:Lutein: OR 0.82; 95% CI 0.48–1.41; $P = 0.48$ Zeaxanthin: OR 0.57; 95% CI 0.34–0.95; $P = 0.03$ Beta-carotene: OR 0.69; 95% CI 0.40–1.19; $P = 0.17$ Alpha-carotene: OR 0.69; 95% CI 0.4–1.19; $P = 0.17$ Beta-cryptoxanthin: OR 0.71; 95% CI 0.42–1.20; $P = 0.20$ Lycopene: OR 1.17; 95% CI 0.68–2.01; $P = 0.58$	Plasma levels:Lutein: OR 0.60; 95% CI 0.24–1.47; $P = 0.26$ Zeaxanthin: OR 0.25; 95% CI 0.08–0.71; $P = 0.004$ Beta-carotene: OR 0.42; 95% CI 0.16–1.12; $P = 0.07$ Alpha-carotene: OR 0.76; 95% CI 0.29–2.03; $P = 0.52$ Beat-cryptoxanthin: OR 0.70; 95% CI 0.29–1.68; $P = 0.40$ Lycopene: OR 1.01; 95% CI 0.41–2.51; $P = 0.99$	Plasma levels:Lutein: OR 0.75; 95% CI 0.23–2.47; $P = 0.63$ Zeaxanthin: OR 1.09; 95% CI 0.37–3.26; $P = 0.83$ Beta-carotene: OR 1.10; 95% CI 0.28–4.26; $P = 0.93$ Alpha-carotene: OR 0.97; 95% CI 0.32–2.91; $P = 0.96$ Beta-cryptoxanthin: OR 1.49; 95% CI 0.48–4.68; $P = 0.49$ Lycopene: OR 0.92; 95% CI 0.27–3.13; $P = 0.59$	Plasma levels:Lutein: OR 1.26; 95% CI 0.52–3.07; $P = 0.60$ Zeaxanthin: OR 0.84; 95% CI 0.34–2.07; $P = 0.68$ Beta-carotene: OR 0.51; 95% CI 0.19–1.36; $P = 0.16$ Alpha-carotene: OR 0.72; 95% CI 0.30–1.73; $P = 0.46$ Beta-cryptoxanthin: OR 0.42; 95% CI 0.15–1.18; $P = 0.09$ Lycopene: OR 1.19; 95% CI 0.49–2.88; $P = 0.70$

Table 1. *Cont.*

Dherani, 2008 [72]	Lutein: OR 0.66; 95% CI 0.43–1.02; P = 0.06 Zeaxanthin: OR 0.66; 95% CI 0.45–0.96; P < 0.03 Beta-cryptoxanthin: OR 0.88; 95% CI 0.63–1.23; P = 0.50 Alpha-carotene: OR 0.69; 95% CI 0.50–0.95; P < 0.05 Beta-carotene: OR 0.77; 95% CI 0.45–1.32; P = 0.30 Lycopene: OR 0.78; 95% CI 0.49–1.23; P = 0.20 Alpha-tocopherol: OR 0.58; 95% CI 0.36–0.94; P = 0.04 Gamma-tocopherol: OR 0.75; 95% CI 0.57–0.98; P = 0.06	Lutein: OR 0.75; 95% CI 0.44–1.31; P = 0.30 Zeaxanthin: OR 0.71; 95% CI 0.43–1.17; P = 0.20 Beta-cryptoxanthin: OR 0.83; 95% CI 0.56–1.22; P = 0.30 Alpha-carotene: OR 0.74; 95% CI 0.48–1.14; P = 0.20 Beta-carotene: OR 0.75; 95% CI 0.44–1.26; P = 0.20 Lycopene: OR 0.83; 95% CI 0.50–1.37; P = 0.40 Alpha-tocopherol: OR 0.60; 95% CI 0.30–1.20; P = 0.10 Gamma-tocopherol: OR 0.99; 95% CI 0.62–1.58; P = 0.90	Lutein: OR 0.53; 95% CI 0.28–1.02; P = 0.10 Zeaxanthin: OR 0.58; 95% CI 0.30–1.12; P = 0.10 Beta-cryptoxanthin: OR 0.93; 95% CI 0.54–1.56; P = 0.90 Alpha-carotene: OR 0.86; 95% CI 0.60–1.22; P = 0.70 Beta-carotene: OR 1.02; 95% CI 0.52–1.99; P = 0.90 Lycopene: OR 1.17; 95% CI 0.54–2.53; P = 0.70 Alpha-tocopherol: OR 0.58; 95% CI 0.25–1.35; P = 0.20 Gamma-tocopherol: OR 0.84; 95% CI 0.44–1.00; P = 0.50	Lutein: OR 0.72; 95% CI 0.30–1.71; P = 0.40 Zeaxanthin: OR 0.84; 95% CI 0.43–1.67; P = 0.60 Beta-cryptoxanthin: OR 1.06; 95% CI 0.54–2.05; P = 0.80 Alpha-carotene: OR 0.80; 95% CI 0.39–1.65; P = 0.60 Beta-carotene: OR 0.71; 95% CI 0.30–1.68; P = 0.40 Lycopene: OR 0.68; 95% CI 0.29–1.61; P = 0.30 Alpha-tocopherol: OR 1.11; 95% CI 0.42–2.91; P = 0.90 Gamma-tocopherol: OR 1.16; 95% CI 0.69–1.93; P = 0.60
Jaques, 2001 [73]	-	Alpha-carotene: OR 0.71; 95% CI 0.37–1.35; P = 0.39 Beta-carotene: OR 0.52; 95% CI 0.28–0.97; P = 0.08 Beta-cryptoxanthin: OR 0.68; 95% CI 0.34–1.35; P = 0.06 Lutein/zeaxanthin: OR 0.52; 95% CI 0.29–0.91; P = 0.08 Lycopene: OR 1.16; 95% CI 0.63–2.16; P = 0.79	-	-
Moeller, 2008 [80]	-	Lutein: OR 0.68; 95% CI 0.48–0.97; P = 0.04 Zeaxanthin: OR 0.68; 95% CI 0.47–0.98; P = 0.01	-	-
Tan, 2008 [76]	-	Beta-carotene: OR 1.09; 95% CI 0.69–1.72; P = 0.715	Beta-carotene: OR 1.06; 95% CI 0.7–1.6; P = 0.854	Beta-carotene: OR 0.76; 95% CI 0.37–1.59; P = 0.317

Table 1. *Cont.*

Theodoropoulou, 2014 [77]	Carotene: OR 0.56; 95% CI 0.45–0.69; P < 0.001	Carotene: OR 0.50; 95% CI 0.39–0.65; P < 0.001	Carotene: OR 0.68; 95% CI 0.43–1.05; P = 0.084	Carotene: OR 0.58; 95% CI 0.40–0.86; P = 0.007
Valero, 2002 [12]	Beta-carotene: OR 0.82; 95% CI 0.51–1.33; P = 0.34 Alpha-carotene: OR 0.64; 95% CI 0.39–1.04; P = 0.07 Beta-cryptoxanthin: OR 0.97; 95% CI 0.61–1.56; P = 0.41 Lutein: OR 1.00; 95% CI 0.64–1.64; P = 0.78 Zeaxanthin: OR 0.99; 95% CI 0.61–1.60; P = 0.83 Lycopene: OR 1.11; 95% CI 0.69–1.78; P = 0.81	Blood lycopene: OR 1.55; 95% CI 1.00–2.38	Blood lycopene: OR 1.20; 95% CI 0.76–1.90	Blood lycopene: OR 1.34; 95% CI 0.87–2.07

Other

Author	Any CAT	NUC	CX	PSC
Mares, 2010 [81]	-	High vs low HEI score: OR 0.63; 95% CI 0.43–0.91	-	-
Rautiainen, 2014 [82]	Highest vs lowest TAC quintile: OR 0.87 95% CI 0.79–0.96; P = 0.03	-	-	-

Table 2. Studies investigating antioxidant rich supplements on cataract incidence.

Author	Sample Size (*n*), Age (years)	Nutrients Examined	Key Findings
Multi-Vitamins			
Mares-Perlman, 2000 [85]	n = 3089	Supplementary Multivitamin, Vitamin C, Vitamin E	The 5-year risk for any CAT was 60% lower for multivitamins or any supplement use containing Vitamin C or E for more than 10 years. 10-year multivitamin use lowered the risk for NUC and CX but not for PSC (OR 0.6, 95% CI 0.4–0.9; OR 0.4, 95% CI 0.2–0.8; and OR, 0.9 95% CI 0.5–1.9; respectively).

Table 2. *Cont.*

Author	Sample Size (*n*), Age (years)	Nutrients Examined	Key Findings
Kuzniarz, 2001 [11]	*n* = 2873, 49–97 years	Supplementary Vitamin A, Thiamine, Riboflavin, Niacin, Pyridoxine, Folate, Vitamin B12	Use of multivitamin supplements was associated with reduced prevalence of NUC, OR 0.6, 95% CI 0.4–1.0, *P* =0.05. For both NUC and CX, longer duration of multivitamin use was associated with reduced prevalence (NUC, trend *P* = 0.02; CX, trend *P* = 0.03). Use of thiamin supplements was associated with reduced prevalence of NUC (OR 0.6, 95% CI 0.4–1.0, *P* = 0.03, dose trend *P* = 0.03) and CX (OR 0.7, 95% CI 0.5–0.9, *P* = 0.01, dose trend *P* = 0.02). Riboflavin (OR 0.8, 95% CI 0.6–1.0, *P* = 0.05) and niacin (OR 0.7, 95% CI 0.6–1.0, *P* = 0.04) supplements exerted a weaker protective influence on CX. Vitamin A supplements were protective against NUC (OR 0.4, 95% CI 0.2–0.8, *P* = 0.01, dose trend *P* = 0.01). Folate (OR 0.4, 95% CI 0.2–0.9, *P* = 0.03) appeared protective for NUC, whereas both folate (NUC 0.6, 95% CI 0.3–0.9, *P* = 0.01, dose trend *P* = 0.04) and Vitamin B12 supplements (OR 0.7, 95% CI 0.5–1.0, *P* = 0.03, dose trend *P* = 0.02) were strongly protective against CX.
Age-Related Eye Disease Study Research Group, 2006 [86]	*n* = 4590, -	Supplementary Centrum^{TM} Multivitamin (Vitamin A, E, C, B1, B2, B12, B6, D, Folic acid, Niacinamide, Biotin, Pantothenic acid, Calcium, Phosphorus, Iodine, Iron, Magnesium, Copper Zinc)	Centrum^{TM} use is associated with a reduction in any lens opacity progression (OR 0.84, 95% CI 0.72–0.98, *P* = 0.025). Also protective for nuclear opacity events (OR 0.75, 95% CI 0.61–0.91, *P* = 0.004).
Zheng Selin, 2013 [83]	*n* = 31120, 45–79 years	Supplemental Vitamin C, Vitamin E, Low dose multivitamins	The multivariable- adjusted HR for Vitamin C supplements only was 1.21 (95% CI 1.04–1.41) in compared to non-users. The HR for long-term Vitamin C users (≥10 years before baseline) was 1.36 (95% CI 1.02–1.81). The HR for Vitamin E use only was 1.59 (95% CI 1.12–2.26). Use of multivitamins only or multiple supplements in addition to Vitamin C or E was not associated with cataract risk.
Single Vitamins			
Christen, 2004 [87]	*n* = 39876, ≥ 45 years	Supplementary Beta-carotene (50mg.d^{-1}, alternate days)	129 CAT in the beta-carotene group and 133 in the placebo group (RR 0.95, 95% CI 0.75–1.21). For cataract extraction, there were 94 cases in the beta-carotene group and 89 cases in the placebo group (RR 1.04, 95% CI 0.78–1.39).
Christen, 2008 [88]	*n* = 39876, ≥ 45 years	Supplementary Vitamin E (600 IU.d^{-1}, alternate days)	No significant difference between the Vitamin E and placebo groups in the incidence of CAT (RR 0.96; 95% CI 0.88–1.04). No significant effects of Vitamin E on the incidence of NUC (RR 0.94; 95% CI 0.87–1.02), CX (RR 0.93; 95% CI 0.81–1.06), or PSC (RR 1.00; 95% CI 0.86–1.16).

Table 2. *Cont.*

Author	Sample Size (*n*), Age (years)	Nutrients Examined	Key Findings
Christen, 2015 [89]	*n* = 35533, ≥50 years	Supplementary Selenium (200 µg.d^{-1} from L-selenomethionine), Vitamin E (400 IU.d^{-1} of all rac-α-tocopheryl acetate)	185 CAT in the selenium group and 204 in placebo (HR 0.91; 95 % CI 0.75–1.11; *P* = 0.37). For Vitamin E, there were 197 cases in the Vitamin E group and 192 in placebo (HR 1.02; 95 % CI 0.84–1.25; *P* = 0.81)
Christen, 2010 [90]	*n* = 11545, ≥50 years	Supplementary Vitamin E (400 IU.d^{-1}, alternate days) Vitamin C (500 mg.d^{-1} alternate days)	579 CAT in the Vitamin E treated group and 595 in the Vitamin E placebo group (HR 0.99; 95% CI 0.88–1.11). For Vitamin C, there were 593 cataracts in the treated group and 581 in the placebo group (HR 1.02; 95% CI 0.91–1.14).
Christen, 2002 [91]	*n* = 22071	Supplementary Beta-carotene (50 mg.d^{-1}, alternate days)	No difference between the beta-carotene and placebo groups in the overall incidence of CAT (998 cases vs 1017 cases; RR 1.00; 95% CI 0.91–1.09) or CAT extraction (584 vs 593; RR 1.00; 95% CI 0.89–1.12).
Ferringo, 2005 [92]	*n* = 1020	Supplementary Vitamin A, Vitamin C, Vitamin E, Beta-carotene	High Vitamin C levels were associated with a protective effect on NUC (OR 0.54; 95% CI 0.30, 0.97) and PSC (OR: 0.37; 95% CI: 0.15–0.93). High Vitamin E levels were associated with increased prevalence of CX (OR 1.99; 95% CI 1.02–3.90), PSC (OR 3.27; 95% CI 1.34–7.96) and of any CAT (OR 1.86; 95% CI 1.08–3.18).
Rautiainen, 2009 [15]	*n* = 24593, 49–83 years	Supplementary Vitamin C	HR of Vitamin C supplement users compared with that for nonusers was 1.25 (95% CI 1.05, 1.50). The HR for the duration of 10 y of use before baseline was 1.46 (95% CI 0.93, 2.31). The HR for the use of multivitamins containing Vitamin C was 1.09 (95% CI 0.94, 1.25). Among women aged 65 y, Vitamin C supplement use increased the risk of CAT by 38% (95% CI 12%, 69%).
The REACT Group, 2002 [84]	*n* = 297	Supplementary Vitamin E (200 mg all-rac alpha-tocopherol acetate), Vitamin C (250 mg ascorbic acid), and b-carotene (6 mg)	After two years of treatment, there was a small positive treatment effect in U.S. patients (*p* = 0.0001); after three years a positive effect (*p* = 0.048) in both the U.S. and the U.K. groups. The positive effect in the U.S. group was even greater after three years: (IPO = 0.389 (Vitamin) vs. IPO = 2.517 (placebo); *p* = 0.0001).

The results of our analysis tend to confirm the belief that nutrients provided as a component of food (Table 1) are superior to mono-nutrient type supplements (Table 2) for the prevention and delay of cataract disease. It appears a good diet is still central to vision-related health, however, small-moderate doses of some antioxidants may still be beneficial in reducing the progression and incidence of specific types of cataract. In general, the literature suggests diets high in fruit and vegetables, Vitamin C, zeaxanthin, lutein and multivitamin-mineral supplements are associated with lower disease rates, while supplemental forms of selenium and Vitamin E had little effect.

Indicators and biomarkers of a healthy diet such as the healthy diet index, plasma antioxidant status and fruit and vegetable intake (Table 1) are associated with slower disease progression in those with moderate disease status, suggesting early intervention with a good diet is warranted [81]. A high intake of fruit and vegetables reduced the incidence of cataracts by 62% when data was appropriately adjusted, suggesting the simple intervention of consuming the recommended daily intake of fruit and vegetables is valid in this population group [74].

Lutein and zeaxanthin are the most abundant carotenoid in lens and have been shown to reduce damage from reactive species in cultured lens epithelial cells [80]. High dietary intake of lutein and zeaxanthin was associated with a 23% lower incidence of nuclear cataracts [80]. Women and men with carotenoid (lutein/zeaxanthin) intakes of approximately 4 to 6 mg/day have reduced rates of cataract extraction [69], demonstrating that dietary intake can not only reduce the incidence but delay disease progression.

Dietary (Table 1) or low dose supplemental (Table 2) Vitamin C reduces the incidence and progression of disease, demonstrated in the majority of the 9 studies. However, a study comparing Vitamin C as part of a multivitamin-mineral (60 mg/day) versus a Vitamin C supplement (1000 mg/day) showed the high dose, single nutrient actually increased the risk of cataracts, while there was no change in the multivitamin-mineral group [15]. Also, Zheng-Selin (2013) [83] reported that a high dose of Vitamin C increased the risk of cataract disease by 21% and Vitamin E increased the risk by 57%. Studies investigating the plasma concentrations of ascorbic acid support findings from the dietary supplement studies that reported higher ascorbic acid concentrations were associated with a lower incidence of cataracts. A small-to-moderate dose of Vitamin C (less than 100 mg/day) appears optimal, while we caution the recommendation of high dose supplemental Vitamin C, deemed to be equal or greater than 500 mg/day, particularly when taken chronically. Intakes of 250 mg/day of Vitamin C did reduce the signs and symptoms of cataracts in a three-month intervention trial [84], suggesting a short-term dose of Vitamin C may benefit those already diagnosed. The value long-term is still in question.

While food contains a complex array of phytochemicals, nutrients and satiety factors proving a range of health benefits, ensuring adequate concentrations of the active ingredients reach the ocular region is a challenge. The intake of a multivitamin and mineral supplement also reduced the incidence of cataract disease, with Centrum™ being the most commonly used brand. Centrum™ contains a large range of nutrients in amounts reflecting a small percentage of the recommended daily allowance, as such is a true dietary supplement rather than an ergogenic aide. One capsule daily reduced the incidence of cataracts by 3% when taken for less than 10 years and by 6% when taken longer, suggesting the benefit for multivitamin intake is small, but useful long-term [83].

Mares-Perlman [85], reported a protective effect of supplementary Vitamin E (Table 2) reducing the incidence of cataracts by 10%, while dietary Vitamin E reduced disease incidence by 80%, confirming the theory that dietary intake is generally more effective than supplements.

In summary, while the literature is mixed, there appears to be a general consensus that a diet high in fruit and vegetables containing Vitamin C, E, A and multivitamin-mineral supplements may be protective against cataracts. While the majority of dietary intake studies show a reduction in the risk of disease and disease progression, the strongest reductions were obtained with diets high in vitamin C, E and A, with lesser effects for diets rich in carotenoids and selenium. Data on supplemental antioxidants provided as low dose multivitamin appear generally positive, while results obtained using single nutrient antioxidants ranged from moderately effective to possibly harmful. Based on current data, a healthy diet and a multivitamin supplement may offer protection against cataracts. If this is the case, it is important to consider how these nutrients are able to reach the lens in sufficient quantities to be effective in protecting the lens from oxidative damage. In order to understand this process, a more thorough understanding of lens structure and function is required so that strategies to enhance the delivery and accumulation of these nutrients particularly in the lens nucleus can be identified, and harnessed to restore depleted antioxidant levels with advancing age.

4. Lens Structure and Function

The transparency of the lens is closely linked to the unique structure and function of its fiber cells. These highly differentiated cells are derived from equatorial epithelial cells, which exit the cell cycle and embark upon a differentiation process that produces extensive cellular elongation, the loss of cellular organelles such as mitochondria and nuclei, and the expression of fiber-specific proteins [93,94]. Fiber cells elongate until fibers from opposite hemispheres meet at the poles and interdigitate to form the lens sutures [95]. Since this process continues throughout life, a gradient of fiber cells at different stages of differentiation is established around an internalized core (nucleus) of mature, anucleate fiber cells that were laid down in the embryo (Figure 3A). While the transparent properties of the lens are a direct result of its highly ordered tissue architecture, the lens is not a purely passive optical element. Maintenance of its architecture requires special mechanisms not only to supply the deeper lying fiber cells with nutrients, but also to control the volume of these cells. Because of its size, the avascular lens cannot rely on passive diffusion alone to transport nutrients and antioxidants from the surrounding humours to deeper lying cells, or to transport waste products back to the surface. Instead, it has been proposed that the lens operates an internal microcirculation system, which maintains lens transparency by delivering nutrients to the lens core faster than would occur by passive diffusion alone [96–98].

A common feature of all vertebrate lenses studied to date is the existence of a standing flow of ionic current that is directed inward at the poles and outward at the equator (Figure 3A). Mathias et al [98] have proposed that these currents measured at the lens surface represent the external portion of a circulating ionic current that drives a unique internal microcirculatory system that maintains fiber cell homeostasis and therefore lens transparency. Briefly, this model of lens transport states that a circulating current of Na^+ ions, primarily enters at the poles and travels into the lens via the extracellular clefts between fiber cells (Figure 3B). Na^+ crosses fiber cell membranes, and returns towards the surface via an intercellular pathway mediated by gap junction channels, where it is actively removed by Na^+ pumps concentrated at the lens equator (Figure 3B, *top panel*). This circulating ionic current creates a net flux of ions that in turn generates fluid flow (Figure 3B, *middle panel*). The accompanying extracellular flow of water convects nutrients towards the deeper lying fiber cells, while the outward intercellular flow, driven by a hydrostatic pressure gradient [99], removes wastes and creates a well-stirred intracellular compartment (Figure 3B, *bottom panel*). Thus active pumping of Na^+ by surface cells is able to regulate the ion composition of inner fiber cells and allows them to maintain the ion gradients necessary to power a variety of secondary active transporters that mediate steady state volume regulation and nutrient/antioxidant uptake [97].

While the evidence in favour of the circulation system has been accumulating over many years, it is not universally accepted [50,100]. The main criticisms have centred around our inability, to date, to actually visualize ion and fluid fluxes within the lens [100] and the perceived need for active metabolism in the lens nucleus. In an effort to measure water fluxes, Candia et al [101], employed Ussing chambers to measure regional water fluxes and showed that water influx at the poles is equal to water efflux at the equator, data which supports the existence of a circulating water fluxes through the lens. Utilizing a novel combination of MRI and confocal microscopy, Vaghefi et al [102,103] have visualized for the first time, the ion and fluid fluxes within the lens that were originally predicted by the microcirculation system. Furthermore, to visualize the delivery of solutes by the microcirculation system Vaghefi et al., have developed MRI imaging protocols that can visualize the rate of penetration of a variety of extracellular contrast agents of varying molecular size into the different regions of the lens [104]. Utilizing this approach, they have shown solute delivery to the lens core is not only dependent on the size of the extracellular tracer molecule, and also occurs at a rate faster than predicted for passive diffusion alone, but also that solute delivery is abolished if the microcirculation system is inhibited [104]. Thus by analogy, it would appear that the delivery of nutrients and antioxidants to the core of the lens is an active process that occurs via an extracellular route which is driven by the lens microcirculation system.

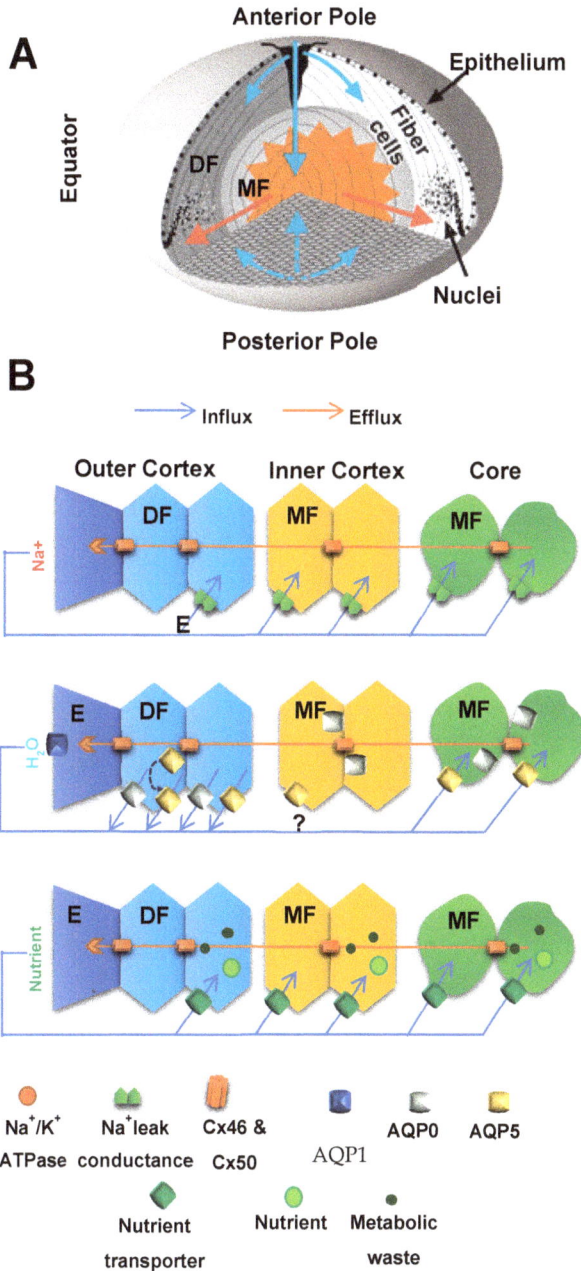

Figure 3. Lens structure and function (**A**) 3-D representation of the microcirculation model, showing ions and fluid fluxes that enter the lens at both poles via the extracellular space (blue arrows) before crossing fiber cell membranes and exiting the lens at the equator via an intracellular pathway (red arrows) mediated by gap junctions. (**B**) Equatorial cross-sections showing how the spatial differences in the distribution of ion channels and transporters between the epithelium (E), differentiating (DF) and mature (MF) fiber cells that generate the circulating flux of Na$^+$ ions (*top*) that drives isotonic fluid fluxes (*middle*) which in turn deliver nutrients to and remove metabolic waste from the MF cells (*bottom*).

To complete the delivery of nutrients and antioxidants to the core of the lens, the mature cells in this region of the lens need to be able to accumulate the solutes convected to them via the microcirculation system. In this regard Lim et al., have shown that mature fiber cells in the lens core differentially express a wide range of Na^+-dependent and independent transporters that are capable of nutrient and antioxidant uptake in this region of the lens [105–110]. Furthermore, it is emerging that distinct metabolic compartments exist in the lens [111], which is not surprising considering that the peripheral differentiating fiber cells contain mitochondria and can perform aerobic metabolism, but mature fiber cells are devoid of mitochondria and rely on anaerobic metabolism to provide their metabolic requirements. This implies that in the different regions of the lens distinct metabolic pathways are utilized to maintain the levels of antioxidants that protect against cataract formation. It follows therefore, that the dysfunction of those region specific pathways is the underlying cause of the different etiologies of cortical and nuclear cataract, which in turn explains the variable efficacy of the different exogenous antioxidant supplements on the two types of cataract observed in epidemiological studies.

5. Conclusions and Future Strategies

With a globally aging population, age related cataracts has grown to epidemic proportions, placing severe pressures on global and local health systems. Since the initiation of lens cataract is strongly associated with oxidative damage, the use of exogenous antioxidant interventions has been advocated as a strategy to slow the progression of lens cataract. However, the majority of these studies are ineffective in slowing down cataract progression. Rather, it appears that a combination and/or a range of endogenous nutrients may be more effective in slowing the progression of the different forms of lens cataract. In this regard, our current understanding of lens structure and function needs to be incorporated into the design of more effective anti-cataract therapies. By designing nutritional strategies that take into account the underlying physiology of how specific nutrients and antioxidants are delivered, taken up, and metabolized to maintain and restore antioxidant levels in the different regions of the lens, appeals as a strategy to prevent both cortical and nuclear cataract, and thereby avert the looming cataract epidemic facing our aging population.

Author Contributions: A.J.B. methodology, data curation, formal analysis, writing—original draft preparation, writing—review and editing; C.I.D.; methodology, data curation, formal analysis J.C.L., writing—original draft preparation, writing—review and editing, visualization, P.J.D. project administration, writing—original draft preparation, writing—review and editing, funding acquisition.

Funding: This research was funded by the Health Research Council of New Zealand and the Auckland Medical Research Foundation.

Acknowledgments: We thank Bo Li for her artistic contributions to Figures 1 and 2.

Conflicts of Interest: The authors declare no conflict of interest.

References

1. Hobbs, R.P.; Bernstein, P.S. Nutrient Supplementation for Age-related Macular Degeneration, Cataract, and Dry Eye. *J. Ophthalmic Vis. Res.* **2014**, *9*, 487–493. [CrossRef]
2. Pascolini, D.; Mariotti, S.P. Global estimates of visual impairment: 2010. *Br. J. Ophthalmol* **2012**, *96*, 614–618. [CrossRef]
3. Weikel, K.A.; Garber, C.; Baburins, A.; Taylor, A. Nutritional modulation of cataract. *Nutr. Rev.* **2014**, *72*, 30–47. [CrossRef] [PubMed]
4. Wild, S.; Roglic, G.; Green, A.; Sicree, R.; King, H. Global prevalence of diabetes: estimates for the year 2000 and projections for 2030. *Diabetes Care* **2004**, *27*, 1047–1053. [CrossRef] [PubMed]
5. Klein, B.E.; Klein, R.; Wang, Q.; Moss, S.E. Older-onset diabetes and lens opacities. The Beaver Dam Eye Study. *Ophthalm. Epid.* **1995**, *2*, 49–55. [CrossRef]
6. United Nations Department of Economic and Social Affairs, Population Division. *World Population Ageing 2017-Highlights (ST/ESA/SER.A/397)*; United Nations: New York, NY, USA, 2017.
7. McCarty, C.A.; Taylor, H.R. The Genetics of Cataract. *Investig. Ophthalmol. Vis. Sci.* **2001**, *42*, 1677–1678.

8. Access Economics. Clear focus—The economic impact of vision loss in New Zealand in 2009. *A report prepared for RANZCO and the RNZFB*, 2010.

9. Brian, G.; Taylor, H. Cataract blindness–challenges for the 21st century. *Bull. World Health Organ.* **2001**, *79*, 249–256.

10. Cumming, R.G.; Mitchell, P.; Smith, W. Diet and cataract: The blue mountains eye study. *Ophthalmology* **2000**, *107*, 450–456. [CrossRef]

11. Kuzniarz, M.; Mitchell, P.; Cumming, R.G.; Flood, V.M. Use of vitamin supplements and cataract: The blue mountains eye study. *Am. J. Ophthalmol.* **2001**, *132*, 19–26. [CrossRef]

12. Valero, M.P.; Fletcher, A.E.; de Stavola, B.L.; Vioque, J.S.; Alepuz, V.C.S. Vitamin C Is Associated with Reduced Risk of Cataract in a Mediterranean Population. *J. Nutr.* **2002**, *132*, 1299–1306. [CrossRef]

13. McNeil, J.J.; Robman, L.; Tikellis, G.; Sinclair, M.I.; McCarty, C.A.; Taylor, H.R. Vitamin E supplementation and cataract: Randomized controlled trial. *Ophthalmology* **2004**, *111*, 75–84. [CrossRef] [PubMed]

14. Age-Related Eye Disease Study Research Group. A randomized, placebo-controlled, clinical trial of high-dose supplementation with vitamins C and E and beta carotene for age-related cataract and vision loss: AREDS report no. 9. *Arch Ophthalmol* **2001**, *119*, 1439–1452. [CrossRef]

15. Rautiainen, S.; Lindblad, B.E.; Morgenstern, R.; Wolk, A. Vitamin C supplements and the risk of age-related cataract: A population-based prospective cohort study in women. *Am. J. Clin. Nutr.* **2009**, *91*, 487–493. [CrossRef]

16. Cheng, R.; Feng, Q.; Ortwerth, B.J. LC-MS display of the total modified amino acids in cataract lens proteins and in lens proteins glycated by ascorbic acid in vitro. *Biochim. Biophys. Acta (Bba) Mol. Basis Dis.* **2006**, *1762*, 533–543. [CrossRef]

17. Linetsky, M.; Shipova, E.; Cheng, R.; Ortwerth, B.J. Glycation by ascorbic acid oxidation products leads to the aggregation of lens proteins. *Biochim. Et Biophys. Acta (Bba) Mol. Basis Dis.* **2008**, *1782*, 22–34. [CrossRef]

18. Christen, W.G.; Glynn, R.J.; Chew, E.Y.; Albert, C.M.; Manson, J.E. Folic Acid, Vitamin B6, and Vitamin B12 in Combination and Age-Related Cataract in a Randomized Trial of Women. *Ophthalmic Epidemiol.* **2016**, *23*, 32–39. [CrossRef]

19. Camacho-Barcia, M.L.; Bulló, M.; Garcia-Gavilán, J.F.; Ruiz-Canela, M.; Corella, D.; Estruch, R.; Fitó, M.; García-Layana, A.; Arós, F.; Fiol, M.; et al. Association of Dietary Vitamin K1 Intake with the Incidence of Cataract Surgery in an Adult Mediterranean Population: A Secondary Analysis of a Randomized Clinical Trial. *JAMA Ophthalmol.* **2017**, *135*, 657–661. [CrossRef]

20. Bourne, R.R; Stevens, G.A.; White, R.A.; Smith, J.L.; Flaxman, S.R.; Price, H.; Jonas, J.B.; Keeffe, J.; Leasher, J.; Naidoo, K.; et al. Causes of vision loss worldwide, 1990–2010: A systematic analysis. *Lancet Glob Health* **2013**, *1*, e339–e349.

21. Michael, R.; Bron, A.J. The ageing lens and cataract: A model of normal and pathological ageing. *Philos. Trans. R. Soc. Lond. Ser. Biol. Sci.* **2011**, *366*, 1278–1292. [CrossRef] [PubMed]

22. Michael, R. Cortical cataract. In *Encyclopedia of the Eye*; Darlene, A.D., Ed.; Academic Press: Oxford, UK, 2010; Volume 1, pp. 532–536.

23. Pesudovs, K.; Elliott, D.B. Refractive error changes in cortical, nuclear, and posterior subcapsular cataracts. *Br. J. Ophthalmol.* **2003**, *87*, 964–967. [CrossRef]

24. Planten, J.T. Changes of refraction in the adult eye due to changing refractive indices of the layers of the lens. *Ophthalmol. J. Int. D'ophtalmologie Int. J. Ophthalmol. Z. Fur Augenheilkd.* **1981**, *183*, 86–90. [CrossRef]

25. Al-Ghoul, K.J.; Costello, M.J. Morphological changes in human nuclear cataracts of late-onset diabetics. *Exp. Eye Res.* **1993**, *57*, 469–486. [CrossRef] [PubMed]

26. Kinoshita, J.H. Pathways of glucose metabolism in the lens. *Investig. Ophthalmol.* **1965**, *4*, 619–628.

27. Kinoshita, J.H. Cataracts in galactosemia: The Jonas, S. Friedenwald Memorial Lecture. *Investig. Ophthalmol.* **1965**, *4*, 786–799.

28. Kinoshita, J.H. Mechanisms initiating cataract formation. Proctor Lecture. *Investig. Ophthalmol.* **1974**, *13*, 713–724.

29. Kador, P.F.; Wyman, M.; Oates, P.J. Aldose reductase, ocular diabetic complications and the development of topical Kinostat®. *Prog. Retin. Eye Res.* **2016**. [CrossRef]

30. Chung, S.S.; Ho, E.C.; Lam, K.S.; Chung, S.K. Contribution of polyol pathway to diabetes-induced oxidative stress. *J. Am. Soc. Nephrol. JASN* **2003**, *14*, S233–S236. [CrossRef]

31. Chan, A.W.; Ho, Y.S.; Chung, S.K.; Chung, S.S. Synergistic effect of osmotic and oxidative stress in slow-developing cataract formation. *Exp. Eye Res.* **2008**, *87*, 454–461. [CrossRef] [PubMed]
32. Pirie, A. Color and solubility of the proteins of human cataracts. *Investig. Ophthalmol.* **1968**, *7*, 634–650.
33. Al-Ghoul, K.J.; Costello, M.J. Fiber cell morphology and cytoplasmic texture in cataractous and normal human lens nuclei. *Curr. Eye Res.* **1996**, *15*, 533–542. [CrossRef]
34. Costello, M.J.; Oliver, T.N.; Cobo, L.M. Cellular architecture in age-related human nuclear cataracts. *Invest Ophthalmol. Vis. Sci.* **1992**, *33*, 3209–3227. [PubMed]
35. Garner, M.H.; Spector, A. Sulfur oxidation in selected human cortical cataracts and nuclear cataracts. *Exp. Eye Res.* **1980**, *31*, 361–369. [CrossRef]
36. Garner, M.H.; Spector, A. Selective oxidation of cysteine and methionine in normal and senile cataractous lenses. *Proc. Natl. Acad. Sci. USA* **1980**, *77*, 1274–1277. [CrossRef] [PubMed]
37. Spector, A.; Roy, D. Disulfide-linked high molecular weight protein associated with human cataract. *Proc. Natl. Acad. Sci. USA* **1978**, *75*, 3244–3248. [CrossRef] [PubMed]
38. Truscott, R.J.; Augusteyn, R.C. The state of sulphydryl groups in normal and cataractous human lenses. *Exp. Eye Res.* **1977**, *25*, 139–148. [CrossRef]
39. Lou, M.F.; Dickerson, J.E., Jr.; Garadi, R. The role of protein-thiol mixed disulfides in cataractogenesis. *Exp. Eye Res.* **1990**, *50*, 819–826. [CrossRef]
40. Lou, M.F.; Dickerson, J.E., Jr.; Tung, W.H.; Wolfe, J.K.; Chylack, L.T., Jr. Correlation of nuclear color and opalescence with protein S-thiolation in human lenses. *Exp. Eye Res.* **1999**, *68*, 547–552. [CrossRef]
41. Lim, J.C.; Umapathy, A.; Donaldson, P.J. Tools to fight the cataract epidemic: A review of experimental animal models that mimic age related nuclear cataract. *Exp. Eye Res.* **2016**, *145*, 432–443. [CrossRef]
42. Lou, M.F. Redox regulation in the lens. *Prog. Retin. Eye Res.* **2003**, *22*, 657–682. [CrossRef]
43. Truscott, R.J. Age-related nuclear cataract-oxidation is the key. *Exp. Eye Res.* **2005**, *80*, 709–725. [CrossRef]
44. Reddy, V.N. Glutathione and its function in the lens–an overview. *Exp. Eye Res.* **1990**, *50*, 771–778. [CrossRef]
45. Rathbun, W.B. *Glutathione Biosynthesis in the Lens and Erythrocyte*; Elsevier/North Holland: Amsterdam, The Netherlands, 1980; pp. 169–173.
46. Kannan, R.; Yi, J.R.; Zlokovic, B.V.; Kaplowitz, N. Molecular characterization of a reduced glutathione transporter in the lens. *Investig. Ophthalmol. Vis. Sci.* **1995**, *36*, 1785–1792.
47. Mackic, J.B.; Jinagouda, S.; McComb, G.J.; Weiss, M.H.; Kannan, R.A.M.; Kaplowitz, N.; Zlokovic, B.V. Transport of Circulating Reduced Glutathione at the Basolateral Side of the Anterior Lens Epithelium: Physiologic Importance and Manipulations. *Exp. Eye Res.* **1996**, *62*, 29–38. [CrossRef] [PubMed]
48. Giblin, F.J. Glutathione: A Vital Lens Antioxidant. *J. Ocul. Pharmacol. Ther.* **2000**, *16*, 121. [CrossRef] [PubMed]
49. Fan, X.; Monnier, V.M.; Whitson, J. Lens glutathione homeostasis: Discrepancies and gaps in knowledge standing in the way of novel therapeutic approaches. *Exp. Eye Res.* **2017**, *156*, 103–111. [CrossRef]
50. Donaldson, P.J.; Musil, L.S.; Mathias, R.T. Point: A Critical Appraisal of the Lens Circulation Model—An Experimental Paradigm for Understanding the Maintenance of Lens Transparency? *Investig. Ophthalmol. Vis. Sci.* **2010**, *51*, 2303. [CrossRef]
51. Rathbun, W.B.; Murray, D.L. Age-related cysteine uptake as rate-limiting in glutathione synthesis and glutathione half-life in the cultured human lens. *Exp. Eye Res.* **1991**, *53*, 205–212. [CrossRef]
52. Sweeney, M.H.J.; Truscott, R.J.W. An impediment to glutathione diffusion in older normal human lenses: A possible precondition for nuclear cataract. *Exp. Eye Res.* **1998**, *67*, 587–595. [CrossRef]
53. Truscott, R.J. Age-related nuclear cataract: A lens transport problem. *Ophthalmic Res.* **2000**, *32*, 185–194. [CrossRef]
54. Zhang, H.; Jay Forman, H.; Choi, J. Υ-Glutamyl Transpeptidase in Glutathione Biosynthesis. In *Methods in Enzymology*; Sies, H., Packer, L., Eds.; Academic Press: Waltham, MA, USA, 2005; Volume 401, pp. 468–483.
55. Bouayed, J.; Bohn, T. Exogenous antioxidants—Double-edged swords in cellular redox state Health beneficial effects at physiologic doses versus deleterious effects at high doses. *Oxid Med. Cell Longev.* **2010**, *3*, 228–237. [CrossRef]
56. Finkel, T. Signal transduction by reactive oxygen species. *J. Cell Biol.* **2011**, *194*, 7–15. [CrossRef]
57. Schieber, M.; Chandel, N.S. ROS function in redox signaling and oxidative stress. *Curr. Biol.* **2014**, *24*, R453–R462. [CrossRef]

58. Sena, L.A.; Chandel, N.S. Physiological roles of mitochondrial reactive oxygen species. *Mol. Cell* **2012**, *48*, 158–167. [CrossRef]

59. Agte, V.; Tarwadi, K. The Importance of Nutrition in the Prevention of Ocular Disease with Special Reference to Cataract. *Ophthalmic Res.* **2010**, *44*, 166–172. [CrossRef]

60. Cui, Y.-H.; Jing, C.-X.; Pan, H.-W. Association of blood antioxidants and vitamins with risk of age-related cataract: A meta-analysis of observational studies. *Am. J. Clin. Nutr.* **2013**, *98*, 778–786. [CrossRef]

61. Fernandez, M.M.; Afshari, N.A. Nutrition and the prevention of cataracts. *Curr. Opin. Ophthalmol.* **2008**, *19*, 66–70. [CrossRef]

62. Fletcher, A.E. Free Radicals, Antioxidants and Eye Diseases: Evidence from Epidemiological Studies on Cataract and Age-Related Macular Degeneration. *Ophthalmic Res.* **2010**, *44*, 191–198. [CrossRef] [PubMed]

63. Mathew, M.C.; Ervin, A.M.; Tao, J.; Davis, R.M. Antioxidant vitamin supplementation for preventing and slowing the progression of age-related cataract. *Cochrane Database Syst. Rev.* **2012**. [CrossRef] [PubMed]

64. McCusker, M.M.; Durrani, K.; Payette, M.J.; Suchecki, J. An eye on nutrition: The role of vitamins, essential fatty acids, and antioxidants in age-related macular degeneration, dry eye syndrome, and cataract. *Clin. Dermatol.* **2016**, *34*, 276–285. [CrossRef]

65. Moeller, S.M.; Jacques, P.F.; Blumberg, J.B. The Potential Role of Dietary Xanthophylls in Cataract and Age-Related Macular Degeneration. *J. Am. Coll. Nutr.* **2000**, *19*, 522S–527S. [CrossRef]

66. Raman, R.; Vaghefi, E.; Braakhuis, A.J. Food components and ocular pathophysiology: A critical appraisal of the role of oxidative mechanisms. *Asia Pac. J. Clin. Nutr.* **2017**, *26*, 572–585. [CrossRef]

67. Seddon, J.M. Multivitamin-multimineral supplements and eye disease: Age-related macular degeneration and cataract. *Am. J. Clin. Nutr.* **2007**, *85*, 304S–307S. [CrossRef] [PubMed]

68. Siegal, M.; Chiu, C.-J.; Taylor, A. Antioxidant Status and Risk for Cataract. In *Preventive Nutrition: The Comprehensive Guide for Health Professionals*; Bendich, A., Deckelbaum, R.J., Eds.; Humana Press: Totowa, NJ, USA, 2005; pp. 463–503.

69. Trumbo, P.R.; Ellwood, K.C. Lutein and zeaxanthin intakes and risk of age-related macular degeneration and cataracts: An evaluation using the Food and Drug Administration's evidence-based review system for health claims. *Am. J. Clin. Nutr.* **2006**, *84*, 971–974. [CrossRef] [PubMed]

70. Zhao, L.-Q.; Li, L.-M.; Zhu, H. The Effect of Multivitamin/Mineral Supplements on Age-Related Cataracts: A Systematic Review and Meta-Analysis. *Nutrients* **2014**, *6*, 931. [CrossRef] [PubMed]

71. Christen, W.G.; Liu, S.; Glynn, R.J.; Gaziano, J.M.; Buring, J.E. Dietary carotenoids, vitamins C and E, and risk of cataract in women: A prospective study. *Arch. Ophthalmol.* **2008**, *126*, 102–109. [CrossRef]

72. Dherani, M.; Murthy, G.V.S.; Gupta, S.K.; Young, I.S.; Maraini, G.; Camparini, M.; Price, G.M.; John, N.; Chakravarthy, U.; Fletcher, A.E. Blood Levels of Vitamin C, Carotenoids and Retinol Are Inversely Associated with Cataract in a North Indian Population. *Investig. Ophthalmol. Vis. Sci.* **2008**, *49*, 3328–3335. [CrossRef] [PubMed]

73. Jacques, P.F.; Chylack, L.T., Jr.; Hankinson, S.E.; Khu, P.M.; Rogers, G.; Friend, J.; Tung, W.; Wolfe, J.K.; Padhye, N.; Willett, W.C.; et al. Long-term Nutrient Intake and Early Age-Related Nuclear Lens Opacities. *Arch. Ophthalmol.* **2001**, *119*, 1009–1019. [CrossRef]

74. Pastor-Valero, M. Fruit and vegetable intake and vitamins C and E are associated with a reduced prevalence of cataract in a Spanish Mediterranean population. *BMC Ophthalmol.* **2013**, *13*, 52. [CrossRef]

75. Ravindran, R.D.; Vashist, P.; Gupta, S.K.; Young, I.S.; Maraini, G.; Camparini, M.; Jayanthi, R.; John, N.; Fitzpatrick, K.E.; Chakravarthy, U.; et al. Inverse Association of Vitamin C with Cataract in Older People in India. *Ophthalmology* **2011**, *118*, 1958–1965.e1952. [CrossRef] [PubMed]

76. Tan, A.G.; Rochtchina, E.; Burlutsky, G.; Mitchell, P.; Cumming, R.G.; Flood, V.M.; Wang, J.J. Antioxidant nutrient intake and the long-term incidence of age-related cataract: The Blue Mountains Eye Study. *Am. J. Clin. Nutr.* **2008**, *87*, 1899–1905. [CrossRef]

77. Theodoropoulou, S.; Samoli, E.; Theodossiadis, P.G.; Papathanassiou, M.; Lagiou, A.; Lagiou, P.; Tzonou, A. Diet and cataract: A case–control study. *Int. Ophthalmol.* **2014**, *34*, 59–68. [CrossRef]

78. Yoshida, M.; Takashima, Y.; Inoue, M.; Iwasaki, M.; Otani, T.; Sasaki, S.; Tsugane, S.; JPHC Study Group. Prospective study showing that dietary vitamin C reduced the risk of age-related cataracts in a middle-aged Japanese population. *Eur. J. Nutr.* **2007**, *46*, 118–124. [CrossRef]

79. Delcourt, C.C.; Carriére, I.; Delage, M.; Barberger-Gateau, P.; Schalch, W.; POLA Study Group. Plasma Lutein and Zeaxanthin and Other Carotenoids as Modifiable Risk Factors for Age-Related Maculopathy and Cataract: The POLA Study. *Investig. Ophthalmol. Vis. Sci.* **2006**, *47*, 2329–2335. [CrossRef]

80. Moeller, S.M.; Voland, R.; Tinker, L.; Blodi, B.A.; Klein, M.L.; Gehrs, K.M.; Johnson, E.J.; Snodderly, D.M.; Wallace, R.B.; Chappell, R.J.; et al. Associations Between Age-Related Nuclear Cataract and Lutein and Zeaxanthin in the Diet and Serum in the Carotenoids in the Age-Related Eye Disease Study (CAREDS), an Ancillary Study of the Women's Health Initiative. *Arch. Ophthalmol.* **2008**, *126*, 354–364. [CrossRef]

81. Mares, J.A.; Voland, R.; Adler, R.; Tinker, L.; Millen, A.E.; Moeller, S.M.; Blodi, B.; Gehrs, K.M.; Wallace, R.B.; Chappell, R.J.; et al. Healthy Diets and the Subsequent Prevalence of Nuclear Cataract in WomenHealthy Diets and Nuclear Cataract in Women. *Arch. Ophthalmol.* **2010**, *128*, 738–749. [CrossRef]

82. Rautiainen, S.; Lindblad, B.E.; Morgenstern, R.; Wolk, A. Total Antioxidant Capacity of the Diet and Risk of Age-Related Cataract: A Population-Based Prospective Cohort of WomenAntioxidant Capacity of Diet and Risk of CataractsAntioxidant Capacity of Diet and Risk of Cataracts. *JAMA Ophthalmol.* **2014**, *132*, 247–252. [CrossRef]

83. Zheng Selin, J.; Rautiainen, S.; Lindblad, B.E.; Morgenstern, R.; Wolk, A. High-Dose Supplements of Vitamins C and E, Low-Dose Multivitamins, and the Risk of Age-related Cataract: A Population-based Prospective Cohort Study of Men. *Am. J. Epidemiol.* **2013**, *177*, 548–555. [CrossRef]

84. Group, T.R.; Chylack, L.T.; Brown, N.P.; Bron, A.; Hurst, M.; Köpcke, W.; Thien, U.; Schalch, W. The Roche European American Cataract Trial (REACT): A randomized clinical trial to investigate the efficacy of an oral antioxidant micronutrient mixture to slow progression of age-related cataract. *Ophthalmic Epidemiol* **2002**, *9*, 49–80. [CrossRef]

85. Mares-Perlman, J.A.; Lyle, B.J.; Klein, R.; Fisher, A.I.; Brady, W.E.; VandenLangenberg, G.M.; Trabulsi, J.N.; Palta, M. Vitamin Supplement Use and Incident Cataracts in a Population-Based Study. *Arch. Ophthalmol.* **2000**, *118*, 1556–1563. [CrossRef]

86. Group, A.-R.E.D.S.R. Centrum Use and Progression of Age-Related Cataract in the Age-Related Eye Disease Study: A Propensity Score Approach. AREDS Report No. 21. *Ophthalmology* **2006**, *113*, 1264–1270. [CrossRef]

87. Christen, W.; Glynn, R.; Sperduto, R.; Chew, E.; Buring, J. Age-related cataract in a randomized trial of beta-carotene in women. *Ophthalmic Epidemiol.* **2004**, *11*, 401–412. [CrossRef] [PubMed]

88. Christen, W.G.; Glynn, R.J.; Chew, E.Y.; Buring, J.E. Vitamin E and Age-Related Cataract in a Randomized Trial of Women. *Ophthalmology* **2008**, *115*, 822–829.e821. [CrossRef]

89. Christen, W.G.; Glynn, R.J.; Gaziano, J.M.; Darke, A.K.; Crowley, J.J.; Goodman, P.J.; Lippman, S.M.; Lad, T.E.; Bearden, J.D.; Goodman, G.E.; et al. Age-Related Cataract in Men in the Selenium and Vitamin E Cancer Prevention Trial Eye Endpoints Study: A Randomized Clinical TrialAge-Related Cataract and Selenium and Vitamin E in MenAge-Related Cataract and Selenium and Vitamin E in Men. *JAMA Ophthalmol.* **2015**, *133*, 17–24. [CrossRef] [PubMed]

90. Christen, W.G.; Glynn, R.J.; Sesso, H.D.; Kurth, T.; MacFadyen, J.; Bubes, V.; Buring, J.E.; Manson, J.E.; Gaziano, J.M. Age-Related Cataract in a Randomized Trial of Vitamins E and C in Men Vitamins E and C and Cataract. *Arch. Ophthalmol.* **2010**, *128*, 1397–1405. [CrossRef] [PubMed]

91. Christen, W.G.; Manson, J.E.; Glynn, R.J.; Gaziano, J.M.; Sperduto, R.D.; Buring, J.E.; Hennekens, C.H. A Randomized Trial of Beta Carotene and Age-Related Cataract in US Physicians. *Arch. Ophthalmol.* **2002**, *121*, 372–378. [CrossRef]

92. Ferrigno, L.; Aldigeri, R.; Rosmini, F.; Sperduto, R.D.; Maraini, G. Associations Between Plasma Levels of Vitamins and Cataract in the Italian-American Clinical Trial of Nutritional Supplements and Age-Related Cataract (CTNS): CTNS Report #2. *Ophthalmic Epidemiol.* **2005**, *12*, 71–80. [CrossRef]

93. Bassnett, S. Lens organelle degradation. *Exp. Eye Res.* **2002**, *74*, 1–6. [CrossRef]

94. Menko, A.S. Lens epithelial cell differentiation. *Exp. Eye Res.* **2002**, *75*, 485–490. [CrossRef]

95. Kuszak, J.R.; Zoltoski, R.K.; Tiedemann, C.E. Development of lens sutures. *Int. J. Dev. Biol.* **2004**, *48*, 889–902. [CrossRef]

96. Donaldson, P.; Kistler, J.; Mathias, R.T. Molecular solutions to mammalian lens transparency. *News Physiol. Sci.* **2001**, *16*, 118–123. [CrossRef]

97. Mathias, R.T.; Kistler, J.; Donaldson, P. The lens circulation. *J. Membr Biol* **2007**, *216*, 1–16. [CrossRef]

98. Mathias, R.T.; Rae, J.L.; Baldo, G.J. Physiological properties of the normal lens. *Physiol. Rev.* **1997**, *77*, 21–50. [CrossRef]

Nutrients **2019**, *11*, 1186

99. Gao, J.; Sun, X.; Moore, L.C.; White, T.W.; Brink, P.R.; Mathias, R.T. Lens intracellular hydrostatic pressure is generated by the circulation of sodium and modulated by gap junction coupling. *J. Gen. Physiol.* **2011**, *137*, 507–520. [CrossRef]

100. Beebe, D.C.; Truscott, R.J.W. Counterpoint: The Lens Fluid Circulation Model—A Critical Appraisal. *Investig. Ophthalmol. Vis. Sci.* **2010**, *51*, 2306–2310. [CrossRef]

101. Candia, O.A.; Mathias, R.; Gerometta, R. Fluid circulation determined in the isolated bovine lens. *Investig. Ophthalmol. Vis. Sci.* **2012**, *53*, 7087–7096. [CrossRef]

102. Vaghefi, E.; Pontre, B.P.; Jacobs, M.D.; Donaldson, P.J. Visualizing ocular lens fluid dynamics using MRI: Manipulation of steady state water content and water fluxes. *Am. J. Physiol.-Regul. Integr. Comp. Physiol.* **2011**, *301*, R335–R342. [CrossRef]

103. Vaghefi, E.; Walker, K.; Pontre, B.P.; Jacobs, M.D.; Donaldson, P.J. Magnetic resonance and confocal imaging of solute penetration into the lens reveals a zone of restricted extracellular space diffusion. *Am. J. Physiol.-Regul. Integr. Compar. Physiol.* **2012**, *302*, R1250–R1259. [CrossRef]

104. Vaghefi, E.; Donaldson, P.J. The lens internal microcirculation system delivers solutes to the lens core faster than would be predicted by passive diffusion. *Am. J. Physiol. Regul. Integr. Comp. Physiol.* **2018**, *315*, R994–R1002. [CrossRef]

105. Li, L.; Lim, J.; Jacobs, M.D.; Kistler, J.; Donaldson, P.J. Regional differences in cystine accumulation point to a sutural delivery pathway to the lens core. *Investig. Ophthalmol. Vis. Sci.* **2007**, *48*, 1253–1260. [CrossRef]

106. Lim, J.; Lam, Y.C.; Kistler, J.; Donaldson, P.J. Molecular characterization of the cystine/glutamate exchanger and the excitatory amino acid transporters in the rat lens. *Investig. Ophthalmol. Vis. Sci.* **2005**, *46*, 2869–2877. [CrossRef]

107. Lim, J.; Li, L.; Jacobs, M.D.; Kistler, J.; Donaldson, P.J. Mapping of glutathione and its precursor amino acids reveals a role for GLYT2 in glycine uptake in the lens core. *Investig. Ophthalmol. Vis. Sci.* **2007**, *48*, 5142. [CrossRef] [PubMed]

108. Lim, J.; Lorentzen, K.A.; Kistler, J.; Donaldson, P.J. Molecular identification and characterisation of the glycine transporter (GLYT1) and the glutamine/glutamate transporter (ASCT2) in the rat lens. *Exp. Eye Res.* **2006**, *83*, 447–455. [CrossRef]

109. Merriman-Smith, B.R.; Krushinsky, A.; Kistler, J.; Donaldson, P.J. Expression patterns for glucose transporters glut1 and glut3 in the normal rat lens and in models of diabetic cataract. *Investig. Ophthalmol. Vis. Sci.* **2003**, *44*, 3458–3466. [CrossRef]

110. Merriman-Smith, R.; Donaldson, P.; Kistler, J. Differential expression of facilitative glucose transporters GLUT1 and GLUT3 in the lens. *Investig. Ophthalmol. Vis. Sci.* **1999**, *40*, 3224.

111. Nye-Wood, M.G.; Spraggins, J.M.; Caprioli, R.M.; Schey, K.L.; Donaldson, P.J.; Grey, A.C. Spatial distributions of glutathione and its endogenous conjugates in normal bovine lens and a model of lens aging. *Exp. Eye Res.* **2017**, *154*, 70–78. [CrossRef]

nutrients

MDPI

Review

Nutraceuticals for the Treatment of Diabetic Retinopathy

Maria Grazia Rossino [1] and Giovanni Casini [1,2,*]

[1]   Department of Biology, University of Pisa, via San Zeno 31, 56127 Pisa, Italy; rossinomariagrazia1@gmail.com
[2]   Interdepartmental Research Center Nutrafood "Nutraceuticals and Food for Health", University of Pisa, 56124 Pisa, Italy
*   Correspondence: giovanni.casini@unipi.it; Tel.: +39-050-2211423

Received: 11 February 2019; Accepted: 28 March 2019; Published: 2 April 2019

Abstract: Diabetic retinopathy (DR) is one of the most common complications of diabetes mellitus and is characterized by degeneration of retinal neurons and neoangiogenesis, causing a severe threat to vision. Nowadays, the principal treatment options for DR are laser photocoagulation, vitreoretinal surgery, or intravitreal injection of drugs targeting vascular endothelial growth factor. However, these treatments only act at advanced stages of DR, have short term efficacy, and cause side effects. Treatment with nutraceuticals (foods providing medical or health benefits) at early stages of DR may represent a reasonable alternative to act upstream of the disease, preventing its progression. In particular, in vitro and in vivo studies have revealed that a variety of nutraceuticals have significant antioxidant and anti-inflammatory properties that may inhibit the early diabetes-driven molecular mechanisms that induce DR, reducing both the neural and vascular damage typical of DR. Although most studies are limited to animal models and there is the problem of low bioavailability for many nutraceuticals, the use of these compounds may represent a natural alternative method to standard DR treatments.

Keywords: retina; oxidative stress; inflammation; microvascular lesions; neoangiogenesis; polyphenols; flavonoids; carotenoids; saponins

---

1. Introduction

Diabetic retinopathy (DR) is a retinal disease representing one of the main causes of vision loss in developed countries. It has been classically considered a microvascular disease of the retina and is characterized, in its later stages, by abnormal growth of retinal vessels, which causes hemorrhages and tractional retinal detachment, leading to vision loss [1]. The understanding of DR has evolved over time and has clarified the role of the neuronal component of the retina in the progression of the disease. Indeed, growing experimental evidence suggests that suffering and death of retinal neurons occur before overt vascular changes [2–4]. For this reason, nowadays DR can be described not only as a microvascular but also as a neurodegenerative disease of the retina [5].

DR is a multifactorial disease but, to date, the exact pathophysiological mechanisms underlying neuro-vascular damage are not thoroughly understood. Nevertheless, different pathways and molecular mechanisms that may cause DR onset have been studied. For instance, the increase in advanced glycation end-products (AGEs) acting at their receptors (RAGE), the formation and activation of protein kinase C (PKC), or the increased flux in the polyol or hexosamine pathway have been examined [6,7]. All of these pathways, along with lower levels of glutathione (GSH), are associated with an increase in oxidative stress. The latter in turn causes different alterations in the diabetic retina as a consequence of severe lipid peroxidation, protein oxidation, oxidative DNA damage, induction of inflammation, and upregulation of growth factors, such as vascular endothelial growth factor (VEGF) [8].

VEGF is a proangiogenic factor that plays a key role in the late vasculopathy. For this reason, current DR treatments consist of the intraocular delivery of anti-VEGF molecules whose action induces restriction or inhibition of abnormal vessel growth. Nevertheless, the administration of anti-VEGF drugs has limitations and may generate different side effects. In addition, the effects are not long-lasting and frequent intravitreal injections are necessary [9–11].

Recent studies have highlighted the neuroprotective role of VEGF that can be noticed in the early phases of DR [12,13]. According to these studies, retinal neurons stressed by diabetes are likely to trigger the release of VEGF as a survival strategy. However, the persistence of the upstream stress conditions determines the accumulation of VEGF, leading to disruption of the blood-retina barrier (BRB) and, in the long term, to neoangiogenesis [14]. It would therefore be appropriate to plan new therapeutic strategies acting upstream of the disease and to prevent its progression by reducing neuronal stress and favoring neuroprotection. Moreover, considering the side effects caused by therapeutic agents administered via intraocular injections, there is a need to develop compounds with antioxidant and/or anti-inflammatory activity that can be administered through alternative delivery modalities. For this reason, in the last few years, several studies have focused on the potential benefits of nutraceuticals.

The term "nutraceutical" was coined by Dr. Stephen De Felice in 1989 and indicates "a food (or part of a food) that provides medical or health benefits, including the prevention and/or treatment of a disease" [15]. Nutraceuticals are effective antioxidants. They may induce the expression of antioxidant enzymes, act as scavengers of reactive oxygen species (ROS), or display singlet oxygen-quenching activity, as in the case of carotenoids [16]. Nutraceuticals may also exert anti-inflammatory effects by reducing the expression or nuclear translocation of nuclear factor kappa-light-chain-enhancer of activated B cells (NF-κB) [17]. Nutraceuticals can be used as natural dietary supplements and therefore can be easily administered, are readily available, and are affordable. A further advantage of nutraceuticals is that they are not likely to induce collateral side effects (if, of course, delivered at the appropriate dosage) such as hypoglycemia, liver injury, or gastric complains, which are characteristic of well-known and popular drugs [18,19].

In this review, we focus our attention on different classes of nutraceuticals, such as polyphenols, carotenoids, saponins, and others (Figure 1), explaining how these substances might counteract DR pathological changes. In particular, we highlight how nutraceuticals may reduce (i) oxidative stress; (ii) inflammation; (iii) neurodegeneration; and (iv) vascular changes. Reviewed literature includes in vitro studies, in vivo studies on animal models, and also clinical studies. Finally, we also consider how the low bioavailability of several nutraceuticals may limit their use.

2. Nutraceuticals and Oxidative Stress

Oxidative stress is caused by an imbalance in the production of ROS and the activity of the biological detoxifying systems. ROS are produced in normal metabolic conditions to support normal cellular functions and modulate a variety of biological processes including cell proliferation, differentiation, and migration, signal transduction, and programmed cell death [8]. However, because of ROS' high reactivity, their accumulation compromises the cell structure and functionality through alterations and degradation of molecules such as DNA, lipids, and proteins [20]. Oxidative stress and ROS production are contrasted by endogenous antioxidant defense enzymes including superoxide dismutase (SOD), catalase (CAT), glutathione peroxidase (GSH-P), and glutathione reductase (GSH-R). In addition to these endogenous enzymatic systems, endogenous non-enzymatic factors also exist and they include GSH (which is regulated by GSH-P and GSH-R), vitamin C, and vitamin E [21]. Besides endogenous antioxidant defenses, exogenous antioxidants of natural origin may be used to preserve redox homeostasis. They may act directly as scavengers of free radicals, indirectly by interrupting free radical chain reactions, or both. They may also decrease oxidative stress by inducing the expression of endogenous antioxidant enzymes [22,23]. For these reasons, it has been recently proposed that therapies based on natural, non-enzymatic antioxidants such as nutraceuticals could relieve the decrease in endogenous antioxidant defenses [23,24].

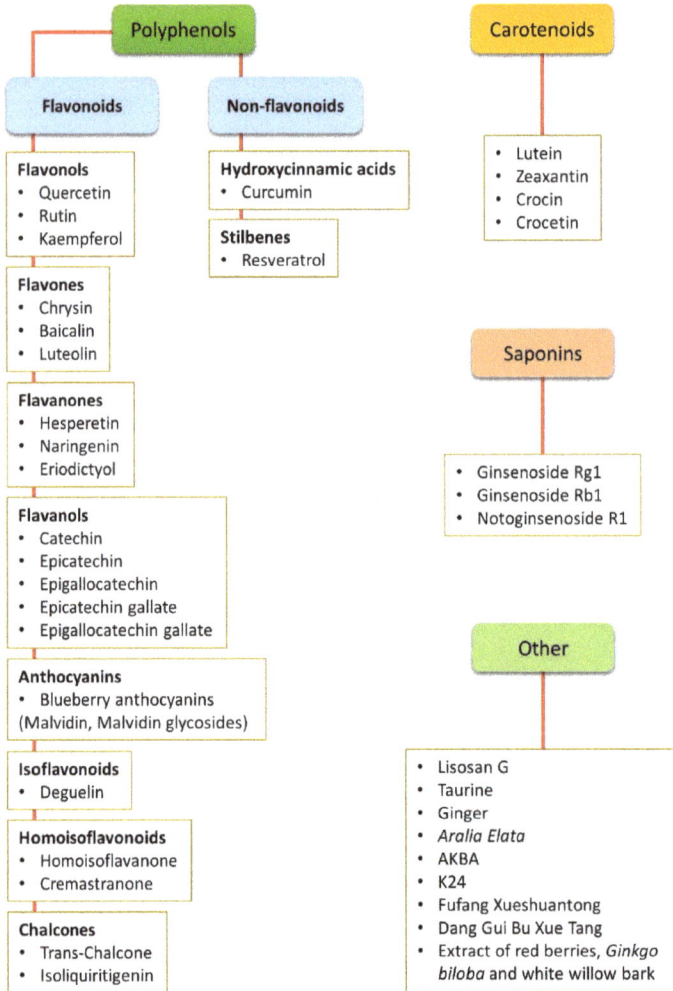

Figure 1. Summary of all the nutraceuticals cited in the present review. The compounds are listed according to their chemical classes, including polyphenols (both flavonoids and non-flavonoids), carotenoids, and saponins. Other compounds that do not belong to any of these classes or that are mixtures of different chemicals are classified as "other". AKBA: Acetyl-11-keto-β-boswellic acid.

The retina is highly susceptible to oxidative stress, which is due principally to the high content of polyunsaturated fatty acids, high oxygen uptake, glucose oxidation, and prolonged exposure to light. In particular, high glucose levels trigger a set of processes, such as AGE accumulation, PKC activation, and increased flux in the polyol and hexosamine pathways, which provoke oxidative stress (see [25] for detail). In turn, an increase in ROS is likely to cause DNA fragmentation resulting in poly-ADP ribose polymerase activation and glyceraldehyde 3-phosphate dehydrogenase inhibition [26]. This causes accumulation of glycolytic metabolites that may induce AGE formation and activation of PKC and of the polyol as well as of the hexosamine pathways, which are known to contribute to DR pathogenesis [6,7]. In summary, oxidative stress creates a propagating cycle, causing a continuous increase in ROS and consequent activation of pathways closely related to the progression of DR [8] (Figure 2).

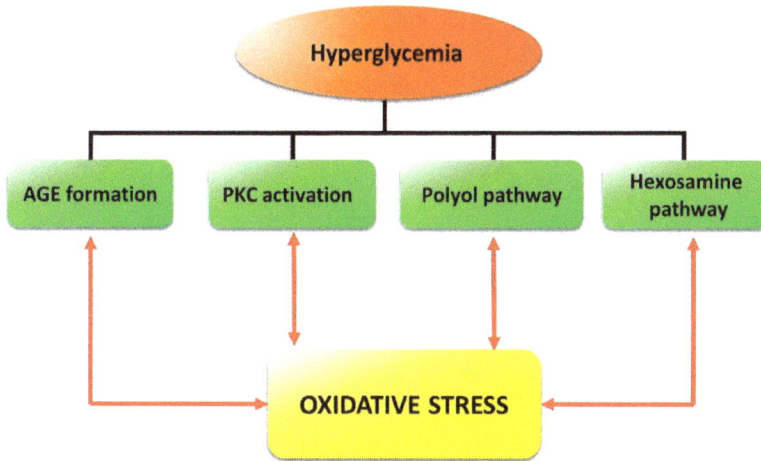

Figure 2. Schematic reconstruction of the events triggered in the retina by hyperglycemia and reinforced by oxidative stress in a vicious cycle. Formation of advanced glycation end-products (AGE) as well as the activation of protein kinase C (PKC), of the polyol pathway, and of the hexosamine pathway, are the main diabetes-induced abnormalities related to diabetic retinopathy.

Different natural dietary compounds have been investigated as possible treatments or adjuvants to counteract retinal oxidative stress typical of DR. They include polyphenols, carotenoids, and saponins, as well as other compounds (Figure 1). They are common in different fruits, vegetables, herbs, and beverages, and are very efficient in strengthening the endogenous antioxidant defenses through a direct scavenger activity and/or through the stimulation of antioxidant enzyme expression. Several classes of these compounds have been tested in vitro and in in vivo animal models. A summary of the effects of nutraceuticals against oxidative stress in models of DR is given in Figure 3.

2.1. Non-Flavonoid Polyphenols

Curcumin, a yellowish polyphenolic substance constituting the major active compound of *Curcuma longa*, is widely known for its antioxidant and anti-inflammatory properties [27–29]. The strong antioxidant power of curcumin has been shown in different studies. In human retinal endothelial cells (HRECs) exposed to high glucose and treated with 10 µM curcumin, intracellular ROS production has been observed to be significantly reduced [30], and similar results have been obtained with the retinal pigment epithelial cell line ARPE-19 [31,32]. The decrease in ROS levels is concomitant with an increased expression of heme oxygenase-1 (HO-1) [31], a redox-sensitive inducible stress protein that, once activated, protects the cell from different types of stress. This observation suggests that curcumin not only generates direct antioxidant activity but it may also act indirectly by enhancing the expression of antioxidant enzymes. This effect is likely to be induced through activation of the transcription nuclear factor erythroid-2-related factor-2 (Nrf2). Once activated, Nrf2 translocates into the nucleus and promotes the transcription of genes that encode antioxidant enzymes (known as phase II antioxidant enzymes), including HO-1 [33,34]. Another recent investigation into high glucose-stressed ARPE-19 cells showed that curcumin-induced inhibition of ROS formation prevents alterations of DNA methyltransferase activity [35]. In in vivo studies with rats with streptozotocin (STZ)-induced diabetes, curcumin has been observed to prevent the retinal increase of malondialdehyde (a marker of oxidative stress) and the decrease in GSH [36]. In the same model, curcumin also inhibited a decrease in total antioxidant capacity by increasing SOD, CAT, and GSH levels [37,38] and prevented an increase in the levels of retinal nitrotyrosine, a marker of oxidative protein damage, and in 8-hydroxy-2′-deoxyguanosine, a marker of oxidative DNA damage [38].

Among other non-flavonoid polyphenols, resveratrol, found in different plants such as grapes, peanuts, and berries, has been described as being able to decrease oxidative stress in retinas of diabetic rats by reducing lipid peroxidation, oxidized to reduced GSH ratio, and superoxide dismutase activity [39]. Recent data also show that resveratrol may reduce the adverse effects of hyperglycemia-induced oxidative stress on retinoic acid metabolism, which is involved in the recycling of 11-cis-retinal in the visual cycle in the retinal pigment epithelium [40].

2.2. Flavonoid Polyphenols

Flavonoids, a class of polyphenols, constitute a variegated group of natural substances characterized by strong antioxidant power. These natural products are present in fruits, vegetables, grains, roots, tea, and wine [41]. In STZ diabetic rats, treatment with different flavonoids ameliorates retinal redox status favoring an increase in GSH and a decrease in lipid peroxidation. It has also been observed that flavonoids are able to increase the levels of antioxidant enzymes such as SOD and CAT. In particular, these findings have been recorded in retinas of diabetic rats treated with quercetin, a common flavonol found in vegetables and fruits [42], with hesperetin, a flavanone commonly present in citrus fruits [43], or with green tea [44]. Green tea is a popular beverage rich in catechin, epicatechin, epigallocatechin, epicatechin gallate, and epigallocatechin gallate. Among these, epigallocatechin gallate is the most abundant catechin in green tea and is widely known for its antioxidant activity. The antioxidant effect of epigallocatechin gallate seems to be associated with a decrease in aldose reductase activity, which catalyzes the rate limiting step in the polyol pathway, and a decrease in AGE accumulation [45].

Eriodictyol, a flavonoid extracted from yerba santa (*Eriodictyon californicum*), a plant native to North America, has been found to reduce ROS production and increase the activity of SOD, GSH-P, and CAT. In addition, it has been shown to enhance the nuclear translocation of Nrf2 and elevate the expression of antioxidant enzyme HO-1 in RGC5 cells treated with high glucose [46].

Anthocyanins constitute another class of flavonoids which are responsible for the red or blue color of plants, fruits, and flowers. In vitro studies with HRECs subjected to high glucose treatment have shown that the blueberry anthocyanins malvidin and malvidin glycoside may produce an antioxidant effect through reduction of ROS levels and an increase in both CAT and SOD activity [47]. In addition, blueberry anthocyanins added to the food of and administered to diabetic rats for 12 weeks have been described as being able to prevent retinal oxidative stress favoring an increase in antioxidant capacity, as demonstrated by an increase in GSH and decrease in ROS levels. This antioxidant activity of blueberry anthocyanins is mediated by activation of Nrf2 and a consequent increase in HO-1 expression [48].

2.3. Carotenoids

Lutein and zeaxantin are the principal constituents of oranges, yellow fruits, and dark green leafy vegetables. Together with *meso*-zeaxanthin, they form the macular pigment of primate eyes [49] and prevent oxidative damage to the retina [50,51]. Their potential role in protecting against visual disorders has been recently reviewed [52–54]. Regarding the effects of carotenoids in experimental models of DR, a decrease in lipid peroxidation, nitrotyrosine levels, and oxidatively modified DNA was observed in the retinas of diabetic rats that had received supplementation with zeaxantin for two months. These effects were accompanied by inhibition of the diabetes-induced decrease in retinal SOD expression and activity, although no effects of zeaxantin were observed on GSH levels [55]. Similarly, lutein administration to one-month-diabetic mice has been observed to prevent retinal oxidative stress and restore retinal ROS levels to normal [56].

Crocetin and crocin are two additional compounds belonging to the class of carotenoids. They can be considered as the active ingredients of saffron, a spice classically used in traditional medicine for its beneficial qualities [57]. Crocetin and crocin, similar to lutein and zeaxantin, are known for their antioxidant and protective actions against ROS. For instance, crocetin has been reported to protect cells of the RGC5 cell line from oxidative stress [58], while treatment with crocin has been shown to prevent upregulation of ROS and nitric oxide in microglia cells cultured in high glucose [59].

2.4. Saponins

Panax notoginseng saponins (PNS), including ginsenoside Rg1, ginsenoside Rb1, and notoginsenoside R1, may generate a protective effect against oxidative stress-induced damage, as observed in STZ diabetic mice treated for two months with North American Ginseng (*Panax quinquefolius*) [60]. In addition, a significant decrease in ROS levels has been recorded in rat retinal capillary endothelial cells exposed to high glucose and treated with 100 μg/mL of PNS [61]. This decrease in ROS levels has been associated with an increase in antioxidant enzymes, including SOD, CAT, and, consequently, GSH. In addition, notoginsenoside R1 was also observed to induce a decrease in the activity of NADPH oxidase, the major enzyme implicated in oxygen radical generation [62].

2.5. Other Compounds

Lisosan G is a fermented powder obtained from organic whole grains (*Triticum aestivum*). It is enriched in bioactive substances such as phenolic components, flavonoids, alpha-lipoic acid, tocopherols, and polyunsaturated fatty acids (see [63] for detail). In mouse retinal explants, Lisosan G has been shown to inhibit an oxidative stress-induced increase in phase II antioxidant enzymes such as HO-1, SOD, and glutamate-cysteine ligase catalytic subunit mRNA expression, while in STZ rats it was observed to seem to inhibit the nuclear translocation of Nrf2, indicating that in these systems Lisosan G is likely to exert antioxidant effects through direct radical scavenging and not through activation of antioxidant enzyme expression [63].

3. Nutraceuticals and Inflammation

Inflammation is a nonspecific response to injury that includes a variety of functional mediators, such as cytokines, chemokines, acute phase proteins, and other pro-inflammatory molecules. Many of these mediators have been detected in the retina of diabetic animals or patients, suggesting that inflammation has a role in the development of DR [64–66]. Reactive gliosis, characterized by increased glial fibrillary acidic protein (GFAP) expression in both Müller cells and astrocytes [4,67], is typically observed in DR [68,69], resulting in the release from these cells of inflammatory cytokines, such as tumor necrosis factor alpha (TNFα), interleukin 1 beta (IL-1β), and others [70,71].

The transcription of inflammatory proteins is regulated by the activation of pro-inflammatory transcription factors, among which NF-κB plays a prominent role. This factor, once activated, translocates into the nucleus, binds to nuclear DNA, and acts as a master switch that promotes the expression of pro-inflammatory cytokines such as IL-1β, interleukin 6, interleukin 8 (IL-8), and, at least in part, TNFα [72]. There is ample evidence suggesting that NF-κB is involved in the pathogenesis of the early phases of DR. In fact, the inhibition of proteins whose expression is regulated by NF-κB decreases capillary degeneration, while direct NF-κB blockade inhibits DR development and progression [64–66,73]. The potential efficacy of some nutraceuticals for the treatment of DR is that they may inhibit NF-κB activation. A summary of the effects of nutraceuticals against inflammation in models of DR is given in Figure 3.

3.1. Non-Flavonoid Polyphenols

Treatment with curcumin prevents TNFα release in HRECs cultured with high glucose [32]. Curcumin also reduces retinal diabetic damage in diabetic rats through inactivation of NF-κB and a decrease in IL-1β levels [38]. There is some indication that curcumin may influence NF-κB by preventing the diabetes-induced retinal activation of calcium/calmodulin-dependent protein kinase II (CAMKII) [27,74–76], a ubiquitous multifunctional protein kinase implicated in the regulation of the transcriptional activity of NF-κB [77]. Curcumin has also been observed to reverse the diabetes-induced upregulation of retinal GFAP in Müller cells of STZ rats [36].

Recent observations have shown that inflammatory markers are reduced in the retinas of STZ rats after administrations of resveratrol via tail vein injections [78]. Similar to curcumin, resveratrol, or an ethanol extract of the root of *Polygonum cuspidatum*, which is rich in resveratrol, attenuates inflammation in the retinas of diabetic rats by reducing NF-κB activity [39,79]. In addition, resveratrol has been described as being able to reduce NF-κB nuclear translocation in the retinas of mice with experimental uveitis [80]. Resveratrol is likely to promote the inhibition of NF-κB through AMP-activated protein kinase (AMPK) activation. Indeed, data obtained from the retinas of mice with STZ-induced diabetes has shown that resveratrol-induced AMPK activation leads to significant suppression of NF-κB phosphorylation and reverses diabetes-induced sirtuin-1 (SIRT1) deactivation [81]. This SIRT1 activation promoted by resveratrol is likely to mediate an inhibition of NF-κB stimulation of DNA transcription, since SIRT1 deacetylates both NF-κB p65 and histone 3, with the effect of decreasing DNA binding by NF-κB [82]. Similarly to curcumin, a mechanism by which resveratrol may negatively modulate NF-κB is the inhibition of retinal CAMKII activation [83].

3.2. Flavonoid Polyphenols

Quercetin displays both antioxidant and anti-inflammatory properties in the retina. In particular, it has been reported to reduce VEGF-induced inflammation by inactivating NF-κB through inhibition of both mitogen-activated protein kinase and Akt in 661W cells [84]. In STZ diabetic rats, quercetin inhibits an increase in retinal GFAP expression and induces a decrease in NF-κB protein expression in specific retinal layers, namely the nerve fiber layer, the inner plexiform layer (IPL), and the inner nuclear layer (INL). This effect of quercetin on NF-κB is also associated with decreased levels of TNFα and IL-1β [42].

Hesperetin is another flavonoid that has been reported to exert antioxidant effects in diabetic retinas, as reported above. This compound has also been observed to inhibit the diabetes-induced over-expression of GFAP and of the pro-inflammatory cytokines TNFα and IL-1β in retinas of diabetic rats [43]. Eriodictyol, a flavonoid of the same class of hesperetin, has been reported to also reduce TNFα in STZ rat retinas [85] or both TNFα and IL-8 in high glucose-stressed RGC-5 cells [46].

Catechin has been observed to increase heat shock protein 27 levels and decrease the production of associated inflammatory factors in retinas of STZ rats [86]. Diabetes induced glial activation in the retina, characterized by increased GFAP expression in Müller cells, has also been found to be inhibited by green tea or by epicatechin [87,88].

3.3. Carotenoids

Among the carotenoids, crocin has been observed not only as being able to protect from oxidative stress, but also to block the pro-inflammatory response in microglial cells challenged with high glucose and free fatty acids. In both the antioxidant and the anti-inflammatory action of crocin, activation of the phosphoinositide 3-kinase (PI3K)/Akt signaling seems to play a significant role [59].

3.4. Other Compounds

The compounds 6-gingerol and the sesquiterpene zerumbone are abundantly present in rhizomes of the plants of the ginger family *Zingiber officinale* and *Zingiber zerumbet*, respectively. They are able to ameliorate retinal damage induced by hyperglycemia by inhibiting NF-κB expression/activation and reducing the expression of pro-inflammatory cytokines [89,90]. In particular, the effect of zerumbone is likely to be due to the blockading of the AGE/RAGE/NF-κB pathway [90].

Oxidative stress

- Malondialdehyde
- ROS levels
- NO levels
- 8-OHdG levels
- Aldose reductase activity
- AGE accumulation
- Lipid peroxidation
- Nitrotyrosine levels
- NADPH activity

- Nrf2 activation
- HO-1 levels
- SOD activity
- CAT activity
- GSH levels

Neurodegeneration

- Bcl-2 upregulation
- Bax downregulation
- Caspase-3 activation
- Cytochrome c levels
- TUNEL staining
- Glutamate excitotoxicity
- pro-NGF levels

- Retinal thickness
- Autophagic flux
- AKT phosphorylation
- BDNF expression
- Trk-B expression
- NGF expression
- Synaptophysin expression
- Erk 1/2 activation
- Glutamate uptake
- Retinal functionality

Nutraceuticals

Inflammation

- NF-κB activation
- NF-κB O-GlcNAcylation
- TNFα expression
- IL-1β expression
- IL-8 expression
- GFAP expression (reactive gliosis)

Vascular changes

- VEGF expression
- VEGFR2 expression and activation
- HIF-1α activation
- Endothelial cell proliferation, migration, and tube formation
- Thickening of basement membrane
- BRB breakdown
- Vascular leakage and permeability
- ICAM-1 expression
- Pericytes loss
- Leukostasis
- MMP-9 expression
- Neovascularization

- ZO-1 expression
- Occludin expression
- Claudin-5 expression

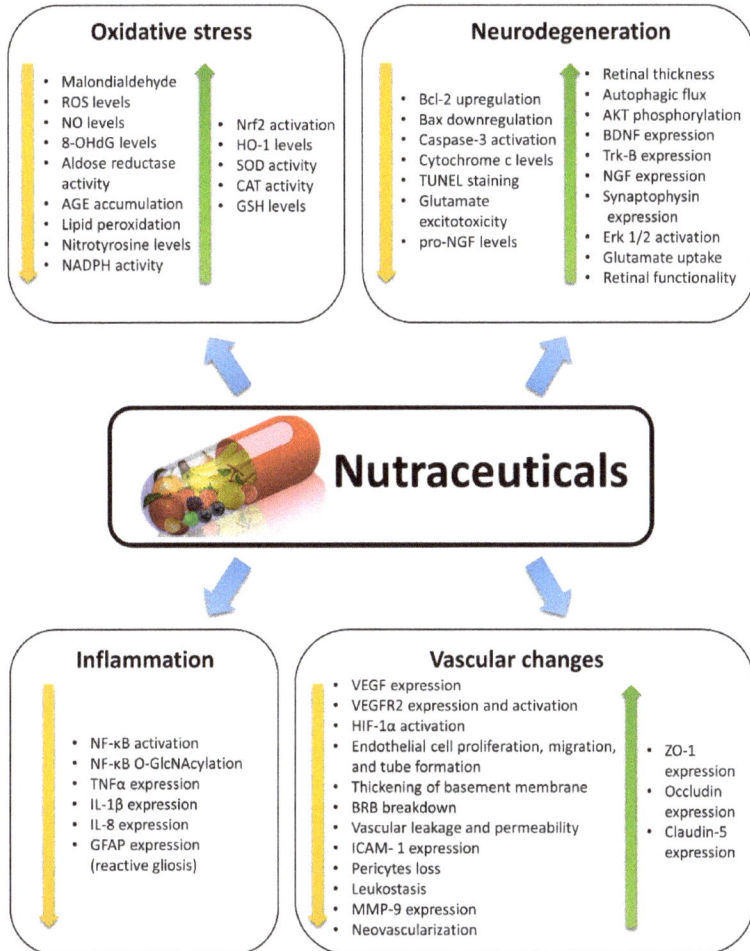

Figure 3. Summary of the effects induced by nutraceuticals as described in the studies reviewed herein. Nutraceuticals exert positive effects in diabetic retinopathy, counteracting the diabetes-induced changes by decreasing (yellow arrows) or increasing (green arrows) the expression/activation of specific factors or the occurrence of some events. 8-OHdG: 8-hydroxy-2'-deoxyguanosine; AGE: Advanced glycation end-products; AKT: Protein kinase B; Bax: Bcl-2-associated X protein; Bcl-2: B cell lymphoma 2; BDNF: Brain-derived neurotrophic factor; BRB: Blood-retina barrier; CAT: Catalase; Erk 1/2: Extracellular signal-regulated kinase 1/2; GFAP: Glial fibrillary acidic protein; GSH: Glutathione; HIF-1α: Hypoxia inducible factor 1α; HO-1: Heme oxygenase-1; ICAM-1: Intercellular cell adhesion molecule 1; IL-1β: Interleukin 1 beta; MMP-9: Matrix metalloproteinase-9; NADPH: Nicotinamide adenine dinucleotide phosphate; Nf-kB: Nuclear factor kappa-light-chain-enhancer of activated B cells; O-GlcNAc: O-linked β-N-acetylglucosamine; NGF: Nerve growth factor; NO: Nitric oxide; Nrf2: Transcription nuclear factor erythroid-2-related factor-2; ROS: Reactive oxygen species; SOD: Superoxide dismutase; TNFα: tumor necrosis factor alpha; Trk-B: Tyrosine receptor kinase B; TUNEL: Terminal deoxynucleotidyl transferase-mediated dUTP nick end labelling; VEGF: Vascular endothelial growth factor; VEGFR2: Vascular endothelial growth factor receptor 2; ZO-1: Zonula occludens 1.

Lisosan G has been reported to block increases in GFAP mRNA expression, indicating reactive gliosis, induced by diabetes in the retinas of STZ rats [63]. In addition, Lisosan G has been shown to exert important anti-inflammatory effects that can be associated with a reduction in NF-κB nuclear translocation, as observed in hepatocytes or in human endothelial progenitor cells [91,92] and has recently been hypothesized in an in vivo rat model of DR, where a Lisosan G-induced reduction of NF-κB phosphorylation was reported [63].

A post-translational modification (O-GlcNAcylation) of NF-κB has been observed in several pathologies, including DR [93]. An extract of *Aralia elata* (a plant traditionally used to treat diabetes in Eastern countries) containing phenolic compounds (3, 4-dihydroxybenzoic acid, chlorogenic acid, and caffeic acid) has been recently shown to reduce glial activation, to suppress NF-κB expression, and to decrease its O-GlcNAcylation in the retinas of STZ diabetic mice [94]. Finally, a fortified extract of red berries, *Ginkgo biloba*, and white willow bark containing carnosine and α-lipoic acid have been reported to attenuate the increase in TNFα levels in the retinas of STZ rats [95].

3.5. Relationships between Inflammation and Oxidative Stress

It is interesting to observe that most of the compounds cited above display both antioxidant and anti-inflammatory properties, as has been reported in different in vitro and in vivo experimental models. In fact, oxidative stress has been recognized as playing a pivotal role in the development of inflammation [96,97]. Accordingly, ROS production is likely to promote activation of NF-κB, an oxidant-sensitive factor and a crosslink between inflammation and oxidative stress [98–100]. Recent evidence in ARPE-19 cells also suggests that high glucose-induced ROS may promote the secretion of inflammatory cytokines through PI3K/Akt/mTOR, and curcumin has been found to inhibit this signaling pathway [101]. In summary, in DR inflammation is likely to be secondary to increased oxidative stress, and the use of appropriate antioxidant compounds may prevent the establishment of an inflammatory state.

4. Nutraceuticals and Neurodegeneration

DR is characterized by an extended loss of neurons due to an increase in apoptosis likely paralleled by a decrease in autophagic capabilities [102]. Neuronal cell vulnerability is evident very early in DR, and it is detectable before any sign of vascular damage [2–4]. This early neuronal impairment leads to retinal functional deficits that can be recorded with electroretinography (ERG) and that are associated with different morphological changes, these mostly including a decrease in thickness of retinal layers, with INL and IPL affected in particular. In retinas of diabetic rodents, an increase in terminal deoxynucleotidyl transferase-mediated dUTP nick end labelling (TUNEL) positive cells can be recorded together with a decrease in anti-apoptotic markers (e.g., B cell lymphoma 2 (Bcl-2)) and an increase in pro-apoptotic markers (e.g., active caspase-3 and Bcl-2-associated X protein (Bax)) [103–105]. Neurodegeneration in DR is likely caused by high glucose-induced oxidative stress and inflammation, but there is evidence that dysregulation of neurotrophic factor expression may also play a role. Neurotrophin nerve growth factor (NGF) and brain-derived neurotrophic factor (BDNF) are expressed by retinal neurons and glia, and are principally involved in cell survival and synaptic modulation [106,107]. A reduction in neurotrophin expression or an imbalance between the mature neurotrophin and its precursor (as in the case of proNGF/NGF) may lead to neuronal damage and neurodegeneration [107,108]. A further cause of neuronal death in DR is represented by increased glutamate levels causing excitotoxicity. This condition is likely to be due to oxidative stress in Müller cells resulting in decreased activity of glutamate-aspartate transporters and down-regulation of glutamine synthetase (GS), which converts glutamate into non-toxic glutamine [109].

Several natural compounds are known for their neuroprotective properties and for their positive effects within the central nervous system. In particular, nutraceuticals rich in flavonoids have been proposed for the treatment and prevention of a variety of neurodegenerative diseases [110,111]. A summary of the effects of nutraceuticals against neurodegeneration in models of DR is given in Figure 3.

4.1. Non-Flavonoid Polyphenols

In diabetic rat retinas, curcumin has been reported to exert antiapoptotic effects by upregulating the expression of Bcl-2 and downregulating the expression of Bax, with reduction of apoptosis of retinal ganglion cells and of cells in the INL and preservation of normal retinal thickness [112]. In addition, curcumin reverses diabetes-induced down-regulation of retinal GS, which may aid glutamate clearance and reduce the risk of excitotoxicity [36]. Interestingly, curcumin may also contribute to inhibition of apoptosis by promoting autophagic flux in retinal neurons. Indeed, curcumin has been reported to stimulate autophagy and exert protective effects in different models of central nervous system neurodegeneration [113].

Resveratrol has been shown to reduce retinal apoptotic levels and attenuate retinal thinning in rats with STZ-induced diabetes [39,78]. The neuroprotective action of resveratrol is likely to be associated with its anti-inflammatory action [83]. Similarly to curcumin, resveratrol may inhibit apoptosis by stimulating autophagy. Indeed, it has been reported to induce autophagy and reduce cell death both in the human retinal pigment epithelial ARPE-19 and in mouse photoreceptor 661W cells exposed to cytotoxic stress [114].

4.2. Flavonoid Polyphenols

The beneficial effects of flavonoids on retinal neurodegeneration in DR have been the subject of numerous studies. Treatment of STZ rats with quercetin protects from diabetes-induced retinal ganglion cell loss, mitigates thinning of retinal layers, reduces caspase-3 expression/activation and the levels of cytochrome c, while increasing Bcl-2 [42,115]. In addition, quercetin improves the expression of neurotrophic factors, of their receptors, and of their downstream signaling molecules. In particular, quercetin treatment favors an increase in Akt phosphorylation and in the expression of BDNF, its receptor Trk-B, and in synaptophysin. These data suggest that the neuroprotective action of quercetin is mediated by the BDNF-Trk-B/Akt-synaptophysin pathway [115]. The possibility that quercetin may affect apoptosis through the promotion of autophagy is supported by observations reporting potent simulation of autophagy by quercetin in Schwann cells with high glucose [116] and quercetin protection from Aβ-induced neurotoxicity through the induction of autophagy in *C. elegans* [117]. Similarly to quercetin, the flavonol kaempferol, which is found in tea, broccoli, apples, strawberries, and beans [118], could also be a stimulator of autophagy, as demonstrated in the human neuroblastoma SH-SY5Y cell line [119].

Another flavonoid that could represent a good choice with which to counteract neurodegeneration in DR is rutin. It is the main glycoside form of quercetin and is abundant in foods such as onions, apples, tea, and red wine [120]. Its neuroprotective effects have been tested in rat retinal ganglion cells subjected to oxidative stress, where treatment with rutin was observed to increase cell survival rate and reduce caspase-3 activation [121]. Rutin anti-apoptotic action has been confirmed in diabetic rats, in which treatment with this compound was observed to cause a decrease in caspase-3 activity and expression, with a concomitant increase in Bcl-2 and preservation of the levels of both BDNF and NGF [122].

Chrysin, a natural flavonoid found in herbs and honeycomb, has been recently shown to protect retinal photoreceptors by maintaining robust retinoid visual cycle-related components in glucose-stimulated human retinal pigment epithelial cells or in the retinal pigment epithelium of diabetic rats [123].

The strong antioxidant power of hesperetin correlates with the neuroprotective actions of this flavonoid. In diabetic retinas, it inhibits neuronal death, reducing caspase-3 expression [43], and prevents retinal thinning, favoring protection of ganglion cells and of cells in the INL [124]. Naringenin, also found in citrus fruits together with hesperetin, exerts similar neuroprotective actions in diabetic retinas, favoring an increase in BDNF and synaptophysin together with reduction in apoptotic levels, as indicated by increases in Bcl-2 and decreases in both Bax and caspase-3 expression [125]. The naringenin-promoted decrease in pro-apoptotic molecules is likely to be due to activation of Akt and Erk 1/2, as shown in hippocampal cells subjected to excitotoxic stress and treated with different concentrations of naringenin [126]. Anti-apoptotic effects of the flavanone eriodictyol have also been reported in high glucose-stressed RGC-5 cells [46].

Other studies have shown that treatment with green tea may prevent neurodegeneration in diabetic retinas. Indeed, the oral administration of green tea to diabetic rats generates a neuroprotective action in the retina characterized by a reduction in neuronal death, restoration of glutamate uptake, and improvement of retinal functionality as recorded with ERG [87]. The neuroprotective effect of epicatechin in retinas of diabetic rats has been proposed to be related to the reduction of pro-NGF production [88].

4.3. Carotenoids

Lutein is the carotenoid with the most recognized neuroprotective effects in the diabetic retina. Its constant intake induces an evident functional improvement, as highlighted by an ERG analysis of oscillatory potentials in the retinas of diabetic mice, which indicates prevention of inner retinal damage [127]. Moreover, lutein treatment restores retinal layer thickness, reduces retinal apoptosis, and preserves both BDNF and synaptophysin levels [128]. Lutein and zeaxantin are present in *Lycium barbarum*, a shrub member of the family Solanaceae which is widely recognized for its beneficial properties and is used in Chinese herbal medicine. *Lycium barbarum* administered to STZ diabetic rats for eight weeks was observed to result in amelioration of retinal ERG [129], which was likely to be related to the strong anti-apoptotic activity of this herb as reported in a retinal ischemia/reperfusion model [130]. The anti-apoptotic action of lutein may be related to autophagy promoting effects of this carotenoid, as reported for both human retinal pigment epithelial ARPE-19 and mouse photoreceptor 661W cells exposed to cytotoxic stress [114].

4.4. Other Compounds

Treatment with Lisosan G restores expression of caspase 3 to control levels in ex vivo mouse retinal explants subjected to oxidative stress. In STZ diabetic rats, Lisosan G reduces neuronal death and favors an improvement in retinal functionality, as evaluated by ERG. This result indicates that treatment with Lisosan G is able to protect both the inner and outer retina from diabetes-induced alterations [63]. Other compounds with documented neuroprotective effects in models of DR include zerumbone, whose anti-apoptotic effects correlate with improvement of retinal histological alterations and reduction of retinal thickness in diabetic rats [90], and *Aralia elata*, which protects mouse retinas from diabetes-induced decreases in retinal thickness, increases in TUNEL labeled ganglion cells, and increases in active caspase-3 [94]. Finally, an anti-apoptotic function, although not a direct neuroprotective effect, has been be attributed to taurine, a non-essential free aminoacid found in *Lycium barbarum* which has been reported to inhibit high glucose-promoted caspase-3 expression and activity in ARPE-19 cells [131].

4.5. Relationships between Oxidative Stress, Inflammation, and Neurodegeneration

Most of the nutraceuticals cited above possess antioxidant, anti-inflammatory, and neuroprotective properties at the same time. It is unlikely that these capacities are expressed independently from each other. Rather, the evidence suggests that they are intimately correlated. Indeed, oxidative stress and ROS toxicity may lead directly to DNA and protein damage, but, as mentioned above, oxidative stress is also linked to inflammation. Both oxidative stress and inflammation, then, would be able to cause neurodegeneration. Treatments with antioxidant compounds in early phases of DR may represent an efficacious way to preserve the retina from further damage due to inflammation and from extensive neurodegeneration. In this sense, nutraceutical antioxidants may represent a novel class of compounds with interesting potential therapeutic value for DR [132].

5. Nutraceuticals and Vascular Changes

On the basis of vascular changes, DR is classified as a non-proliferative diabetic retinopathy (NPDR) or proliferative diabetic retinopathy (PDR). NPDR is characterized by microvascular damage including BRB breakdown, pericyte loss, acellular capillaries, capillary occlusion, and thickening of the basement membrane. In PDR, neoangiogenesis phenomena are observed and new blood vessels are generated. These vessels create a deleterious action in the retina because of their mechanic traction, which, in the end, causes retinal detachment and consequent blindness [133]. As outlined below, VEGF, acting at its main receptor vascular endothelial growth factor receptor-2 (VEGFR2), plays prominent roles in both phases of DR.

The BRB represents a filter allowing selective passage of substances from the bloodstream to the retina, thereby regulating osmotic equilibrium, ionic concentrations, and transport of nutrients. These functions are based on the presence of tight and adherens junctions between adjacent cells. Tight junctions are composed of proteins like occludin, claudin, and zonula occludens 1 (ZO-1). These proteins are the principal compounds implicated in BRB functionality, creating a strong bond between endothelial cells and regulating the transport of solutes and molecules through prevention of the unchecked diffusion of substances between the bloodstream and neuroretina [134]. In DR, oxidative stress and inflammation result in complex changes causing upregulation of cytokines and growth factors, among which VEGF is the most implicated in BRB dysfunctions [135,136]. Indeed, VEGF upregulation is correlated with alterations of the tight junction structure caused by VEGF-induced phosphorylation and downregulation of tight junction proteins (i.e., ZO-1 and occludin) [137,138]. In addition, overexpressed VEGF also induces phosphorylation of the adherens junction protein VE-cadherin, further favoring increased BRB permeability [139]. VEGF upregulation in DR also correlates with increased expression of intercellular cell adhesion molecule 1 (ICAM-1), which in turn promotes leucocyte adhesion and capillary occlusion [13]. Other cytokines and chemokines are implicated in BRB impairment. For instance, TNFα overexpression is associated with decreases in occludin, claudin, and ZO-1 expression, while IL-1β induces barrier dysfunction through leukocyte recruitment and release of the vasoactive amine histamine [140,141]. Matrix metalloproteinases (MMPs) play important roles both in the early stages of DR, when MMP-2 and MMP-9 promote the apoptosis of retinal capillary cells, and in the later phase, when they facilitate neovascularization by degrading the extracellular matrix [142].

Other early vascular pathological changes in NPDR include loss of pericytes and thickening of the basement membrane. Pericytes are contractile cells located at the surface of capillaries, implicated in blood vessel stability, blood flow regulation, and formation of the BRB. In NPDR, pericyte loss occurs even before endothelial injury and is directly correlated with accumulation of AGEs, impairment of the BRB, and vascular leakage [143,144]. Apoptosis of pericytes in NPDR also leads to formation of microaneurysms and acellular capillaries [145]. Thickening of the basement membrane, due to the increase in vascular basal membrane compounds such as laminin and collagen IV [136], may contribute to the disruption of the tight link between pericytes and endothelial cells, causing pericyte apoptosis, whereas the endothelium, deprived of proliferation control, can give rise to new vessels [146].

PDR is characterized by neovascularization coupled with fibrotic responses at the vitreoretinal interface, and subsequent blindness due to vitreous hemorrhage, retinal fibrosis, tractional retinal detachment, and neovascular glaucoma [147–149]. Out of all the angiogenesis regulators, VEGF has been most extensively studied and provides the basis for current anti-angiogenic therapy [150]. VEGF plays a crucial role in PDR pathogenesis by promoting neovascularization through binding to VEGFR2 expressed on endothelial cells, inducing endothelial cell proliferation and sprouting angiogenesis [151].

The protective actions of nutraceuticals against microvascular changes typical of NPDR have been investigated in a variety of DR models. However, these models do not reproduce the neoangiogenesis characterizing PDR, and evidence of possible antiangiogenic properties of nutraceuticals has been found in other experimental models favoring the growth of new retinal vessels, mainly rodents with oxygen induced retinopathy (OIR) or experimental choroidal neovascularization (CNV). Other indications of the possible antiangiogenic effects of nutraceuticals have been derived from observations of their efficacy in inhibiting endothelial cell proliferation, migration, and tube formation. A summary of the effects of nutraceuticals against vascular changes in models of DR or of neoangiogenesis is given in Figure 3.

5.1. Non-Flavonoid Polyphenols

The vasoprotective potential of curcumin has been tested in vitro and in vivo. Treatment with curcumin prevents increases in glucose-induced VEGF expression as well as cellular proliferation in HRECs [30]. In addition, pre-treatment with curcumin has been shown to prevent capillary degeneration in rat retinas after ischemia reperfusion injury [152]. In diabetic rodents, curcumin has also been observed to protect pericytes from structural degeneration and to reduce VEGF expression, retinal vascular leakage, thickening of the basement membrane, vessel diameter, and vessel tortuosity [27,29,112,153]. Finally, curcumin has been reported to suppress experimental CNV and activation of hypoxia inducible factor 1α (HIF-1α, a transcription factor promoting VEGF expression and release) in mice [154].

In hypoxic ARPE-19 cells, resveratrol has been found to significantly inhibit HIF-1α and VEGF by blocking the PI3K/Akt/mTOR signaling pathway and by promoting proteasomal HIF-1α degradation [155]. Resveratrol also reduces diabetes-induced VEGF and ICAM-1 expression, leukocyte adhesion, pericytes loss, and prevents BRB breakdown as well as vascular leakage in the retinas of diabetic mice and rats [78,80,156,157]. In addition, extracts of *Polygonum cuspidatum*, containing resveratrol, have been shown to inhibit retinal vascular permeability and the loosening of the tight junctions in diabetic rats [79]. In mice with CNV induced by laser photocoagulation, resveratrol has been observed to significantly inhibit CNV growth [155,158] and reduce retinal neovascular lesions in very low-density lipoprotein receptor mutant mice, which are characterized by retinal neovascularization, by inhibiting VEGF expression as well as endothelial cell proliferation and migration [159]. The potential antiangiogenic effects of resveratrol and its possible use in DR treatments have been recently reviewed [160].

5.2. Flavonoid Polyphenols

Many of the microvascular changes and angiogenesis processes that occur in DR are inhibited by treatment with different flavonoids. Quercetin has been reported to reduce VEGF and MMP-9 expression in the retinas of diabetic rats [161]. In experiments with the rhesus choroids-retina endothelial cell line RF/6A, quercetin has also been reported to inhibit VEGF-induced endothelial cell proliferation, migration, and tube formation, suggesting that it may efficiently inhibit choroidal or retinal neovascularization [162,163].

Chrysin has been found to ameliorate diabetes-mediated microvascular and neovascular abnormalities in studies with HRECs and with retinas of *db/db* mice. Indeed, it increases the stability between endothelial cells by increasing ZO-1 and VE-cadherin expression and reduced vascular permeability and vasoregression. Chrysin also restricts the phenomena of neovascularization and prevents the onset of neovascular tufts. Its actions are likely to be mediated by inhibition of the upregulated HIF-1α-VEGF-VEGFR2 axis [164]. In addition, intravitreally injected chrysin has been found to exert an inhibitory effect on CNV in an experimental rat model [165].

Among the green tea catechins, epigallocatechin gallate treatment of ARPE-19 cells reduces VEGF, VEGFR2, and MMP-9 mRNA expression and inhibits proliferation, vascular permeability, and tube formation in VEGF-induced human retinal microvascular endothelial cells (HRMECs). In addition, it also reduces BRB breakdown in VEGF-induced animal models [166]. Epicatechin has been reported to reduce apoptosis and AGE accumulation in retinal vascular cells of intravenously AGE injected rats [167]. Interestingly, green tea fractions have been reported to decrease neovascularization in the OIR rat model; however, the active components of green tea displaying such effects do not seem to contain catechins [168].

Hesperetin in diabetic rats has been found to inhibit VEGF expression, decrease vascular permeability and leakage, and restore the normal thickness of the basement membrane [169]. Another flavanone compound, naringenin, has been reported to attenuate laser-induced CNV in rats [170], an effect that is increased if naringenin is complexed with β-cyclodextrin, which improves naringenin water solubility [171]. Belonging to the same class of flavonoids as hesperetin and naringenin, eriodictyol has been described as being able to lower the retinal levels of VEGF, ICAM-1, and endothelial nitric oxide synthase, which is involved in BRB breakdown, in STZ rat retinas [85].

Both the flavone glycoside baicalin, found in several plant species of the genus *Scutellaria*, and the natural flavone luteolin, abundantly present in several plant products, including broccoli, pepper, thyme, and celery, display antiangiogenic properties in models of retinal neovascularization. Indeed, intravitreally-injected baicalin inhibits the growth of CNV in rats [172], while intravitreal luteolin has been reported to inhibit retinal neovascularization in the mouse OIR model and to suppress hypoxia-induced VEGF expression (via inhibition of HIF-1α) as well as VEGF-induced migration and tube formation in HRMECs [173]. Similarly to baicalin and luteolin, deguelin, a derivative of the isoflavonoid rotenone and a naturally occurring insecticide isolated from plants of the *Mundulea sericea* family, effectively reduces both CNV and OIR neovascularization [174,175]. It has also been shown to inhibit tube formation of human umbilical vein endothelial cells (HUVECs) and in vivo angiogenesis of chick chorioallantoic membrane [174], which is consistent with deguelin antiangiogenic activity. In addition, deguelin analogs have been recently produced which inhibit HIF-1α and reduce both in vitro angiogenesis and neovascularization in the OIR model [176].

In line with the other flavonoids cited above, the naturally occurring homoisoflavonoids cremastranone and homoisoflavanone, which are both found in *Cremastra appendiculata*, traditionally known as a medicinal plant in East Asia, have been observed to reduce both CNV and neovascularization in the OIR model, and to inhibit HMREC or HUVEC proliferation, migration, and tube formation [177,178].

Chalcones are natural compounds which are present in edible plants. Intraperitoneal administration of trans-chalcone in a mouse OIR model has been shown to significantly inhibit neovascularization and VEGF as well as ICAM-1 upregulation [179]. In addition, intravitreal administrations of isoliquiritigenin, from licorice root, have been observed to alleviate neoangiogenesis in both the CNV and the OIR models, and suppress neovascularization in the corneal neovascularization assay and VEGF-induced vessel growth in an *ex ovo* chick chorioallantoic membrane assay [180].

Blueberry anthocyanins are very effective in preventing the onset of microvascular damage. In the retinas of diabetic rats treated with *Vaccinium myrtillus* extracts, VEGF levels have been seen to be reduced, the expression of the tight junction proteins claudin-5, occludin, and ZO-1 is restored, and BRB breakdown is prevented [181].

5.3. Carotenoids

Dietary lutein has been shown recently to promote a decrease in the extent of CNV induced by laser photocoagulation in mice. This effect increases in an additive manner when lutein is administered together with ω-3 long-chain polyunsaturated fatty acids and it is accompanied by reductions in oxidative stress and in inflammatory mediators [182].

5.4. Saponins

Rk1 ginsenoside, a derivative of natural ginseng, has been implicated in the prevention of pathological loss of vascular integrity thanks to its strong anti-vascular permeability action. Rk1 ginsenoside reduces leakage of retinal vessels in diabetic mice and, in HRMECs, inhibits endothelial permeability caused by VEGF and other vasoactive factors such as thrombin and histamine [183]. Ginsenoside Re has also been reported to exert protective effects against vascular damage in the retinas of diabetic rats [184].

5.5. Other Compounds

In retinas of STZ rats, Lisosan G prevents VEGF upregulation and VEGFR2 stimulation, as demonstrated by reduction of VEGFR2 phosphorylation. Consequently, the diabetes-induced reduction of occludin and ZO-1 expression is also inhibited by Lisosan G. These effects result in protection of the BRB, as evidenced by a dramatic reduction in vascular leakage in the retinas of STZ rats treated with Lisosan G with respect to the retinas of control STZ rats [63].

Acetyl-11-keto-β-boswellic acid (AKBA) is an active principle derived from the plant *Boswellia serrata*. It has been found to efficiently inhibit pathologic neovascularization in a mouse OIR model. AKBA inhibits upregulation of VEGF expression, which is typical of OIR, likely by affecting the Src homology region 2 domain-containing phosphatase 1/signal transducer and activator of transcription 3/VEGF axis [185].

Osteomeles schwerinae C. K. Schneid (Rosaceae) is a native plant in Asia. An ethanolic extract of this plant, referred to as K24, has an inhibitory effect on AGE-induced retinal vascular leakage by suppressing the expression of VEGF and decreasing occludin downregulation. In addition, K24 inhibits neovascular growth in retinas of OIR mice [186].

Extracts of *Zingiber officinale* orally administered to diabetic rats result in the normalization of the retinal vessel diameter and reduction of basement membrane thickness [89]. Diabetes-induced BRB breakdown is prevented with extracts of *Zingiber zerumbet* rhizome, containing principally kaempferol, quercetin, curcumin, and zerumbone. An ethanol extract of the rhizome administered to diabetic rats reduces vascular permeability and vessel dilation, favors an increase in tight junction protein expression, reduces VEGF and pro inflammatory molecule expression, causes a decrease in adhesion molecules such as ICAM-1, and alleviates leukostasis [187]. The vasoprotective effect of ginsenosides is also observed when they are in combination with other compounds. For instance, *Panax notoginseng* may be combined with other Chinese herbs, such as *Salvia miltiorrhiza*, *Astragalus membranaceus*, and *Scrophularia ningpoensis*, to generate a compound called Fufang Xueshuantong, which causes an improvement in microvascular lesions, induces decreases in VEGF and ICAM-1 expression and BRB breakdown together with an increase in occludin expression [188,189]. Similarly, adding *Panax notoginsen* to Dang Gui Bu Xue Tang, an aqueous extract of *Radix Astragali* and *Radix Angelica sinensis* used in traditional Chinese medicine, reduces VEGF levels, occludin expression, vascular permeability, leukostasis, and the number of acellular capillaries in the retinas of diabetic rats [190].

Another extract that may reduce vascular damage in DR is the fortified extract of red berries, *Ginkgo biloba*, and white willow bark, as cited above. Indeed, in addition to inhibition of TNFα levels, it also induces attenuation of VEGF upregulation in the retinas of STZ rats [95].

5.6. Relationships between Oxidative Stress, Inflammation, Neurodegeneration, and Vascular Damage

As discussed above, nutraceuticals display neuroprotective effects due to their antioxidant and anti-inflammatory properties, as demonstrated in different experimental models of DR. These same compounds, or compounds that have been demonstrated to possess antioxidant and/or anti-inflammatory properties in other models, also protect the retina from the vascular damage and vascular proliferation typical of DR. It is interesting to note that in studies analyzing VEGF in DR models after treatment with neuroprotectants, decrease in apoptotic markers is always associated with a decrease in VEGF expression and/or release (see for instance [12,63]). These observations can be explained by assuming that those compounds also exert an independent regulation of the VEGF biosynthetic pathways or of the cell response to VEGF, as suggested by the observed effects of nutraceuticals on VEGF-induced endothelial cell proliferation, migration, and tube formation, or in models of retinal neoangiogenesis. However, the existence of a causal relationship between neuronal damage and vascular responses is a more likely hypothesis. Therefore, the effects of diabetes in the retina may include an initial high glucose-induced oxidative stress that elicits an inflammatory response and provokes damage of neurons and of other retinal cells. Neuronal suffering then would trigger expression and release of VEGF, mainly from Müller cells, which would act as a neuroprotective factor.

Indeed, VEGF has recognized neuroprotective properties, but its prolonged upregulation will induce microvascular damage, BRB breakdown, and, in the long term, neoangiogenesis [3]. The assumption of nutraceuticals from the earliest evidence of diabetes will strengthen the antioxidant power in the retina, reducing oxidative stress and inflammation, with consequent protection from cell death, absence of VEGF upregulation, and no induction of vascular changes (Figure 4).

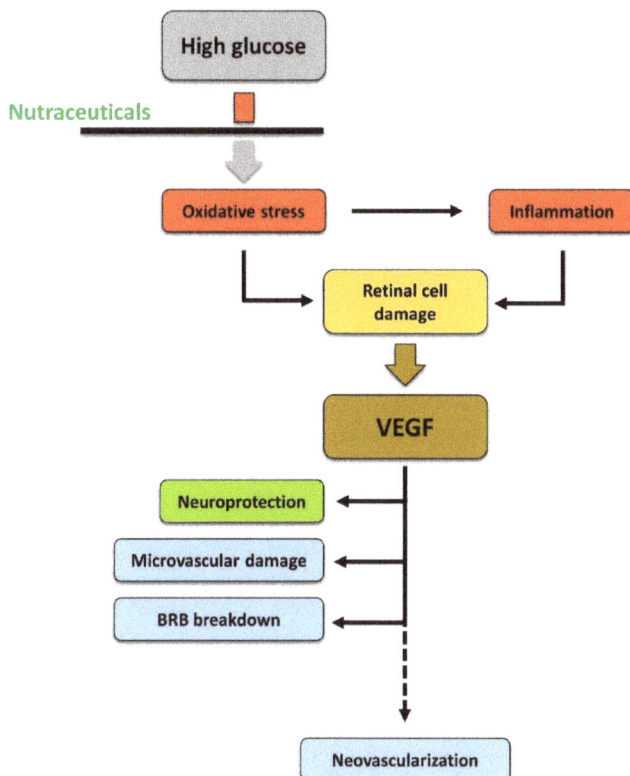

Figure 4. Hypothetic cascade of events induced by high glucose in the retina leading to diabetic retinopathy and the effects of nutraceuticals. See text for explanation.

6. Clinical Studies

There are only a few clinical studies investigating the possible use of nutraceuticals for the treatment of DR, and most of them have been focused on carotenoids. Randomized clinical trials in patients with NPDR have shown that supplementation with lutein for three or for nine months results in increased visual acuity and contrast sensitivity, while foveal thickness decreases, indicating an alleviation of macular edema [191,192]. Similar results have been obtained in a placebo-controlled randomized clinical trial with patients affected by diabetic maculopathy refractory to conventional therapy, in which administration of 15 mg crocin tablets per day for three months caused a significant improvement of both best-corrected visual acuity and central macular thickness [193]. In addition, Type 2 diabetes patients having a higher ratio of serum non-pro-vitamin A carotenoids (lutein, zeaxanthin, lycopene) to pro-vitamin A carotenoids (α-carotene, β-carotene and β-cryptoxanthin) have shown a 66% reduction in risk for DR [194]. Moreover, the optical density of the macular pigment, which comprises the carotenoids lutein and zeaxanthin [49], has been reported to be lower in patients with Type 2 diabetes than in age-matched controls, and still lower in patients with Type 2 diabetes and

DR [195]. Finally, a retrospective study with Type 2 diabetic patients after two years of carotenoid supplementation has suggested that carotenoids may have a beneficial effect on the macular function of diabetic patients [196].

In addition to clinical studies on carotenoids, there are also a few papers which have reported the use of other nutraceuticals in patients suffering from DR. For instance, a standardized phytosomal curcuminoid mixture (Meriva®) greatly improves curcumin absorption [197], and in one study 38 diabetic patients treated with Meriva® showed improvements in diabetic microangiopathy and retinopathy at four weeks post-treatment [198]. In addition, a recent study has investigated potential beneficial effects of green tea. Indeed, a clinic-based, case-control study performed on diabetic patients with Type 2 diabetes showed that those who regularly drank Chinese green tea every week for at least one year in their lives had a DR risk reduction of about 50% compared with those who had not [199].

7. Bioavailability of Nutraceuticals

Bioavailability is a pharmacokinetic term referring to the fraction of bioactive compound that reaches the blood circulation without undergoing alterations. The index of bioavailability of nutraceuticals is important because it allows for the calculation of the right dose of nutraceutical to ingest. For this reason, understanding the oral bioavailability of a nutraceutical compound is as important as understanding its therapeutic potential. After ingestion, botanical compounds must overcome a series of threats that may alter their structure before they can reach systemic circulation, for instance, the environment of the gastrointestinal tract and the intestinal as well as the hepatic metabolism. Unfortunately, many nutraceuticals have low oral bioavailability, and therefore investigations to improve this aspect are of fundamental importance. Recently, significant steps forward have been made to develop new technologies using analogous compounds, nanoformulations, or nanoparticles, which may protect the nutraceutical from enteric adverse conditions [200–203].

Curcumin is characterized by poor bioavailability mainly due to low solubility, rapid metabolism and poor absorption, which, despite its medical efficacy, limits its clinical applications [204]. Conjugation of curcumin to metal oxide nanoparticles or encapsulation in lipid nanoparticles, dendrimers, nanogels, or polymeric nanoparticles, improves the water solubility and bioavailability of curcumin, thus increasing its pharmacological effectiveness [76]. The encapsulation of curcumin in the calix [4] arene nanoassembly limits curcumin degradation and increases its solubility, enhancing the effect of the compound on antioxidant and anti-inflammatory markers in both in vivo and in vitro models [205]. Similar results have been obtained using a different nanocarrier formulation comprising Pluronic-F127 stabilized D-α-Tocopherol polyethene glycol 1000 succinate nanoparticles [206]. A recent study has reported that, among different tested curcumin formulations, only that containing a hydrophilic carrier may provide therapeutic levels of curcumin in rabbit retinas [32].

Resveratrol, similarly to curcumin, is known for its poor oral bioavailability and scarce pharmacokinetic properties due to low aqueous solubility and low photostability, which compromise its great potential. In fact, as shown by pharmacokinetic studies, the levels of unmetabolized resveratrol after oral administration are reduced to about 1% due to its high intestinal and hepatic metabolism [207]. To solve this problem, different resveratrol nanoformulations have been tested, including liposomes, solid lipid nanoparticles, polymeric nanoparticles, and cyclodextrins. The use of these alternative administration methods generates different advantages because they improve solubility, bioavailability, and physical chemical stability, and favor a controlled drug release [208–210]. The use of resveratrol analogs could be another alternative choice for administration of this nutraceutical. The pharmacokinetic profiles of resveratrol and its analog perolstilbene have been analyzed in rats, showing that the bioavailability of perolstilbene was 80% and that of resveratrol 20% [211]. A summary of oral delivery systems for resveratrol has recently been published [160].

Nanoparticles can also be used to increase the bioavailability of epigallocatechin gallate, another nutraceutical characterized by low solubility and stability. Different nanosystems have been used for epigallocatechin gallate delivery, including liposomes, gold nanoparticles, inorganic nanocarriers, and lipid as well as polymeric nanoparticles [212,213].

A recent study has reported that the distribution of an orally administered nutraceutical may vary substantially depending on tissue type. Indeed, in a pilot study, ^{13}C-lutein was detected in a variety of tissues in a rhesus macaque after a single oral administration, but not in the retina [214]. Some improvement in lutein delivery to ocular tissues may derive from lutein encapsulation into hyaluronic acid-coated PLGA nanoparticles, which have been demonstrated to efficiently bind ARPE-19 cells and improve the physicochemical properties of lutein [215].

8. Conclusions

A review of the effects produced by the administration of nutraceuticals in DR-related models indicates that all of the pathologic conditions seen in DR, including oxidative stress, inflammation, neurodegeneration, and vascular lesions can be alleviated by many of these natural compounds. There is evidence suggesting that oxidative stress, induced by diabetes through different pathways, might promote inflammation and cause neurodegeneration. Neuronal suffering, in turn, would trigger VEGF upregulation, causing subsequent vascular damage. Therefore, it appears that an increased antioxidant defense, if established before extended neuronal and vascular lesions, could reduce the subsequent pathological changes. A continuous supplementation of nutraceuticals with diet could afford a sufficient antioxidant power, and nutraceutical-based approaches may be the most efficacious, economic, and sustainable treatments to limit or even prevent the development of DR in diabetic subjects.

Despite this attractive perspective, however, clinical studies examining the real potential of nutraceuticals to ameliorate DR are still very limited in number. This is probably due the fact that it is not totally clear whether nutraceuticals should be tested to treat DR or to prevent DR. The evidence reported in this review has led us to hypothesize a chain of events (see Figure 4) that could be prevented by nutraceuticals; nutraceuticals may not be as efficient in treating the disease once it has been established. When investigating the preventative value of nutraceuticals, clinical studies are probably more difficult to organize and would require considerably long time periods.

Another reason for limiting clinical studies is likely the poor bioavailability of most nutraceuticals. As long as efficient delivery methods are not available for nutraceuticals to exert significant biological action in the retina, it will be difficult to design meaningful clinical investigations. Studies investigating new strategies for nutraceutical delivery, mainly based on nanoformulations, are very recent (they have appeared in the last ten years), and hopefully in the near future new research may fill this gap and promote new clinical experimentation of nutraceuticals.

Author Contributions: M.G.R. and G.C. collected literature and wrote the manuscript. G.C. had primary responsibility for the final content. All authors read and approved the manuscript.

Funding: This research was funded by the Italian Ministry of University and Research (FFABR 2017).

Acknowledgments: We thank Rosario Amato, Massimo dal Monte, and Alessandro Massolo for their critical reading of the manuscript.

Conflicts of Interest: The authors declare no conflict of interest.

References

1. Curtis, T.; Gardiner, T.; Stitt, A. Microvascular lesions of diabetic retinopathy: Clues towards understanding pathogenesis? *Eye* **2009**, *23*, 1496. [CrossRef] [PubMed]
2. Harrison, W.W.; Bearse, M.A.; Ng, J.S.; Jewell, N.P.; Barez, S.; Burger, D.; Schneck, M.E.; Adams, A.J. Multifocal electroretinograms predict onset of diabetic retinopathy in adult patients with diabetes. *Investig. Ophthalmol. Vis. Sci.* **2011**, *52*, 772–777. [CrossRef]
3. Hernández, C.; Dal Monte, M.; Simó, R.; Casini, G. Neuroprotection as a therapeutic target for diabetic retinopathy. *J. Diabetes Res.* **2016**. [CrossRef] [PubMed]

4. Simo, R.; Hernandez, C. Neurodegeneration in the diabetic eye: New insights and therapeutic perspectives. *Trends Endocrinol. Metab.* **2014**, *25*, 23–33. [CrossRef] [PubMed]

5. Simó, R.; Stitt, A.W.; Gardner, T.W. Neurodegeneration in diabetic retinopathy: Does it really matter? *Diabetologia* **2018**. [CrossRef]

6. Tarr, J.M.; Kaul, K.; Chopra, M.; Kohner, E.M.; Chibber, R. Pathophysiology of diabetic retinopathy. *ISRN Ophthalmol.* **2013**. [CrossRef]

7. Wang, W.; Lo, A. Diabetic retinopathy: Pathophysiology and treatments. *Int. J. Mol. Sci.* **2018**, *19*, 1816. [CrossRef]

8. Kowluru, R.A.; Chan, P.-S. Oxidative stress and diabetic retinopathy. *Exp. Diabetes Res.* **2007**, *2007*. [CrossRef] [PubMed]

9. Duh, E.J.; Sun, J.K.; Stitt, A.W. Diabetic retinopathy: Current understanding, mechanisms, and treatment strategies. *JCI Insight* **2017**, *2*. [CrossRef] [PubMed]

10. Simó, R.; Sundstrom, J.M.; Antonetti, D.A. Ocular anti-VEGF therapy for diabetic retinopathy: The role of VEGF in the pathogenesis of diabetic retinopathy. *Diabetes Care* **2014**, *37*, 893–899. [CrossRef]

11. Zhao, Y.; Singh, R.P. The role of anti-vascular endothelial growth factor (anti-VEGF) in the management of proliferative diabetic retinopathy. *Drugs Context* **2018**, *7*. [CrossRef]

12. Amato, R.; Biagioni, M.; Cammalleri, M.; Dal Monte, M.; Casini, G. VEGF as a survival factor in ex vivo models of early diabetic retinopathy. *Investig. Ophthalmol. Vis. Sci.* **2016**, *57*, 3066–3076. [CrossRef]

13. Behl, T.; Kotwani, A. Exploring the various aspects of the pathological role of vascular endothelial growth factor (VEGF) in diabetic retinopathy. *Pharmacol. Res.* **2015**, *99*, 137–148. [CrossRef]

14. Witmer, A.; Vrensen, G.; Van Noorden, C.; Schlingemann, R. Vascular endothelial growth factors and angiogenesis in eye disease. *Prog. Retin. Eye Res.* **2003**, *22*, 1–29. [CrossRef]

15. Brower, V. Nutraceuticals: Poised for a healthy slice of the healthcare market? *Nat. Biotechnol.* **1998**, *16*, 728. [CrossRef]

16. Milatovic, D.; Zaja-Milatovic, S.; Gupta, R.C. Oxidative Stress and Excitotoxicity: Antioxidants from Nutraceuticals. In *Nutraceuticals*; Elsevier: Amsterdam, The Netherlands, 2016; pp. 401–413.

17. Aggarwal, B.B.; Van Kuiken, M.E.; Iyer, L.H.; Harikumar, K.B.; Sung, B. Molecular targets of nutraceuticals derived from dietary spices: Potential role in suppression of inflammation and tumorigenesis. *Exp. Biol. Med.* **2009**, *234*, 825–849. [CrossRef]

18. Chauhan, B.; Kumar, G.; Kalam, N.; Ansari, S.H. Current concepts and prospects of herbal nutraceutical: A review. *J. Adv. Pharm. Technol. Res.* **2013**, *4*, 4. [CrossRef]

19. Kalra, E.K. Nutraceutical-definition and introduction. *AAPS Pharmsci* **2003**, *5*, 27–28. [CrossRef]

20. Calderon, G.; Juarez, O.; Hernandez, G.; Punzo, S.; De la Cruz, Z. Oxidative stress and diabetic retinopathy: Development and treatment. *Eye* **2017**, *31*, 1122. [CrossRef]

21. Tokarz, P.; Kaarniranta, K.; Blasiak, J. Role of antioxidant enzymes and small molecular weight antioxidants in the pathogenesis of age-related macular degeneration (AMD). *Biogerontology* **2013**, *14*, 461–482. [CrossRef]

22. Ahmadinejad, F.; Geir Møller, S.; Hashemzadeh-Chaleshtori, M.; Bidkhori, G.; Jami, M.-S. Molecular mechanisms behind free radical scavengers function against oxidative stress. *Antioxidants* **2017**, *6*, 51. [CrossRef] [PubMed]

23. Nimse, S.B.; Pal, D. Free radicals, natural antioxidants, and their reaction mechanisms. *RSC Adv.* **2015**, *5*, 27986–28006. [CrossRef]

24. Cui, K.; Luo, X.; Xu, K.; Murthy, M.V. Role of oxidative stress in neurodegeneration: Recent developments in assay methods for oxidative stress and nutraceutical antioxidants. *Prog. Neuro-Psychopharmacol. Biol. Psychiatry* **2004**, *28*, 771–799. [CrossRef]

25. Wu, M.Y.; Yiang, G.T.; Lai, T.T.; Li, C.J. The Oxidative Stress and Mitochondrial Dysfunction during the Pathogenesis of Diabetic Retinopathy. *Oxid. Med. Cell. Longev.* **2018**, *5*. [CrossRef] [PubMed]

26. Brownlee, M. The pathobiology of diabetic complications: A unifying mechanism. *Diabetes* **2005**, *54*, 1615–1625. [CrossRef]

27. Parsamanesh, N.; Moossavi, M.; Bahrami, A.; Butler, A.E.; Sahebkar, A. Therapeutic potential of curcumin in diabetic complications. *Pharmacol. Res.* **2018**. [CrossRef]

28. Peddada, K.V.; Verma, V.; Nebbioso, M. Therapeutic potential of curcumin in major retinal pathologies. *Int. Ophthalmol.* **2018**. [CrossRef] [PubMed]

29. Wang, L.L.; Sun, Y.; Huang, K.; Zheng, L. Curcumin, a potential therapeutic candidate for retinal diseases. *Mol. Nutr. Food Res.* **2013**, *57*, 1557–1568. [CrossRef] [PubMed]

30. Premanand, C.; Rema, M.; Sameer, M.Z.; Sujatha, M.; Balasubramanyam, M. Effect of curcumin on proliferation of human retinal endothelial cells under in vitro conditions. *Investig. Ophthalmol. Vis. Sci.* **2006**, *47*, 2179–2184. [CrossRef]

31. Woo, J.M.; Shin, D.-Y.; Lee, S.J.; Joe, Y.; Zheng, M.; Yim, J.H.; Callaway, Z.; Chung, H.T. Curcumin protects retinal pigment epithelial cells against oxidative stress via induction of heme oxygenase-1 expression and reduction of reactive oxygen. *Mol. Vis.* **2012**, *18*, 901.

32. Platania, C.B.M.; Fidilio, A.; Lazzara, F.; Piazza, C.; Geraci, F.; Giurdanella, G.; Leggio, G.M.; Salomone, S.; Drago, F.; Bucolo, C. Retinal Protection and Distribution of Curcumin in Vitro and in Vivo. *Front. Pharm.* **2018**, *9*. [CrossRef] [PubMed]

33. Li, Y.; Zou, X.; Cao, K.; Xu, J.; Yue, T.; Dai, F.; Zhou, B.; Lu, W.; Feng, Z.; Liu, J. Curcumin analog 1, 5-bis (2-trifluoromethylphenyl)-1, 4-pentadien-3-one exhibits enhanced ability on Nrf2 activation and protection against acrolein-induced ARPE-19 cell toxicity. *Toxicol. Appl. Pharmacol.* **2013**, *272*, 726–735. [CrossRef] [PubMed]

34. Yang, C.; Zhang, X.; Fan, H.; Liu, Y. Curcumin upregulates transcription factor Nrf2, HO-1 expression and protects rat brains against focal ischemia. *Brain Res.* **2009**, *1282*, 133–141. [CrossRef] [PubMed]

35. Maugeri, A.; Mazzone, M.G.; Giuliano, F.; Vinciguerra, M.; Basile, G.; Barchitta, M.; Agodi, A. Curcumin Modulates DNA Methyltransferase Functions in a Cellular Model of Diabetic Retinopathy. *Oxid. Med. Cell. Longev.* **2018**, *2*. [CrossRef] [PubMed]

36. Zuo, Z.-F.; Zhang, Q.; Liu, X.-Z. Protective effects of curcumin on retinal Müller cell in early diabetic rats. *Int. J. Ophthalmol.* **2013**, *6*, 422. [CrossRef] [PubMed]

37. Gupta, S.K.; Kumar, B.; Nag, T.C.; Agrawal, S.S.; Agrawal, R.; Agrawal, P.; Saxena, R.; Srivastava, S. Curcumin prevents experimental diabetic retinopathy in rats through its hypoglycemic, antioxidant, and anti-inflammatory mechanisms. *J. Ocul. Pharmacol. Ther.* **2011**, *27*, 123–130. [CrossRef]

38. Kowluru, R.A.; Kanwar, M. Effects of curcumin on retinal oxidative stress and inflammation in diabetes. *Nutr. Metab.* **2007**, *4*, 8. [CrossRef]

39. Soufi, F.G.; Mohammad-nejad, D.; Ahmadieh, H. Resveratrol improves diabetic retinopathy possibly through oxidative stress–nuclear factor κB–apoptosis pathway. *Pharmacol. Rep.* **2012**, *64*, 1505–1514. [CrossRef]

40. Al-Hussaini, H.; Kilarkaje, N. Effects of trans-resveratrol on type 1 diabetes-induced inhibition of retinoic acid metabolism pathway in retinal pigment epithelium of Dark Agouti rats. *Eur. J. Pharm.* **2018**, *834*, 142–151. [CrossRef]

41. Panche, A.; Diwan, A.; Chandra, S. Flavonoids: An overview. *J. Nutr. Sci.* **2016**, *5*. [CrossRef]

42. Kumar, B.; Gupta, S.K.; Nag, T.C.; Srivastava, S.; Saxena, R.; Jha, K.A.; Srinivasan, B.P. Retinal neuroprotective effects of quercetin in streptozotocin-induced diabetic rats. *Exp. Eye Res.* **2014**, *125*, 193–202. [CrossRef]

43. Kumar, B.; Gupta, S.K.; Srinivasan, B.; Nag, T.C.; Srivastava, S.; Saxena, R.; Jha, K.A. Hesperetin rescues retinal oxidative stress, neuroinflammation and apoptosis in diabetic rats. *Microvasc. Res.* **2013**, *87*, 65–74. [CrossRef]

44. Kumar, B.; Gupta, S.K.; Nag, T.C.; Srivastava, S.; Saxena, R. Green tea prevents hyperglycemia-induced retinal oxidative stress and inflammation in streptozotocin-induced diabetic rats. *Ophthalmic Res.* **2012**, *47*, 103–108. [CrossRef]

45. Sampath, C.; Sang, S.; Ahmedna, M. In vitro and in vivo inhibition of aldose reductase and advanced glycation end products by phloretin, epigallocatechin 3-gallate and [6]-gingerol. *Biomed. Pharmacother.* **2016**, *84*, 502–513. [CrossRef]

46. Lv, P.; Yu, J.; Xu, X.; Lu, T.; Xu, F. Eriodictyol inhibits high glucose-induced oxidative stress and inflammation in retinal ganglial cells. *J. Cell. Biochem.* **2018**, *14*, 27848. [CrossRef] [PubMed]

47. Huang, W.; Yan, Z.; Li, D.; Ma, Y.; Zhou, J.; Sui, Z. Antioxidant and Anti-Inflammatory Effects of Blueberry Anthocyanins on High Glucose-Induced Human Retinal Capillary Endothelial Cells. *Oxid. Med. Cell. Longev.* **2018**, *2018*. [CrossRef] [PubMed]

48. Song, Y.; Huang, L.; Yu, J. Effects of blueberry anthocyanins on retinal oxidative stress and inflammation in diabetes through Nrf2/HO-1 signaling. *J. Neuroimmunol.* **2016**, *301*, 1–6. [CrossRef] [PubMed]

49. Bone, R.A.; Landrum, J.T.; Hime, G.W.; Cains, A.; Zamor, J. Stereochemistry of the human macular carotenoids. *Investig. Ophthalmol. Vis. Sci.* **1993**, *34*, 2033–2040.

50. Khachik, F.; Bernstein, P.S.; Garland, D.L. Identification of lutein and zeaxanthin oxidation products in human and monkey retinas. *Investig. Ophthalmol. Vis. Sci.* **1997**, *38*, 1802–1811.

51. Scripsema, N.K.; Hu, D.-N.; Rosen, R.B. Lutein, zeaxanthin, and meso-zeaxanthin in the clinical management of eye disease. *J. Ophthalmol.* **2015**. [CrossRef]

52. Jia, Y.-P.; Sun, L.; Yu, H.-S.; Liang, L.-P.; Li, W.; Ding, H.; Song, X.-B.; Zhang, L.-J. The pharmacological effects of lutein and zeaxanthin on visual disorders and cognition diseases. *Molecules* **2017**, *22*, 610. [CrossRef]

53. Li, B.; Ahmed, F.; Bernstein, P.S. Studies on the singlet oxygen scavenging mechanism of human macular pigment. *Arch. Biochem. Biophys.* **2010**, *504*, 56–60. [CrossRef]

54. Neelam, K.; Goenadi, C.J.; Lun, K.; Yip, C.C.; Eong, K.-G.A. Putative protective role of lutein and zeaxanthin in diabetic retinopathy. *Br. J. Ophthalmol.* **2017**, *101*, 551–558. [CrossRef]

55. Kowluru, R.A.; Menon, B.; Gierhart, D.L. Beneficial effect of zeaxanthin on retinal metabolic abnormalities in diabetic rats. *Investig. Ophthalmol. Vis. Sci.* **2008**, *49*, 1645–1651. [CrossRef]

56. Sasaki, M.; Ozawa, Y.; Kurihara, T.; Kubota, S.; Yuki, K.; Noda, K.; Kobayashi, S.; Ishida, S.; Tsubota, K. Neurodegenerative influence of oxidative stress in the retina of a murine model of diabetes. *Diabetologia* **2010**, *53*, 971–979. [CrossRef]

57. José Bagur, M.; Alonso Salinas, G.; Jiménez-Monreal, A.; Chaouqi, S.; Llorens, S.; Martínez-Tomé, M.; Alonso, G. Saffron: An Old Medicinal Plant and a Potential Novel Functional Food. *Molecules* **2018**, *23*, 30. [CrossRef]

58. Yamauchi, M.; Tsuruma, K.; Imai, S.; Nakanishi, T.; Umigai, N.; Shimazawa, M.; Hara, H. Crocetin prevents retinal degeneration induced by oxidative and endoplasmic reticulum stresses via inhibition of caspase activity. *Eur. J. Pharmacol.* **2011**, *650*, 110–119. [CrossRef]

59. Yang, X.; Huo, F.; Liu, B.; Liu, J.; Chen, T.; Li, J.; Zhu, Z.; Lv, B. Crocin inhibits oxidative stress and pro-inflammatory response of microglial cells associated with diabetic retinopathy through the activation of PI3K/Akt signaling pathway. *J. Mol. Neurosci.* **2017**, *61*, 581–589. [CrossRef] [PubMed]

60. Sen, S.; Querques, M.A.; Chakrabarti, S. North American Ginseng (Panax quinquefolius) prevents hyperglycemia and associated pancreatic abnormalities in diabetes. *J. Med. Food* **2013**, *16*, 587–592. [CrossRef]

61. Fan, Y.; Qiao, Y.; Huang, J.; Tang, M. Protective effects of Panax notoginseng saponins against high glucose-induced oxidative injury in rat retinal capillary endothelial cells. *Evid.-Based Complement. Altern. Med.* **2016**, *2016*. [CrossRef]

62. Fan, C.; Qiao, Y.; Tang, M. Notoginsenoside R1 attenuates high glucose-induced endothelial damage in rat retinal capillary endothelial cells by modulating the intracellular redox state. *Drug Des. Dev. Ther.* **2017**, *11*, 3343. [CrossRef] [PubMed]

63. Amato, R.; Rossino, M.G.; Cammalleri, M.; Locri, F.; Pucci, L.; Monte, M.D.; Casini, G. Lisosan G Protects the Retina from Neurovascular Damage in Experimental Diabetic Retinopathy. *Nutrients* **2018**, *10*, 1932. [CrossRef] [PubMed]

64. Joussen, A.M.; Poulaki, V.; Le, M.L.; Koizumi, K.; Esser, C.; Janicki, H.; Schraermeyer, U.; Kociok, N.; Fauser, S.; Kirchhof, B. A central role for inflammation in the pathogenesis of diabetic retinopathy. *FASEB J.* **2004**, *18*, 1450–1452. [CrossRef] [PubMed]

65. Kern, T.S. Contributions of inflammatory processes to the development of the early stages of diabetic retinopathy. *J. Diabetes Res.* **2007**. [CrossRef] [PubMed]

66. Tang, J.; Kern, T.S. Inflammation in diabetic retinopathy. *Prog. Retin. Eye Res.* **2011**, *30*, 343–358. [CrossRef] [PubMed]

67. Lieth, E.; Gardner, T.W.; Barber, A.J.; Antonetti, D.A. Retinal neurodegeneration: Early pathology in diabetes. *Clin. Exp. Ophthalmol. Viewp.* **2000**, *28*, 3–8. [CrossRef]

68. Lieth, E.; Barber, A.J.; Xu, B.; Dice, C.; Ratz, M.J.; Tanase, D.; Strother, J.M. Glial reactivity and impaired glutamate metabolism in short-term experimental diabetic retinopathy. Penn State Retina Research Group. *Diabetes* **1998**, *47*, 815–820. [CrossRef] [PubMed]

69. Mizutani, M.; Gerhardinger, C.; Lorenzi, M. Muller cell changes in human diabetic retinopathy. *Diabetes* **1998**, *47*, 445–449. [CrossRef]

70. Bringmann, A.; Iandiev, I.; Pannicke, T.; Wurm, A.; Hollborn, M.; Wiedemann, P.; Osborne, N.N.; Reichenbach, A. Cellular signaling and factors involved in Muller cell gliosis: Neuroprotective and detrimental effects. *Prog. Retin. Eye Res.* **2009**, *28*, 423–451. [CrossRef] [PubMed]

71. Lee, H.J.; Suk, J.E.; Patrick, C.; Bae, E.J.; Cho, J.H.; Rho, S.; Hwang, D.; Masliah, E.; Lee, S.J. Direct transfer of alpha-synuclein from neuron to astroglia causes inflammatory responses in synucleinopathies. *J. Biol. Chem.* **2010**, *285*, 9262–9272. [CrossRef]

72. Liu, T.; Zhang, L.; Joo, D.; Sun, S.C. NF-kappaB signaling in inflammation. *Signal Transduct. Target* **2017**, *2*, 14. [CrossRef]

73. Suryavanshi, S.V.; Kulkarni, Y.A. NF-kappabeta: A Potential Target in the Management of Vascular Complications of Diabetes. *Front. Pharm.* **2017**, *8*. [CrossRef]

74. Li, J.; Wang, P.; Ying, J.; Chen, Z.; Yu, S. Curcumin Attenuates Retinal Vascular Leakage by Inhibiting Calcium/Calmodulin-Dependent Protein Kinase II Activity in Streptozotocin-Induced Diabetes. *Cell Physiol Biochem* **2016**, *39*, 1196–1208. [CrossRef] [PubMed]

75. Pradhan, D.D.T.; Tripathy, G. Pharmacognostic evaluation of curcumin on diabetic retinopathy in alloxan-induced diabetes through NF-KB and Brn3a related mechanism. *Pharmacogn. J.* **2018**, *10*, 324–332. [CrossRef]

76. Shome, S.; Talukdar, A.D.; Choudhury, M.D.; Bhattacharya, M.K.; Upadhyaya, H. Curcumin as potential therapeutic natural product: A nanobiotechnological perspective. *J Pharm Pharm.* **2016**, *68*, 1481–1500. [CrossRef] [PubMed]

77. Fan, W.; Cooper, N.G. Glutamate-induced NFkappaB activation in the retina. *Investig. Ophthalmol. Vis. Sci.* **2009**, *50*, 917–925. [CrossRef]

78. Chen, Y.; Meng, J.; Li, H.; Wei, H.; Bi, F.; Liu, S.; Tang, K.; Guo, H.; Liu, W. Resveratrol exhibits an effect on attenuating retina inflammatory condition and damage of diabetic retinopathy via PON1. *Exp Eye Res* **2019**, *181*, 356–366. [CrossRef]

79. Sohn, E.; Kim, J.; Kim, C.S.; Lee, Y.M.; Kim, J.S. Extract of Polygonum cuspidatum Attenuates Diabetic Retinopathy by Inhibiting the High-Mobility Group Box-1 (HMGB1) Signaling Pathway in Streptozotocin-Induced Diabetic Rats. *Nutrients* **2016**, *8*, 140. [CrossRef]

80. Kubota, S.; Kurihara, T.; Mochimaru, H.; Satofuka, S.; Noda, K.; Ozawa, Y.; Oike, Y.; Ishida, S.; Tsubota, K. Prevention of ocular inflammation in endotoxin-induced uveitis with resveratrol by inhibiting oxidative damage and nuclear factor-kappaB activation. *Investig. Ophthalmol. Vis. Sci.* **2009**, *50*, 3512–3519. [CrossRef] [PubMed]

81. Kubota, S.; Ozawa, Y.; Kurihara, T.; Sasaki, M.; Yuki, K.; Miyake, S.; Noda, K.; Ishida, S.; Tsubota, K. Roles of AMP-activated protein kinase in diabetes-induced retinal inflammation. *Investig. Ophthalmol. Vis. Sci.* **2011**, *52*, 9142–9148. [CrossRef] [PubMed]

82. Bagul, P.K.; Deepthi, N.; Sultana, R.; Banerjee, S.K. Resveratrol ameliorates cardiac oxidative stress in diabetes through deacetylation of NFkB-p65 and histone 3. *J. Nutr. Biochem.* **2015**, *26*, 1298–1307. [CrossRef] [PubMed]

83. Kim, Y.H.; Kim, Y.S.; Kang, S.S.; Cho, G.J.; Choi, W.S. Resveratrol inhibits neuronal apoptosis and elevated Ca2+/calmodulin-dependent protein kinase II activity in diabetic mouse retina. *Diabetes* **2010**, *59*, 1825–1835. [CrossRef]

84. Lee, M.; Yun, S.; Lee, H.; Yang, J. Quercetin Mitigates Inflammatory Responses Induced by Vascular Endothelial Growth Factor in Mouse Retinal Photoreceptor Cells through Suppression of Nuclear Factor Kappa B. *Int. J. Mol. Sci.* **2017**, *18*, 2497. [CrossRef]

85. Bucolo, C.; Leggio, G.M.; Drago, F.; Salomone, S. Eriodictyol prevents early retinal and plasma abnormalities in streptozotocin-induced diabetic rats. *Biochem Pharm.* **2012**, *84*, 88–92. [CrossRef]

86. Wang, W.; Zhang, Y.; Jin, W.; Xing, Y.; Yang, A. Catechin Weakens Diabetic Retinopathy by Inhibiting the Expression of NF-kappaB Signaling Pathway-Mediated Inflammatory Factors. *Ann. Clin. Lab. Sci.* **2018**, *48*, 594–600. [PubMed]

87. Silva, K.C.; Rosales, M.A.; Hamassaki, D.E.; Saito, K.C.; Faria, A.M.; Ribeiro, P.A.; Faria, J.B.; Faria, J.M. Green tea is neuroprotective in diabetic retinopathy. *Investig. Ophthalmol. Vis. Sci.* **2013**, *54*, 1325–1336. [CrossRef]

88. Al-Gayyar, M.M.; Matragoon, S.; Pillai, B.A.; Ali, T.K.; Abdelsaid, M.A.; El-Remessy, A.B. Epicatechin blocks pro-nerve growth factor (proNGF)-mediated retinal neurodegeneration via inhibition of p75 neurotrophin receptor expression in a rat model of diabetes [corrected]. *Diabetologia* **2011**, *54*, 669–680. [CrossRef] [PubMed]

89. Dongare, S.; Gupta, S.K.; Mathur, R.; Saxena, R.; Mathur, S.; Agarwal, R.; Nag, T.C.; Srivastava, S.; Kumar, P. Zingiber officinale attenuates retinal microvascular changes in diabetic rats via anti-inflammatory and antiangiogenic mechanisms. *Mol. Vis.* **2016**, *22*, 599–609.

90. Tzeng, T.F.; Liou, S.S.; Tzeng, Y.C.; Liu, I.M. Zerumbone, a Phytochemical of Subtropical Ginger, Protects against Hyperglycemia-Induced Retinal Damage in Experimental Diabetic Rats. *Nutrients* **2016**, *8*, 449. [CrossRef] [PubMed]

91. Giusti, L.; Gabriele, M.; Penno, G.; Garofolo, M.; Longo, V.; Del Prato, S.; Lucchesi, D.; Pucci, L. A Fermented Whole Grain Prevents Lipopolysaccharides-Induced Dysfunction in Human Endothelial Progenitor Cells. *Oxid. Med. Cell. Longev.* **2017**. [CrossRef] [PubMed]

92. La Marca, M.; Beffy, P.; Pugliese, A.; Longo, V. Fermented wheat powder induces the antioxidant and detoxifying system in primary rat hepatocytes. *PLoS ONE* **2013**, *8*. [CrossRef] [PubMed]

93. Kim, S.J.; Yoo, W.S.; Choi, M.; Chung, I.; Yoo, J.M.; Choi, W.S. Increased O-GlcNAcylation of NF-kappaB Enhances Retinal Ganglion Cell Death in Streptozotocin-induced Diabetic Retinopathy. *Curr. Eye Res.* **2016**, *41*, 249–257. [CrossRef] [PubMed]

94. Kim, S.J.; Kim, M.J.; Choi, M.Y.; Kim, Y.S.; Yoo, J.M.; Hong, E.K.; Ju, S.; Choi, W.S. Aralia elata inhibits neurodegeneration by downregulating O-GlcNAcylation of NF-kappaB in diabetic mice. *Int. J. Ophthalmol.* **2017**, *10*, 1203–1211. [CrossRef] [PubMed]

95. Bucolo, C.; Marrazzo, G.; Platania, C.B.; Drago, F.; Leggio, G.M.; Salomone, S. Fortified extract of red berry, Ginkgo biloba, and white willow bark in experimental early diabetic retinopathy. *J. Diabetes Res.* **2013**. [CrossRef]

96. Gill, R.; Tsung, A.; Billiar, T. Linking oxidative stress to inflammation: Toll-like receptors. *Free Radic. Biol. Med.* **2010**, *48*, 1121–1132. [CrossRef]

97. Reuter, S.; Gupta, S.C.; Chaturvedi, M.M.; Aggarwal, B.B. Oxidative stress, inflammation, and cancer: How are they linked? *Free Radic. Biol. Med.* **2010**, *49*, 1603–1616. [CrossRef]

98. Lugrin, J.; Rosenblatt-Velin, N.; Parapanov, R.; Liaudet, L. The role of oxidative stress during inflammatory processes. *Biol. Chem.* **2014**, *395*, 203–230. [CrossRef]

99. Morgan, M.J.; Liu, Z.G. Crosstalk of reactive oxygen species and NF-kappaB signaling. *Cell Res.* **2011**, *21*, 103–115. [CrossRef] [PubMed]

100. Forrester, S.J.; Kikuchi, D.S.; Hernandes, M.S.; Xu, Q.; Griendling, K.K. Reactive Oxygen Species in Metabolic and Inflammatory Signaling. *Circ. Res.* **2018**, *122*, 877–902. [CrossRef]

101. Ran, Z.; Zhang, Y.; Wen, X.; Ma, J. Curcumin inhibits high glucoseinduced inflammatory injury in human retinal pigment epithelial cells through the ROSPI3K/AKT/mTOR signaling pathway. *Mol. Med. Rep.* **2018**, *12*. [CrossRef]

102. Amato, R.; Catalani, E.; Dal Monte, M.; Cammalleri, M.; Di Renzo, I.; Perrotta, C.; Cervia, D.; Casini, G. Autophagy-mediated neuroprotection induced by octreotide in an ex vivo model of early diabetic retinopathy. *Pharm. Res.* **2018**, *128*, 167–178. [CrossRef] [PubMed]

103. Barber, A.J. Diabetic retinopathy: Recent advances towards understanding neurodegeneration and vision loss. *Sci. China Life Sci.* **2015**, *58*, 541–549. [CrossRef] [PubMed]

104. Barber, A.J.; Baccouche, B. Neurodegeneration in diabetic retinopathy: Potential for novel therapies. *Vis. Res.* **2017**, *139*, 82–92. [CrossRef] [PubMed]

105. Ola, M.S.; Nawaz, M.I.; Khan, H.A.; Alhomida, A.S. Neurodegeneration and neuroprotection in diabetic retinopathy. *Int. J. Mol. Sci.* **2013**, *14*, 2559–2572. [CrossRef] [PubMed]

106. Ola, M.S.; Alhomida, A.S. Neurodegeneration in diabetic retina and its potential drug targets. *Curr. Neuropharmacol.* **2014**, *12*, 380–386. [CrossRef] [PubMed]

107. Mohamed, R.; El-Remessy, A.B. Imbalance of the Nerve Growth Factor and Its Precursor: Implication in Diabetic Retinopathy. *J. Clin. Exp. Ophthalmol.* **2015**, *6*, 2155–9570. [CrossRef] [PubMed]

108. Ola, M.S.; Nawaz, M.I.; El-Asrar, A.A.; Abouammoh, M.; Alhomida, A.S. Reduced levels of brain derived neurotrophic factor (BDNF) in the serum of diabetic retinopathy patients and in the retina of diabetic rats. *Cell. Mol. Neurobiol.* **2013**, *33*, 359–367. [CrossRef]

109. Li, Q.; Puro, D.G. Diabetes-induced dysfunction of the glutamate transporter in retinal Muller cells. *Investig. Ophthalmol. Vis. Sci.* **2002**, *43*, 3109–3116.

110. Solanki, I.; Parihar, P.; Mansuri, M.L.; Parihar, M.S. Flavonoid-based therapies in the early management of neurodegenerative diseases. *Adv. Nutr.* **2015**, *6*, 64–72. [CrossRef]

111. Vauzour, D.; Vafeiadou, K.; Rodriguez-Mateos, A.; Rendeiro, C.; Spencer, J.P. The neuroprotective potential of flavonoids: A multiplicity of effects. *Genes Nutr.* **2008**, *3*, 115–126. [CrossRef]

112. Yang, F.; Yu, J.; Ke, F.; Lan, M.; Li, D.; Tan, K.; Ling, J.; Wang, Y.; Wu, K. Curcumin Alleviates Diabetic Retinopathy in Experimental Diabetic Rats. *Ophthalmic Res.* **2018**, *60*, 43–54. [CrossRef]

113. Shakeri, A.; Cicero, A.F.G.; Panahi, Y.; Mohajeri, M.; Sahebkar, A. Curcumin: A naturally occurring autophagy modulator. *J. Cell. Physiol.* **2018**, *21*, 27404. [CrossRef]

114. Sheu, S.J.; Chen, J.L.; Bee, Y.S.; Chen, Y.A.; Lin, S.H.; Shu, C.W. Differential autophagic effects of vital dyes in retinal pigment epithelial ARPE-19 and photoreceptor 661W cells. *PLoS ONE* **2017**, *12*, e0174736. [CrossRef] [PubMed]

115. Ola, M.S.; Ahmed, M.M.; Shams, S.; Al-Rejaie, S.S. Neuroprotective effects of quercetin in diabetic rat retina. *Saudi J. Biol. Sci.* **2017**, *24*, 1186–1194. [CrossRef]

116. Qu, L.; Liang, X.; Gu, B.; Liu, W. Quercetin alleviates high glucose-induced Schwann cell damage by autophagy. *Neural Regen. Res.* **2014**, *9*, 1195–1203. [CrossRef] [PubMed]

117. Regitz, C.; Dussling, L.M.; Wenzel, U. Amyloid-beta (Abeta(1)(-)(4)(2))-induced paralysis in Caenorhabditis elegans is inhibited by the polyphenol quercetin through activation of protein degradation pathways. *Mol. Nutr. Food Res.* **2014**, *58*, 1931–1940. [CrossRef] [PubMed]

118. Somerset, S.M.; Johannot, L. Dietary flavonoid sources in Australian adults. *Nutr. Cancer* **2008**, *60*, 442–449. [CrossRef] [PubMed]

119. Filomeni, G.; Graziani, I.; De Zio, D.; Dini, L.; Centonze, D.; Rotilio, G.; Ciriolo, M.R. Neuroprotection of kaempferol by autophagy in models of rotenone-mediated acute toxicity: Possible implications for Parkinson's disease. *Neurobiol. Aging* **2012**, *33*, 767–785. [CrossRef]

120. Havsteen, B. Flavonoids, a class of natural products of high pharmacological potency. *Biochem. Pharm.* **1983**, *32*, 1141–1148. [CrossRef]

121. Nakayama, M.; Aihara, M.; Chen, Y.N.; Araie, M.; Tomita-Yokotani, K.; Iwashina, T. Neuroprotective effects of flavonoids on hypoxia-, glutamate-, and oxidative stress-induced retinal ganglion cell death. *Mol. Vis.* **2011**, *17*, 1784–1793.

122. Ola, M.S.; Ahmed, M.M.; Ahmad, R.; Abuohashish, H.M.; Al-Rejaie, S.S.; Alhomida, A.S. Neuroprotective Effects of Rutin in Streptozotocin-Induced Diabetic Rat Retina. *J. Mol. Neurosci.* **2015**, *56*, 440–448. [CrossRef] [PubMed]

123. Kang, M.K.; Lee, E.J.; Kim, Y.H.; Kim, D.Y.; Oh, H.; Kim, S.I.; Kang, Y.H. Chrysin Ameliorates Malfunction of Retinoid Visual Cycle through Blocking Activation of AGE-RAGE-ER Stress in Glucose-Stimulated Retinal Pigment Epithelial Cells and Diabetic Eyes. *Nutrients* **2018**, *10*, 1046. [CrossRef]

124. Shimouchi, A.; Yokota, H.; Ono, S.; Matsumoto, C.; Tamai, T.; Takumi, H.; Narayanan, S.P.; Kimura, S.; Kobayashi, H.; Caldwell, R.B.; et al. Neuroprotective effect of water-dispersible hesperetin in retinal ischemia reperfusion injury. *Jpn. J. Ophthalmol.* **2016**, *60*, 51–61. [CrossRef]

125. Al-Dosari, D.I.; Ahmed, M.M.; Al-Rejaie, S.S.; Alhomida, A.S.; Ola, M.S. Flavonoid Naringenin Attenuates Oxidative Stress, Apoptosis and Improves Neurotrophic Effects in the Diabetic Rat Retina. *Nutrients* **2017**, *9*, 1161. [CrossRef] [PubMed]

126. Xu, X.H.; Ma, C.M.; Han, Y.Z.; Li, Y.; Liu, C.; Duan, Z.H.; Wang, H.L.; Liu, D.Q.; Liu, R.H. Protective Effect of Naringenin on Glutamate-Induced Neurotoxicity in Cultured Hippocampal Cells. *Arch. Biol. Sci.* **2015**, *67*, 639–646. [CrossRef]

127. Sasaki, M.; Yuki, K.; Kurihara, T.; Miyake, S.; Noda, K.; Kobayashi, S.; Ishida, S.; Tsubota, K.; Ozawa, Y. Biological role of lutein in the light-induced retinal degeneration. *J. Nutr. Biochem.* **2012**, *23*, 423–429. [CrossRef] [PubMed]

128. Ozawa, Y.; Sasaki, M.; Takahashi, N.; Kamoshita, M.; Miyake, S.; Tsubota, K. Neuroprotective Effects of Lutein in the Retina. *Curr. Pharm. Des.* **2012**, *18*, 51–56. [CrossRef] [PubMed]

129. Hu, C.K.; Lee, Y.J.; Colitz, C.M.; Chang, C.J.; Lin, C.T. The protective effects of Lycium barbarum and Chrysanthemum morifolum on diabetic retinopathies in rats. *Vet. Ophthalmol.* **2012**, *15*, 65–71. [CrossRef] [PubMed]

130. Li, S.Y.; Yang, D.; Yeung, C.M.; Yu, W.Y.; Chang, R.C.C.; So, K.F.; Wong, D.; Lo, A.C.Y. Lycium Barbarum Polysaccharides Reduce Neuronal Damage, Blood-Retinal Barrier Disruption and Oxidative Stress in Retinal Ischemia/Reperfusion Injury. *PLoS ONE* **2011**, *6*. [CrossRef]

131. Song, M.K.; Roufogalis, B.D.; Huang, T.H.W. Reversal of the Caspase-Dependent Apoptotic Cytotoxicity Pathway by Taurine from Lycium barbarum (Goji Berry) in Human Retinal Pigment Epithelial Cells: Potential Benefit in Diabetic Retinopathy. *Evid.-Based Complement. Altern. Med.* **2012**. [CrossRef]

132. Kelsey, N.A.; Wilkins, H.M.; Linseman, D.A. Nutraceutical Antioxidants as Novel Neuroprotective Agents. *Molecules* **2010**, *15*, 7792–7814. [CrossRef]

133. Stitt, A.W.; Lois, N.; Medina, R.J.; Adamson, P.; Curtis, T.M. Advances in our understanding of diabetic retinopathy. *Clin. Sci.* **2013**, *125*. [CrossRef] [PubMed]

134. Frey, T.; Antonetti, D.A. Alterations to the blood-retinal barrier in diabetes: Cytokines and reactive oxygen species. *Antioxid. Redox Signal.* **2011**, *15*, 1271–1284. [CrossRef]

135. Eshaq, R.S.; Aldalati, A.M.Z.; Alexander, J.S.; Harris, N.R. Diabetic retinopathy: Breaking the barrier. *Pathophysiology* **2017**, *24*, 229–241. [CrossRef] [PubMed]

136. Kusuhara, S.; Fukushima, Y.; Ogura, S.; Inoue, N.; Uemura, A. Pathophysiology of Diabetic Retinopathy: The Old and the New. *Diabetes Metab. J.* **2018**, *42*, 364–376. [CrossRef] [PubMed]

137. Antonetti, D.A.; Barber, A.J.; Hollinger, L.A.; Wolpert, E.B.; Gardner, T.W. Vascular endothelial growth factor induces rapid phosphorylation of tight junction proteins occludin and zonula occluden 1. A potential mechanism for vascular permeability in diabetic retinopathy and tumors. *J. Biol. Chem.* **1999**, *274*, 23463–23467. [CrossRef]

138. Murakami, T.; Felinski, E.A.; Antonetti, D.A. Occludin Phosphorylation and Ubiquitination Regulate Tight Junction Trafficking and Vascular Endothelial Growth Factor-induced Permeability. *J. Biol. Chem.* **2009**, *284*, 21036–21046. [CrossRef]

139. Esser, S.; Lampugnani, M.G.; Corada, M.; Dejana, E.; Risau, W. Vascular endothelial growth factor induces VE-cadherin tyrosine phosphorylation in endothelial cells. *J. Cell Sci.* **1998**, *111*, 1853–1865.

140. Aveleira, C.A.; Lin, C.M.; Abcouwer, S.F.; Ambrosio, A.F.; Antonetti, D.A. TNF-alpha signals through PKCzeta/NF-kappaB to alter the tight junction complex and increase retinal endothelial cell permeability. *Diabetes* **2010**, *59*, 2872–2882. [CrossRef]

141. Bamforth, S.D.; Lightman, S.L.; Greenwood, J. Interleukin-1 beta-induced disruption of the retinal vascular barrier of the central nervous system is mediated through leukocyte recruitment and histamine. *Am. J. Pathol.* **1997**, *150*, 329–340. [PubMed]

142. Kowluru, R.A.; Zhong, Q.; Santos, J.M. Matrix metalloproteinases in diabetic retinopathy: Potential role of MMP-9. *Expert Opin. Investig. Drugs* **2012**, *21*, 797–805. [CrossRef]

143. Hammes, H.P.; Lin, J.H.; Renner, O.; Shani, M.; Lundqvist, A.; Betsholtz, C.; Brownlee, M.; Deutsch, U. Pericytes and the pathogenesis of diabetic retinopathy. *Diabetes* **2002**, *51*, 3107–3112. [CrossRef] [PubMed]

144. Xu, J.; Chen, L.J.; Yu, J.; Wang, H.J.; Zhang, F.; Liu, Q.; Wu, J. Involvement of Advanced Glycation End Products in the Pathogenesis of Diabetic Retinopathy. *Cell. Physiol. Biochem.* **2018**, *48*, 705–717. [CrossRef]

145. Ejaz, S. Importance of pericytes and mechanisms of pericyte loss during diabetes retinopathy. *Diabetes Obes. Metab.* **2008**, *10*, 53–63. [CrossRef] [PubMed]

146. Beltramo, E.; Porta, M. Pericyte Loss in Diabetic Retinopathy: Mechanisms and Consequences. *Curr. Med. Chem.* **2013**, *20*, 3218–3225. [CrossRef]

147. Loukovaara, S.; Gucciardo, E.; Repo, P.; Vihinen, H.; Lohi, J.; Jokitalo, E.; Salven, P.; Lehti, K. Indications of lymphatic endothelial differentiation and endothelial progenitor cell activation in the pathology of proliferative diabetic retinopathy. *Acta Ophthalmol.* **2015**, *93*, 512–523. [CrossRef] [PubMed]

148. Roy, S.; Amin, S.; Roy, S. Retinal fibrosis in diabetic retinopathy. *Exp. Eye Res.* **2016**, *142*, 71–75. [CrossRef] [PubMed]

149. Stitt, A.W.; Curtis, T.M.; Chen, M.; Medina, R.J.; McKay, G.J.; Jenkins, A.; Gardiner, T.A.; Lyons, T.J.; Hammes, H.P.; Simo, R.; et al. The progress in understanding and treatment of diabetic retinopathy. *Prog. Retin. Eye Res.* **2016**, *51*, 156–186. [CrossRef] [PubMed]

150. Adamis, A.P.; Miller, J.W.; Bernal, M.T.; Damico, D.J.; Folkman, J.; Yeo, T.K.; Yeo, K.T. Increased Vascular Endothelial Growth-Factor Levels in the Vitreous of Eyes with Proliferative Diabetic-Retinopathy. *Am. J. Ophthalmol.* **1994**, *118*, 445–450. [CrossRef]

151. Osaadon, P.; Fagan, X.J.; Lifshitz, T.; Levy, J. A review of anti-VEGF agents for proliferative diabetic retinopathy. *Eye* **2014**, *28*, 510–520. [CrossRef]

152. Wang, L.L.; Li, C.Z.; Guo, H.; Kern, T.S.; Huang, K.; Zheng, L. Curcumin Inhibits Neuronal and Vascular Degeneration in Retina after Ischemia and Reperfusion Injury. *PLoS ONE* **2011**, *6*, e23194. [CrossRef]

153. Mrudula, T.; Suryanarayana, P.; Srinivas, P.N.B.S.; Reddy, G.B. Effect of curcumin on hyperglycemia-induced vascular endothelial growth factor expression in streptozotocin-induced diabetic rat retina. *Biochem. Biophys. Res. Commun.* **2007**, *361*, 528–532. [CrossRef]

154. Xie, P.; Zhang, W.; Yuan, S.; Chen, Z.; Yang, Q.; Yuan, D.; Wang, F.; Liu, Q. Suppression of experimental choroidal neovascularization by curcumin in mice. *PLoS ONE* **2012**, *7*, 28. [CrossRef] [PubMed]

155. Lee, C.S.; Choi, E.Y.; Lee, S.C.; Koh, H.J.; Lee, J.H.; Chung, J.H. Resveratrol Inhibits Hypoxia-Induced Vascular Endothelial Growth Factor Expression and Pathological Neovascularization. *Yonsei Med. J.* **2015**, *56*, 1678–1685. [CrossRef] [PubMed]

156. Kim, Y.H.; Kim, Y.S.; Roh, G.S.; Choi, W.S.; Cho, G.J. Resveratrol blocks diabetes-induced early vascular lesions and vascular endothelial growth factor induction in mouse retinas. *Acta Ophthalmol.* **2012**, *90*, E31–E37. [CrossRef]

157. Li, W.L.; Jiang, D.Y. Effect of Resveratrol on Bcl-2 and VEGF Expression in Oxygen-Induced Retinopathy of Prematurity. *J. Pediatric Ophthalmol. Strabismus* **2012**, *49*, 230–235. [CrossRef]

158. Nagai, N.; Kubota, S.; Tsubota, K.; Ozawa, Y. Resveratrol prevents the development of choroidal neovascularization by modulating AMP-activated protein kinase in macrophages and other cell types. *J. Nutr. Biochem.* **2014**, *25*, 1218–1225. [CrossRef] [PubMed]

159. Hua, J.; Guerin, K.I.; Chen, J.; Michan, S.; Stahl, A.; Krah, N.M.; Seaward, M.R.; Dennison, R.J.; Juan, A.M.; Hatton, C.J.; et al. Resveratrol Inhibits Pathologic Retinal Neovascularization in Vldlr(-/-) Mice. *Investig. Ophthalmol. Vis. Sci.* **2011**, *52*, 2809–2816. [CrossRef]

160. Popescu, M.; Bogdan, C.; Pintea, A.; Rugina, D.; Ionescu, C. Antiangiogenic cytokines as potential new therapeutic targets for resveratrol in diabetic retinopathy. *Drug Des. Dev.* **2018**, *12*, 1985–1996. [CrossRef]

161. Chen, B.; He, T.; Xing, Y.Q.; Cao, T. Effects of quercetin on the expression of MCP-1, MMP-9 and VEGF in rats with diabetic retinopathy. *Exp. Ther. Med.* **2017**, *14*, 6022–6026. [CrossRef]

162. Li, F.T.; Bai, Y.J.; Zhao, M.; Huang, L.Z.; Li, S.S.; Li, X.X.; Chen, Y. Quercetin Inhibits Vascular Endothelial Growth Factor-Induced Choroidal and Retinal Angiogenesis in vitro. *Ophthalmic Res.* **2015**, *53*, 109–116. [CrossRef] [PubMed]

163. Chen, Y.; Li, X.X.; Xing, N.Z.; Cao, X.G. Quercetin inhibits choroidal and retinal angiogenesis in vitro. *Graefes Arch. Clin. Exp. Ophthalmol.* **2008**, *246*, 373–378. [CrossRef]

164. Kang, M.K.; Park, S.H.; Kim, Y.H.; Lee, E.J.; Antika, L.D.; Kim, D.Y.; Choi, Y.J.; Kang, Y.H. Dietary Compound Chrysin Inhibits Retinal Neovascularization with Abnormal Capillaries in db/db Mice. *Nutrients* **2016**, *8*, 782. [CrossRef]

165. Song, J.H.; Kim, Y.H.; Lee, S.C.; Kim, M.H.; Lee, J.H. Inhibitory Effect of Chrysin (5,7-Dihydroxyflavone) on Experimental Choroidal Neovascularization in Rats. *Ophthalmic Res.* **2016**, *56*, 49–55. [CrossRef] [PubMed]

166. Lee, H.S.; Jun, J.H.; Jung, E.H.; Koo, B.A.; Kim, Y.S. Epigalloccatechin-3-gallate Inhibits Ocular Neovascularization and Vascular Permeability in Human Retinal Pigment Epithelial and Human Retinal Microvascular Endothelial Cells via Suppression of MMP-9 and VEGF Activation. *Molecules* **2014**, *19*, 12150–12172. [CrossRef] [PubMed]

167. Kim, J.; Kim, C.S.; Moon, M.K.; Kim, J.S. Epicatechin breaks preformed glycated serum albumin and reverses the retinal accumulation of advanced glycation end products. *Eur. J. Pharmacol.* **2015**, *748*, 108–114. [CrossRef] [PubMed]

168. Saito, Y.; Hasebe-Takenaka, Y.; Ueda, T.; Nakanishi-Ueda, T.; Kosuge, S.; Aburada, M.; Shimada, T.; Ikeya, Y.; Onda, H.; Ogura, H.; et al. Effects of green tea fractions on oxygen-induced retinal neovascularization in the neonatal rat. *J. Clin. Biochem. Nutr.* **2007**, *41*, 43–49. [CrossRef]

169. Kumar, B.; Gupta, S.K.; Srinivasan, B.P.; Nag, T.C.; Srivastava, S.; Saxena, R. Hesperetin ameliorates hyperglycemia induced retinal vasculopathy via anti-angiogenic effects in experimental diabetic rats. *Vasc. Pharmacol.* **2012**, *57*, 201–207. [CrossRef] [PubMed]

170. Shen, Y.; Zhang, W.Y.; Chiou, G.C.Y. Effect of naringenin on NaIO3-induced retinal pigment epithelium degeneration and laser-induced choroidal neovascularization in rats. *Int. J. Ophthalmol.* **2010**, *3*, 5–8. [CrossRef]

171. Xu, X.R.; Yu, H.T.; Hang, L.; Shao, Y.; Ding, S.H.; Yang, X.W. Preparation of Naringenin/beta-Cyclodextrin Complex and Its More Potent Alleviative Effect on Choroidal Neovascularization in Rats. *Biomed Res. Int.* **2014**. [CrossRef]

172. Yang, S.J.; Jo, H.; Kim, J.G.; Jung, S.H. Baicalin Attenuates Laser-Induced Choroidal Neovascularization. *Curr. Eye Res.* **2014**, *39*, 745–751. [CrossRef]

173. Park, S.W.; Cho, C.S.; Jun, H.O.; Ryu, N.H.; Kim, J.H.; Yu, Y.S.; Kim, J.S.; Kim, J.H. Anti-Angiogenic Effect of Luteolin on Retinal Neovascularization via Blockade of Reactive Oxygen Species Production. *Investig. Ophthalmol. Vis. Sci.* **2012**, *53*, 7718–7726. [CrossRef]

174. Kim, J.H.; Kim, J.H.; Yu, Y.S.; Park, K.H.; Kang, H.J.; Lee, H.Y.; Kim, K.W. Antiangiogenic effect of deguelin on choroidal neovascularization. *J. Pharmacol. Exp. Ther.* **2008**, *324*, 643–647. [CrossRef]

175. Kim, J.H.; Kim, J.H.; Yu, Y.S.; Shin, J.Y.; Lee, H.Y.; Kim, K.W. Deguelin inhibits retinal neovascularization by down-regulation of HIF-1 alpha in oxygen-induced retinopathy. *J. Cell. Mol. Med.* **2008**, *12*, 2407–2415. [CrossRef]

176. An, H.; Lee, S.; Lee, J.M.; Jo, D.H.; Kim, J.; Jeong, Y.S.; Heo, M.J.; Cho, C.S.; Choi, H.; Seo, J.H.; et al. Novel Hypoxia-Inducible Factor 1alpha (HIF-1alpha) Inhibitors for Angiogenesis-Related Ocular Diseases: Discovery of a Novel Scaffold via Ring-Truncation Strategy. *J. Med. Chem.* **2018**, *61*, 9266–9286. [CrossRef]

177. Basavarajappa, H.D.; Lee, B.; Lee, H.; Sulaiman, R.S.; An, H.; Magana, C.; Shadmand, M.; Vayl, A.; Rajashekhar, G.; Kim, E.Y.; et al. Synthesis and Biological Evaluation of Novel Homoisoflavonoids for Retinal Neovascularization. *J. Med. Chem.* **2015**, *58*, 5015–5027. [CrossRef]

178. Kim, J.H.; Kim, J.H.; Yu, Y.S.; Jun, H.O.; Kwon, H.J.; Park, K.H.; Kim, K.W. Inhibition of choroidal neovascularization by homoisoflavanone, a new angiogenesis inhibitor. *Mol. Vis.* **2008**, *14*, 556–561.

179. Lamoke, F.; Labazi, M.; Montemari, A.; Parisi, G.; Varano, M.; Bartoli, M. Trans-Chalcone prevents VEGF expression and retinal neovascularization in the ischemic retina. *Exp. Eye Res.* **2011**, *93*, 350–354. [CrossRef] [PubMed]

180. Jhanji, V.; Liu, H.M.; Law, K.; Lee, V.Y.W.; Huang, S.F.; Pang, C.P.; Yam, G.H.F. Isoliquiritigenin from licorice root suppressed neovascularisation in experimental ocular angiogenesis models. *Br. J. Ophthalmol.* **2011**, *95*, 1309–1315. [CrossRef] [PubMed]

181. Kim, J.; Kim, C.S.; Lee, Y.M.; Sohn, E.; Jo, K.; Kim, J.S. Vaccinium myrtillus extract prevents or delays the onset of diabetes-induced blood-retinal barrier breakdown. *Int. J. Food Sci. Nutr.* **2015**, *66*, 236–242. [CrossRef] [PubMed]

182. Yanai, R.; Chen, S.; Uchi, S.H.; Nanri, T.; Connor, K.M.; Kimura, K. Attenuation of choroidal neovascularization by dietary intake of omega-3 long-chain polyunsaturated fatty acids and lutein in mice. *PLoS ONE* **2018**, *13*. [CrossRef]

183. Maeng, Y.S.; Maharjan, S.; Kim, J.H.; Park, J.H.; Yu, Y.S.; Kim, Y.M.; Kwon, Y.G. Rk1, a Ginsenoside, Is a New Blocker of Vascular Leakage Acting through Actin Structure Remodeling. *PLoS ONE* **2013**, *8*. [CrossRef] [PubMed]

184. Shi, Y.W.; Wan, X.S.; Shao, N.; Ye, R.Y.; Zhang, N.; Zhang, Y.J. Protective and anti-angiopathy effects of ginsenoside Re against diabetes mellitus via the activation of p38 MAPK, ERK1/2 and JNK signaling. *Mol. Med. Rep.* **2016**, *14*, 4849–4856. [CrossRef] [PubMed]

185. Lulli, M.; Cammalleri, M.; Fornaciari, I.; Casini, G.; Dal Monte, M. Acetyl-11-keto-beta-boswellic acid reduces retinal angiogenesis in a mouse model of oxygen-induced retinopathy. *Exp. Eye Res.* **2015**, *135*, 67–80. [CrossRef] [PubMed]

186. Lee, Y.M.; Kim, J.; Kim, C.S.; Jo, K.; Yoo, N.H.; Sohn, E.; Kim, J.S. Anti-glycation and anti-angiogenic activities of 5'-methoxybiphenyl-3,4,3'-triol, a novel phytochemical component of Osteomeles schwerinae. *Eur. J. Pharmacol.* **2015**, *760*, 172–178. [CrossRef]

187. Hong, T.Y.; Tzeng, T.F.; Liou, S.S.; Liu, I.M. The ethanol extract of Zingiber zerumbet rhizomes mitigates vascular lesions in the diabetic retina. *Vasc. Pharmacol.* **2016**, *76*, 18–27. [CrossRef] [PubMed]

188. Duan, H.H.; Huang, J.M.; Li, W.; Tang, M.K. Protective Effects of Fufang Xueshuantong on Diabetic Retinopathy in Rats. *Evid.-Based Complement. Altern. Med.* **2013**. [CrossRef] [PubMed]

189. Jian, W.J.; Yu, S.Y.; Tang, M.K.; Duan, H.H.; Huang, J.M. A combination of the main constituents of Fufang Xueshuantong Capsules shows protective effects against streptozotocin-induced retinal lesions in rats. *J. Ethnopharmacol.* **2016**, *182*, 50–56. [CrossRef] [PubMed]

190. Gao, D.H.; Guo, Y.J.; Li, X.J.; Li, X.M.; Li, Z.P.; Xue, M.; Ou, Z.M.; Liu, M.; Yang, M.X.; Liu, S.H.; et al. An Aqueous Extract of Radix Astragali, Angelica sinensis, and Panax notoginseng Is Effective in Preventing Diabetic Retinopathy. *Evid.-Based Complement. Altern. Med.* **2013**. [CrossRef] [PubMed]

191. Hu, B.J.; Hu, Y.N.; Lin, S.; Ma, W.J.; Li, X.R. Application of Lutein and Zeaxanthin in nonproliferative diabetic retinopathy. *Int. J. Ophthalmol.* **2011**, *4*, 303–306. [CrossRef]

192. Zhang, P.C.; Wu, C.R.; Wang, Z.L.; Wang, L.Y.; Han, Y.; Sun, S.L.; Li, Q.S.; Ma, L. Effect of lutein supplementation on visual function in nonproliferative diabetic retinopathy. *Asia Pac. J. Clin. Nutr.* **2017**, *26*, 406–411. [CrossRef] [PubMed]

193. Sepahi, S.; Mohajeri, S.A.; Hosseini, S.M.; Khodaverdi, E.; Shoeibi, N.; Namdari, M.; Tabassi, S.A.S. Effects of Crocin on Diabetic Maculopathy: A Placebo-Controlled Randomized Clinical Trial. *Am. J. Ophthalmol.* **2018**, *190*, 89–98. [CrossRef] [PubMed]

194. Brazionis, L.; Rowley, K.; Itsiopoulos, C.; O'Dea, K. Plasma carotenoids and diabetic retinopathy. *Br. J. Nutr.* **2009**, *101*, 270–277. [CrossRef] [PubMed]

195. Lima, V.C.; Rosen, R.B.; Maia, M.; Prata, T.S.; Dorairaj, S.; Farah, M.E.; Sallum, J. Macular Pigment Optical Density Measured by Dual-Wavelength Autofluorescence Imaging in Diabetic and Nondiabetic Patients: A Comparative Study. *Investig. Ophthalmol. Vis. Sci.* **2010**, *51*, 5840–5845. [CrossRef] [PubMed]

196. Moschos, M.M.; Dettoraki, M.; Tsatsos, M.; Kitsos, G.; Kalogeropoulos, C. Effect of carotenoids dietary supplementation on macular function in diabetic patients. *Eye Vis.* **2017**, *4*. [CrossRef]

197. Cuomo, J.; Appendino, G.; Dern, A.S.; Schneider, E.; McKinnon, T.P.; Brown, M.J.; Togni, S.; Dixon, B.M. Comparative absorption of a standardized curcuminoid mixture and its lecithin formulation. *J. Nat. Prod.* **2011**, *74*, 664–669. [CrossRef] [PubMed]

198. Steigerwalt, R.; Nebbioso, M.; Appendino, G.; Belcaro, G.; Ciammaichella, G.; Cornelli, U.; Luzzi, R.; Togni, S.; Dugall, M.; Cesarone, M.R.; et al. Meriva (R), a lecithinized curcumin delivery system, in diabetic microangiopathy and retinopathy. *Panminerva Med.* **2012**, *54*, 11–16. [PubMed]

199. Ma, Q.H.; Chen, D.D.; Sun, H.P.; Yan, N.; Xu, Y.; Pan, C.W. Regular Chinese Green Tea Consumption Is Protective for Diabetic Retinopathy: A Clinic-Based Case-Control Study. *J. Diabetes Res.* **2015**. [CrossRef]

200. Acosta, E. Bioavailability of nanoparticles in nutrient and nutraceutical delivery. *Curr. Opin. Colloid Interface Sci.* **2009**, *14*, 3–15. [CrossRef]

201. Huang, Q.R.; Yu, H.L.; Ru, Q.M. Bioavailability and Delivery of Nutraceuticals Using Nanotechnology. *J. Food Sci.* **2010**, *75*, R50–R57. [CrossRef]

202. Ting, Y.W.; Jiang, Y.; Ho, C.T.; Huang, Q.R. Common delivery systems for enhancing in vivo bioavailability and biological efficacy of nutraceuticals. *J. Funct. Foods* **2014**, *7*, 112–128. [CrossRef]

203. Yao, M.F.; McClements, D.J.; Xiao, H. Improving oral bioavailability of nutraceuticals by engineered nanoparticle-based delivery systems. *Curr. Opin. Food Sci.* **2015**, *2*, 14–19. [CrossRef]

204. Anand, P.; Kunnumakkara, A.B.; Newman, R.A.; Aggarwal, B.B. Bioavailability of curcumin: Problems and promises. *Mol. Pharm.* **2007**, *4*, 807–818. [CrossRef]

205. Granata, G.; Paterniti, I.; Geraci, C.; Cunsolo, F.; Esposito, E.; Cordaro, M.; Blanco, A.R.; Cuzzocrea, S.; Consoli, G.M.L. Potential Eye Drop Based on a Calix[4]arene Nanoassembly for Curcumin Delivery: Enhanced Drug Solubility, Stability, and Anti Inflammatory Effect. *Mol. Pharm.* **2017**, *14*, 1610–1622. [CrossRef] [PubMed]

206. Davis, B.M.; Pahlitzsch, M.; Guo, L.; Balendra, S.; Shah, P.; Ravindran, N.; Malaguarnera, G.; Sisa, C.; Shamsher, E.; Hamze, H.; et al. Topical Curcumin Nanocarriers are Neuroprotective in Eye Disease. *Sci. Rep.* **2018**, *8*. [CrossRef] [PubMed]

207. Rotches-Ribalta, M.; Andres-Lacueva, C.; Estruch, R.; Escribano, E.; Urpi-Sarda, M. Pharmacokinetics of resveratrol metabolic profile in healthy humans after moderate consumption of red wine and grape extract tablets. *Pharmacol. Res.* **2012**, *66*, 375–382. [CrossRef] [PubMed]

208. Bonechi, C.; Martini, S.; Ciani, L.; Lamponi, S.; Rebmann, H.; Rossi, C.; Ristori, S. Using Liposomes as Carriers for Polyphenolic Compounds: The Case of Trans-Resveratrol. *PLoS ONE* **2012**, *7*, e41438. [CrossRef]

209. Jung, K.H.; Lee, J.H.; Park, J.W.; Quach, C.H.T.; Moon, S.H.; Cho, Y.S.; Lee, K.H. Resveratrol-loaded polymeric nanoparticles suppress glucose metabolism and tumor growth in vitro and in vivo. *Int. J. Pharm.* **2015**, *478*, 251–257. [CrossRef]

210. Summerlin, N.; Soo, E.; Thakur, S.; Qu, Z.; Jambhrunkar, S.; Popat, A. Resveratrol nanoformulations: Challenges and opportunities. *Int. J. Pharm.* **2015**, *479*, 282–290. [CrossRef] [PubMed]

211. Kapetanovic, I.M.; Muzzio, M.; Huang, Z.H.; Thompson, T.N.; McCormick, D.L. Pharmacokinetics, oral bioavailability, and metabolic profile of resveratrol and its dimethylether analog, pterostilbene, in rats. *Cancer Chemother. Pharmacol.* **2011**, *68*, 593–601. [CrossRef]

212. Granja, A.; Frias, I.; Neves, A.R.; Pinheiro, M.; Reis, S. Therapeutic Potential of Epigallocatechin Gallate Nanodelivery Systems. *Biomed Res. Int.* **2017**. [CrossRef] [PubMed]

213. Minnelli, C.; Moretti, P.; Fulgenzi, G.; Mariani, P.; Laudadio, E.; Armeni, T.; Galeazzi, R.; Mobbili, G. A Poloxamer-407 modified liposome encapsulating epigallocatechin-3-gallate in the presence of magnesium: Characterization and protective effect against oxidative damage. *Int. J. Pharm.* **2018**, *552*, 225–234. [CrossRef] [PubMed]

214. Jeon, S.; Li, Q.; Rubakhin, S.S.; Sweedler, J.V.; Smith, J.W.; Neuringer, M.; Kuchan, M.; Erdman, J.W., Jr. (13)C-lutein is differentially distributed in tissues of an adult female rhesus macaque following a single oral administration: A pilot study. *Nutr. Res.* **2019**, *61*, 102–108. [CrossRef] [PubMed]

215. Chittasupho, C.; Posritong, P.; Ariyawong, P. Stability, Cytotoxicity, and Retinal Pigment Epithelial Cell Binding of Hyaluronic Acid-Coated PLGA Nanoparticles Encapsulating Lutein. *AAPS Pharmscitech* **2018**, *20*. [CrossRef] [PubMed]

Review

Saffron (*Crocus sativus* L.) in Ocular Diseases: A Narrative Review of the Existing Evidence from Clinical Studies

Rebekka Heitmar [1,*], **James Brown** [1] **and Ioannis Kyrou** [2,3,4]

[1] Aston University, School of Life and Health Sciences, Aston Triangle, Birmingham B4 7ET, UK; j.e.p.brown@aston.ac.uk

[2] Aston Medical Research Institute, Aston Medical School, Aston University, Birmingham B4 7ET, UK; I.Kyrou@aston.ac.uk

[3] WISDEM, University Hospitals Coventry and Warwickshire NHS Trust, Coventry CV2 2DX, UK

[4] Translational & Experimental Medicine, Division of Biomedical Sciences, Warwick Medical School, University of Warwick, Coventry CV4 7AL, UK

* Correspondence: R.Heitmar1@aston.ac.uk

Received: 1 March 2019; Accepted: 13 March 2019; Published: 18 March 2019

Abstract: Saffron (*Crocus sativus* L.) and its main constituents, i.e., crocin and crocetin, are natural carotenoid compounds, which have been reported to possess a wide spectrum of properties and induce pleiotropic anti-inflammatory, anti-oxidative, and neuroprotective effects. An increasing number of experimental, animal, and human studies have investigated the effects and mechanistic pathways of these compounds in order to assess their potential therapeutic use in ocular diseases (e.g., in age related macular degeneration, glaucoma, and diabetic maculopathy). This narrative review presents the key findings of published clinical studies that examined the effects of saffron and/or its constituents in the context of ocular disease, as well as an overview of the proposed underlying mechanisms mediating these effects.

Keywords: saffron; *Crocus Sativus* L.; crocin; crocetin; supplements; anti-oxidant; anti-inflammatory; AMD; diabetes; glaucoma

---

1. Introduction

In addition to uncorrected refractive error, age related macular degeneration (AMD), glaucoma, cataract, and other retinal diseases (e.g., diabetic retinopathy and retinitis pigmentosa) are the major causes of blindness worldwide [1–4]. Among these ocular diseases, AMD is currently listed as the leading cause of irreversible vision loss in the developed world [2]. While the introduction of anti-VEGF (vascular endothelial growth factor) treatment has had a positive impact on preserving vision and slowing progression in AMD [5], there is still no cure to date. Furthermore, the increasing prevalence rates of obesity and related cardio-metabolic disease, including type 2 diabetes mellitus (T2DM) and cardiovascular disease (CVD) [6], adds significantly to the imposed health care burden and increases the treatment challenges. Indeed, increasing data link AMD to a number of lipid pathway genes, CVD phenotypes, excess body weight, and central/abdominal obesity [7–9]. Moreover, most of the aforementioned conditions are also associated with aging and exhibit overlapping pathophysiology with common mechanistic pathways, such as inflammation, oxidative stress, apoptosis, and neurodegeneration [6,7,10–13]. As these pathways may be affected by the effects of various nutritional supplements and botanical/herbal compounds (e.g., saffron and its constituents), increasing research interest is now focused on the potential therapeutic use of such natural products [11,14–16].

Saffron is mainly used in cooking as a colouring and flavouring spice that is comprised of the dried stigmas of the *Crocus sativus* L. flower, a stemless herb that belongs to the Iridaceae family [17,18]. Based on phytochemical studies, the pharmacologically active saffron constituents include bitter principles (e.g., picrocrocin), volatile agents (e.g., safranal, which can be obtained by picrocrocin hydrolysis), and dye materials (e.g., crocetin and its glycoside, i.e., crocin, which gives saffron its characteristic colour) [17–19]. In herbal medicine, saffron is traditionally used as a nerve sedative, stress-reliever, anti-depressant, aphrodisiac, expectorant, and anti-spasmodic agent [17,18,20,21]. Growing evidence from pharmacological studies has further shown that saffron or its active constituents may potentially exert neuroprotective, anti-convulsant, anti-depressive, anxiolytic, anti-oxidant, anti-inflammatory, hypolipidemic, anti-atherogenic, anti-hypertensive, and even anti-tumour effects [17,18,20,22,23]. Of note, a recent meta-analysis by Pourmasoumi et al. showed that saffron may be beneficial for several CVD-risk related outcomes (e.g., blood pressure, body weight, waist circumference, and fasting blood glucose levels), suggesting that saffron may have protective effects for multiple systemic conditions related to such CVD risk factors [24]. Furthermore, favourable results have been recently reported from both animal and clinical studies examining the effects of saffron and its constituents on CVD risk [22,24,25], endothelial function [26], inflammatory diseases [23,27], oxidative stress [26,28], and glycaemic factors [24,29], indicating that saffron may have promising potential as adjunct therapy in both systemic conditions and ocular diseases mainly through its anti-inflammatory and anti-oxidative effects [14,17,20,22].

This narrative review presents key findings of clinical studies that investigated the impact of saffron or one of its constituents on vision-related outcome measures, and an overview of the proposed mechanisms mediating these effects. As saffron is a natural product and, hence, saffron supplements are under less tight regulation regarding sale and dosage, this review is also intended to provide concise information on dosage and potential side effects based on the available literature.

2. Methods

Although not a systematic review, in order to ensure the quality and consistency of our approach, in the present narrative review we applied a predefined search strategy and followed, where relevant, the Preferred Reporting Items for Systematic Reviews (PRISMA) guidelines [30,31].

Search strategy: A systematic search of the following databases was conducted: PubMed, Scopus, Google Scholar, Cochrane library, and Web of Science. Studies published on these databases available up to the 31st of January 2019 were included for screening. Using Boolean operators (e.g., AND, OR), the applied search terms included combinations of the following key words: "saffron", "safranal", "crocetin", "crocin", "eye", "ocular", "retina", "diabetes", "macula", "glaucoma", "age related macular degeneration", "AMD" "anti-inflammatory", "anti-oxidant", "neuroprotective", "nutrition", and "supplement". Only publications in the English language were included. To minimise the risk of omitting relevant studies, the reference lists of all eligible papers were also manually checked.

Study selection: After elimination of duplicate records by one author (RH), two authors (RH and IK) independently reviewed the remaining publications to decide which were suitable for inclusion in this review. As a first step to eliminate unsuitable studies, all titles and abstracts of the publications identified through the performed database searches were screened, and studies that were clearly irrelevant were removed. Subsequently, the remaining papers were evaluated by reviewing the full text versions. Finally, all clinical studies in adults with ocular diseases which assessed the impact of saffron and/or its constituents on vision-related outcome measures, such as visual acuity, visual field parameters, contrast sensitivity, electrophysiology parameters (electroretinography (ERG), focal ERG (fERG), and multifocal ERG (mfERG)), macular thickness measures, and intraocular pressure (IOP), were included and reviewed in detail. Any discrepancies regarding inclusion of clinical studies were resolved by consensus and, if necessary, discussion with a third author (JB). Table 1 presents the PICOS (Population/Participants/Problem, Intervention, Comparators, Outcomes, Study Design) criteria that were followed in order to identify clinical studies for the present review.

Table 1. Predefined PICOS (Population/Participants/Problem, Intervention, Comparators, Outcomes, Study Design) criteria that were followed in order to identify and include clinical studies in the present review.

Parameters	Descriptions
Population/Participants/Problem	Adults with ocular disease
Intervention	Any intervention with oral administration of saffron or one of its constituents
Comparison	Studies with any comparator/control that incorporated a non-intervention group or studies with a pre- vs. post-intervention comparison without a comparator/control group
Outcomes	Vision-related outcome measures, such as visual acuity, visual field parameters, contrast sensitivity, electrophysiology parameters (ERG, fERG, mfERG), macular thickness measures, and IOP
Setting	Clinical studies/trials

ERG: electroretinography; fERG: focal electroretinography; mfERG: multifocal electroretinography; IOP: intraocular pressure.

3. Clinical Evidence Regarding the Impact of Oral Supplementation with Saffron or One of Its Constituents on Vision-Related Parameters in Adults with Ocular Diseases

Based on the results of the performed literature searches, there were eight published clinical studies that have assessed the impact of oral supplementation with saffron or one of its constituents on vision-related parameters in adults with ocular diseases [32–39]. Of these, six were randomized controlled trials (RCTs) [32,35–39], while two are longitudinal interventional clinical studies reporting on pre- (baseline) versus post-intervention comparisons without a comparator/control group [33,34]. For the two latter studies that are from the same group and report on different outcome measures, it is also not clear whether their design involves, at least partly, participants of the same cohort, and efforts to contact the corresponding authors in order to clarify this point were not successful [33,34]. The key characteristics and findings of these eight clinical studies are presented in the following sections and are summarized in Table 2.

Table 2. Clinical studies investigating the effects of oral supplementation of saffron or crocin on vision-related parameters in adults with ocular diseases.

Ocular Disease	Number of Subjects	Constituent Dosage	Study Design	Primary Outcome Measures—Findings	Proposed Mechanisms	Reference
AMD [bilateral early AMD]	N = 25	Saffron 20 mg daily	Double-blind, placebo controlled, cross over, RCT three-month period with cross over for another three months	fERG: increased amplitude in saffron, but not in placebo group BCVA: increased (one line) in saffron, but not in placebo group	Anti-oxidant Neuroprotective	Falsini et al. (2010) [32]
AMD [bilateral early AMD]	N = 29	Saffron 20 mg daily	Longitudinal interventional open-label study three monthly follow-ups over a 15-month period	fERG: increased amplitude that stabilized after three months BCVA: increased (two lines) that stabilized after three months	Anti-oxidant Neuroprotective	Piccardi et al. (2012) [33]
AMD [bilateral early AMD]	N = 33	Saffron 20 mg daily	Longitudinal, 3 monthly follow-ups over a 12-month period	fERG: increased amplitude and sensitivity amplitude that stabilized after three months independent of genotype	Anti-oxidant Anti-inflammatory Neuroprotective	Marangoni et al. (2013) [34]
AMD [dry and wet AMD]	N = 40	Saffron 15 mg twice daily	Placebo controlled, RCT six-month period with follow-ups at three and six months	CMT: decreased in saffron and placebo groups in wet AMD, but not in dry AMD ERG: amplitude increased in the saffron group (dry and wet AMD) compared to placebo after three months, but not six months	Neuroprotective Anti-depressant	Lashay et al. (2016) [35]
AMD [mild/moderate dry AMD]	N = 54	Saffron 50 mg daily	Placebo controlled, RCT three months	CMT: unchanged BCVA: increased (one line) in saffron, but not in placebo group CS: increased in saffron, but not in placebo group	Anti-oxidant Hemorheological activity	Riazi et al. (2017) [36]
AMD [mild/moderate AMD]	N = 96	Saffron 20 mg daily	Double-blind, placebo controlled, cross over, RCT three months followed by cross over into the other arm for three months	BCVA: increased in saffron group [and AREDS * + saffron], but not in placebo mfERG response density: increased in AREDS+saffron, but not in the saffron or placebo group mfERG latency: decreased in saffron group, but not in placebo group	Anti-oxidant Neuroprotective	Broadhead et al. (2019) [37]

Nutrients **2019**, *11*, 649

Table 2. *Cont.*

Ocular Disease	Number of Subjects	Constituent Dosage	Study Design	Primary Outcome Measures—Findings	Proposed Mechanisms	Reference
POAG [clinically stable POAG]	*N* = 34	Saffron 30 mg daily	Double-blind, placebo controlled RCT 1 month duration 1 month wash-out	IOP: reduction after three and four weeks compared to placebo IOP returned to pre-intervention levels after a 4-week wash out period	Antioxidant Neuroprotective	Bonyadi et al. (2014) [38]
DME	*N* = 60 (101 DME eyes)	Crocin 5 mg or 15 mg daily	Double-masked, placebo controlled, phase 2 RCT 3 months	CMT: significantly decreased after three months compared to placebo only in the 15 mg group BCVA: significantly improved after three months compared to placebo only in the 15 mg group HbA1c and FBG: significantly decreased after three months compared to placebo only in the 15 mg group	Anti-oxidant Hemorheological activity Anti-inflammatory	Sepahi et al. (2018) [39]

*: participants were requested to continue on any supplements (including AREDS-based therapies) they had been taking prior to this study. AMD: age related macular degeneration; POAG: primary open angle glaucoma; DME: diabetic macular edema; RCT: randomized clinical trial; BCVA: best corrected visual acuity; AREDS: Age-Related Eye Disease Study; CMT: central macular thickness; CS: contrast sensitivity; ERG: electroretinography; fERG: focal electroretinography; mfERG: multifocal electroretinography; IOP: intraocular pressure; HbA1c: haemoglobin A1c; FBG: fasting blood glucose.

Age related macular degeneration: six published clinical studies assessed vision-related parameters in AMD patients under oral saffron supplementation (Table 2) [32–37]. Based on these, both objective (ERG) and subjective measures (Snellen, LogMar, EDTRS charts) of visual acuity were shown to significantly improve with all tested dosages of saffron (daily dose range: 20–50 mg), even after short-term oral supplementation (e.g., three months) [32–37]. Central macular thickness (CMT) [35] and contrast sensitivity (CS) [36] were also assessed in studies (Table 2), with the latter reportedly increasing with saffron supplementation [36], whilst the former was shown to decrease only in wet, but not dry AMD following saffron supplementation [35]. As the formulation, dosage, intervention duration, test methods, and outcome measures varied across these six studies, a direct quantitative comparison was not possible. Of note, longer-term data are currently available only from the two non-RCT studies with oral saffron supplementation (20 mg daily) in patients with bilateral early AMD over 12 and 15 months, respectively (Table 2) [33,34]. In these two longer studies, the noted improvements were achieved within three months of oral saffron supplementation, after which the gained functionality seemed to plateau [33,34].

Glaucoma: the only existing clinical study in patients with primary open angle glaucoma (POAG) reported that oral saffron supplementation (30 mg daily for one month) can significantly reduce the IOP after three weeks (Table 2) [38]. In this pilot double-blind, placebo controlled RCT, all participants had stable POAG (verified by visual field and optic nerve head examinations) and were treated with topical timolol 0.5% twice daily and dorzolamide 2% three times daily. Compared to placebo, the addition of daily oral saffron supplementation for one month resulted in significantly decreased IOP after the third week, but this effect reverted after a 4-week wash-out period [38].

Diabetic maculopathy: despite the alarmingly increasing T2DM prevalence, only one eligible RCT was identified in patients with refractory diabetic maculopathy; it examined the effects of oral crocin supplementation (5 mg or 15 mg daily) (Table 2) [39]. In this study, Sepahi et al. reported a decrease in CMT that appeared dose dependent, showing significant reduction after three months of daily oral crocin supplementation compared to placebo in the higher (15 mg), but not in the lower (5 mg) dosage group [39]. Visual function, as measured by best corrected visual acuity (BCVA), was also significantly improved only in the higher dosage group compared to placebo. Moreover, in addition to the ocular findings, oral crocin supplementation significantly reduced HbA1c and fasting blood glucose levels only with the higher dose. Interestingly, while the lower crocin dose did not induce significant reductions compared to placebo, the study authors noted that this dose could clinically improve CMT, BCVA, HbA1c, and fasting glycaemia [39].

4. Action Time-Course of Oral Saffron Supplementation

To date, there are limited data regarding the time-course of the effects of oral saffron supplementation on visual-related parameters in patients with ocular diseases. From most of the aforementioned published clinical studies (Table 2) [32–39], it appears that saturation of the saffron-induced effect(s) is reached within a three-month period of oral supplementation. It is plausible that this saturation can be reached earlier; however, this remains uncertain, since most of these studies included testing at three-month intervals. Moreover, it is noteworthy that in a study utilizing a light damage model of photoreceptor degeneration in rats, the protection noted with the applied daily dose of saffron (1 mg/kg) was detectable at two days, increasing to 10 days [40].

5. Safety Profile of Oral Saffron Supplementation

A relatively limited number of studies have examined the toxicity of saffron [18,22,41–43]. According to *in vivo* studies, saffron has very low toxicity for doses of up to 1.5 gr per day, while toxic effects have been documented for daily doses ≥ 5 gr, with a lethal dose of approximately 20 gr [44]. The side effects of saffron are partly attributed to its dye/coloured constituents, which can accumulate in the skin, mucosas, and/or sclera mimicking icteric symptoms [44].

Taking into account that toxic effects of saffron appear to require doses of grams per day, clinical data from studies in healthy volunteers suggest that saffron and crocin in doses of milligrams per day can be considered safe [45,46]. Indeed, a randomized, double-blind, placebo-controlled study examining the safety of oral crocin supplementation (20 mg daily compared to placebo) in healthy volunteers concluded that this was relatively safe within the one-month study period [45]. In this short-term study, there were no major adverse events and no changes in the studied haematological, biochemical, hormonal, and urinary parameters, except for decreasing amylase levels, mixed white blood cells, and partial thromboplastin time after one month [45]. Saffron doses of 200 mg and 400 mg for seven days were also shown to be relatively safe in another double-blind, placebo-controlled study in healthy volunteers, with changes in haematological and biochemical parameters that were not considered clinically important [46]. Notably, in this study, one female participant in each of the two saffron groups exhibited abnormal uterine bleeding [46].

Furthermore, systematic review data from RCTs examining the effectiveness of saffron on behavioural and psychological outcomes also support the safety of saffron supplementation, since in all these studies there were no significant differences in adverse events between saffron and placebo groups [47]. Nausea, sedation, appetite fluctuation, and headache were among the most common adverse effects reported in these RCTs, which used saffron daily doses ranging from 20 mg to 400 mg (supplements containing additional compounds with putative synergistic effects were also used in some of these RCTs) [47]. Similarly, a good safety profile has been reported in the available clinical studies in patients with ocular conditions, most of whom had regular follow-up that also included telephone calls and interviews to ensure compliance and monitor possible side effects [32–39]. Indeed, these studies concluded that the applied dose regimens of saffron and crocin were safe without significantly increased safety risk [32–39]. As such, the available clinical evidence supports the safety of oral saffron supplementation; however, longer-term studies are still required in order to comprehensively evaluate the long term safety profile of saffron and its constituents in various conditions.

Overall, caution is currently advised regarding saffron supplementation in patients with renal insufficiency or bleeding disorders, and in those on anti-coagulation treatment due to potential inhibition of platelet aggregation [21,41]. Finally, saffron supplementation has also potential emmenagogue and abortifacient effects [48]. In fact, saffron has been reportedly used for induction of abortion (doses >10 gr) [41,44], with amounts higher than those used in foods (>5 gr) having uterine stimulant effects; thus, it should be avoided during pregnancy [21].

6. Proposed Mechanisms/Pathways Mediating the Effects of Saffron and/or Its Constituents in Ocular Diseases

Taking into account the underlying pathophysiology of AMD, diabetic retinopathy, and glaucoma, improvements in the overall pro-inflammatory and oxidative stress load, as well as in the systemic vascular/endothelial health, have the potential to also positively impact on ocular-related parameters. Therefore, given the favourable effects of saffron and/or its constituents on multiple CVD-risk factors and inflammation [23,24,27,28], it is highly plausible that the same underlying anti-oxidant, anti-apoptotic, and anti-inflammatory mechanisms/pathways are also facilitating the noted positive effects of oral saffron supplementation in ocular diseases (Figure 1). For example, reduced circulating levels of glutathione (GSH; a key radical scavenger) have been shown in patients with POAG, leading to increased oxidative stress [49]. Accordingly, a possible mechanistic pathway mediating the effects of saffron and its constituents (crocin and crocetin; both anti-oxidant carotenoids) is considered to involve increased GSH levels that protect against reactive oxygen species and apoptosis [23]. Similarly, crocin and crocetin may also suppress the activation of pro-inflammatory pathways (e.g., the nuclear factor kappa-B pathway) [23,50–52]. Furthermore, these two saffron constituents appear to enhance oxygen diffusion [53], and improve ocular blood flow [54], factors that play an important role in diseases such as AMD. Of note, pathways targeting not only oxidative stress and inflammation, but also endothelial function are also particularly relevant to glaucoma treatment and may mediate the effects of saffron and/or its constituents in patients with glaucoma. Indeed, the trabecular meshwork, which constitutes the target tissue of glaucoma in the anterior chamber, is comprised by endothelial-like cells, and is sensitive to oxidative damage, since it lacks effective antioxidant mechanisms [55,56]. As such, the trabecular meshwork is altered in most types of glaucoma, consequently increasing IOP [56], and may be a target for the effects of saffron and/or its constituents in patients with glaucoma; however, further research is required to clarify these potential mechanisms/effects. Finally, saffron and its constituents have been shown to improve glycaemia by enhancing insulin sensitivity and preventing pancreatic beta-cell failure [39,57], which in turn may improve diabetic retinopathy/maculopathy [39].

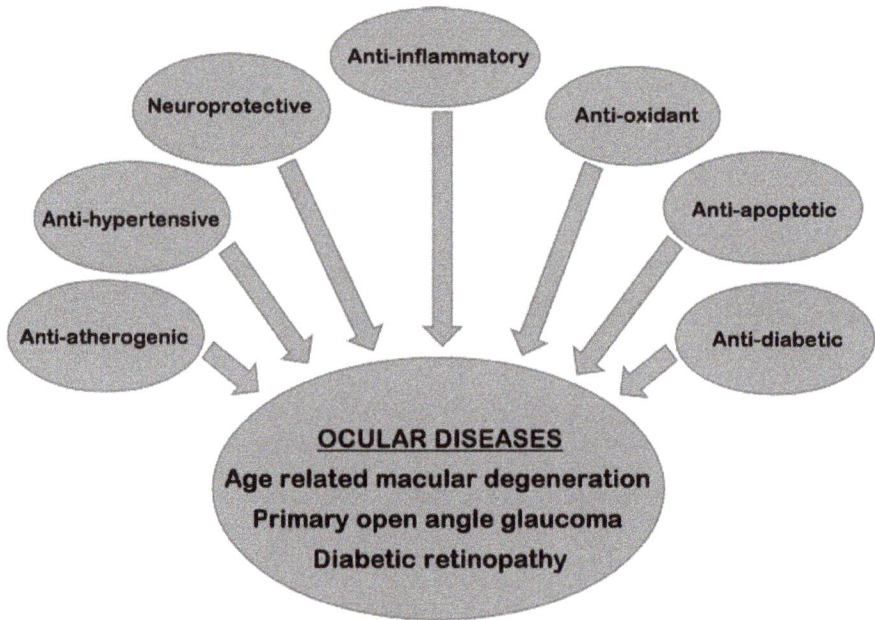

Figure 1. Simplified schematic representation of the potential mechanisms that may mediate the effects of saffron and/or its constituents (e.g., crocin) in various ocular diseases (e.g., age related macular degeneration, primary open angle glaucoma, and diabetic retinopathy), including anti-inflammatory, anti-oxidant, anti-apoptotic, neuroprotective, antidiabetic, anti-atherogenic, and anti-hypertensive effects.

Moreover, a recent study by Corso et al. demonstrated that saffron can decrease ATP-induced retinal cytotoxicity by targeting the P2X7 receptors [58]. As these receptors are found in both inner and outer retinal cells, which are affected in neurodegenerative disorders such as AMD, this suggests another potential mechanistic pathway via which saffron may protect against neurodegeneration [58]. This is also supported by data showing that saffron may protect photoreceptors against retinal stress, maintaining both function and morphology in a mammalian retinal model after exposure to damaging bright continuous light [59]. Laabich et al. have also shown that crocin protects retinal photoreceptors against light-induced cell death in primary retinal cell cultures [60].

It should be also noted that the pharmacokinetic details regarding whether and which saffron constituents/metabolites reach the various tissues after oral intake of saffron or one of its constituents are not fully clarified yet. Therefore, the exact mechanisms by which saffron and/or its constituents act on the retina (particularly, on the photoreceptors and bipolar cells) remain to be further elucidated. Indeed, differences in uptake and tissue distribution may further account for different underlying pathways/mechanisms which can mediate the observed effects of oral saffron supplementation in patients with ocular disease. For example, lutein is considered to exert protective effects against AMD through its anti-oxidative properties and its accumulation in the macula acting as a blue light filter [61]. On the other hand, saffron and its constituents may improve ocular function more indirectly by improving the aforementioned systemic CVD-risk factors through anti-inflammatory and anti-oxidant effects [22,23,26–28,62].

Pharmacokinetic studies in rats have shown that after oral administration, crocin is largely excreted from the gastro-intestinal tract, which also serves as an important site for crocin hydrolysis [63]. As such, orally administered crocin is hydrolysed to crocetin before intestinal absorption [41,64]. Furthermore, crocetin has been shown to be quickly absorbed after oral administration in mice, with a short plasma half-life, resulting in rapid elimination without accumulation in the body [41]. Interestingly, the progress of saffron supplementation has been examined both in an animal model with light-induced retinal degeneration and AMD patients [42]. Indeed, experiments in albino rats under saffron treatment showed that crocetin was detectable in the collected blood samples [42]. However, there were no traces of saffron-related metabolites in any other examined tissue, with the exception of degenerating retinas where modest amounts of crocins were found in 7 of the 15 tested animals [42]. These findings suggest that crocins may be resynthesized from circulating crocetin, which can reach the retina following blood-brain barrier damage, as also supported by the fact that no crocin or crocetin metabolites were found in healthy retinas or other parts of the central nervous system [42]. In addition, based on analyses of blood and urine samples (collected within two hours after the morning saffron dose) from two AMD patients under saffron supplementation for over a year and three healthy volunteers after two weeks under the same saffron dose (20 mg daily), crocetin was consistently quantified in both blood and urine samples from the healthy volunteers, but not in those from the AMD patients, suggesting that these metabolites are immediately absorbed and used [42]. In line with this, data from an open-label, single dose escalation study (single crocetin dose of 7.5, 15, and 22.5 mg in one-week intervals) have also indicated that crocetin is absorbed more quickly than other carotenoids (e.g., lutein) in healthy adults [65].

Finally, an increasing body of RCTs indicate that saffron may improve the effects and symptoms of depression [47,66–68]. As depression may often affect patients with chronic diseases (e.g., with AMD, diabetic retinopathy, or glaucoma) [35,69,70] in whom it may also reduce treatment adherence (e.g., potentially lower adherence to glaucoma treatment, and, hence increase the glaucoma progression risk) [70], it seems plausible to hypothesize that some of the benefits of oral saffron supplementation in patients with ocular diseases may also have a component relating to the potential anti-depressive effects of saffron and/or its constituents [35]. However, currently there is a paucity of clinical data addressing this hypothesis that requires testing in the context of well-designed RCTs.

Overall, the majority of the data regarding the potential mechanisms/pathways mediating the effects of saffron and/or its constituents in ocular diseases originate from either *in vitro* experiments on human cell lines or animal models. These experiments/models allow the study of effects and mechanistic pathways that may be otherwise inaccessible, but have obvious limitations regarding their applicability to humans due to their nature (e.g., models involving light- or drug-induced ocular damage), which also cannot fully account for the multifactorial and progressive nature of most ocular diseases in humans. In addition, findings documented in clinical studies with saffron or its constituents that relate to subjective measures of vision and quality of life indices cannot be replicated and further tested in animal models. Thus, it becomes evident that translating and clarifying the evidence regarding the underlying mechanisms/pathways induced by saffron and/or its constituents in ocular diseases in humans requires additional data from both pre-clinical and clinical studies.

7. Summary of the Literature

The existing clinical evidence suggests that oral supplementation with saffron or crocin may have positive effects on various vision-related parameters in adults with AMD, POAG, and diabetic maculopathy (Table 2) [32–39]. Moreover, the findings from these clinical studies support the good safety profile of oral supplementation with saffron (range of tested daily doses: 20 to 50 mg) and crocin (5 mg and 15 mg daily) in these patients, but long term safety data are still scarce. As such, it is not possible to draw firm conclusions for evidence-based recommendations regarding oral saffron supplementation from the clinical studies conducted so far, since the existing data are considered rather limited, particularly regarding long term outcomes and for conditions other than AMD. In addition, further research is also required to clarify the exact underlying mechanisms that mediate the noted positive outcomes of oral saffron supplementation in ocular diseases. Based on the available data from *in vitro*, animal, and human studies, it is currently considered that pleiotropic effects of saffron and/or its constituents relating to anti-inflammatory and anti-oxidant pathways, as well as to improvements in oxygen diffusion and ocular blood flow are likely to facilitate the documented benefits of saffron supplementation in ocular diseases (Figure 1) [22,23,26–28,53,54,62].

8. Conclusions

Saffron supplementation appears to have promising potential as an effective and safe adjunct therapy in certain ocular diseases [32–39,61]. It is important to highlight that nutritional/dietary supplements (i.e., concentrated sources of vitamins and minerals and/or other substances with a nutritional or physiological effect that are marketed in "dose" form, such as tablets, capsules, pills, or liquids in measured doses) have to comply to certain national/international regulations [71,72]. Indeed, although such supplements are regulated as foods, these are subject to different regulations compared to other foods and from drugs in order to protect consumers against potential health risks from these products and the general public against potentially misleading information [72]. However, such regulations vary from country to country, despite certain shared existing regulatory frameworks (e.g., the European Directive on Food Supplements) [71,72]. Particularly for saffron, it is also noteworthy that its content may vary depending on the source [61]. In this context, it becomes evident that there is still an unmet need for additional bioavailability and pharmacokinetic studies, as well as well-designed, adequately powered and long term RCTs in order to form evidence-based recommendations for the potential therapeutic role of oral saffron supplementation in ocular diseases.

Author Contributions: R.H. and I.K. contributed equally. Conceptualization, R.H. and I.K.; Writing-Original Draft Preparation, R.H. and I.K.; Writing-review and editing, R.H., J.B., and I.K.; All authors reviewed, edited and approved the final version of the manuscript.

Funding: This research received no external funding.

Conflicts of Interest: The authors declare no conflict of interest.

References

1. Bourne, R.R.; Stevens, G.A.; White, R.A.; Smith, J.L.; Flaxman, S.R.; Price, H.; Jonas, J.B.; Keeffe, J.; Leasher, J.; Naidoo, K.; et al. Vision Loss Expert Group. Causes of vision loss worldwide, 1990–2010: A systematic analysis. *Lancet Glob. Health* **2013**, *1*, e339–e349. [CrossRef]

2. Wong, T.Y.; Su, X.; Li, X.; Cheung, C.M.; Klein, R.; Cheng, C.Y.; Wong, T.Y. Global prevalence of age-related macular degeneration and disease burden projection for 2020 and 2040: A systematic review and meta-analysis. *Lancet Glob. Health* **2014**, *2*, e106–e116. [CrossRef]

3. Jonas, J.B.; Aung, T.; Bourne, R.R.; Bron, A.M.; Ritch, R.; Panda-Jonas, S. Glaucoma. *Lancet* **2017**, *390*, 2183–2193. [CrossRef]

4. Wong, T.Y.; Cheung, C.M.; Larsen, M.; Sharma, S.; Simo, R. Diabetic retinopathy. *Nat. Rev. Dis. Primers* **2016**, *17*, 16012. [CrossRef]

5. Hernández-Zimbrón, L.F.; Zamora-Alvarado, R.; Velez-Montoya, R.; Zenteno, E.; Gulias-Cañizo, R.; Quiroz-Mercado, H.; Gonzalez-Salinas, R. Age-Related Macular Degeneration: New Paradigms for Treatment and Management of AMD. *Oxid. Med. Cell. Longev.* **2018**, *2018*, 8374647. [CrossRef] [PubMed]

6. Kyrou, I.; Randeva, H.S.; Tsigos, C.; Kaltsas, G.; Weickert, M.O. Clinical Problems Caused by Obesity. In *Endotext [Internet]*; Feingold, K.R., Anawalt, B., Boyce, A., Chrousos, G., Dungan, K., Grossman, A., Hershman, J.M., Kaltsas, G., Koch, C., Kopp, P., et al., Eds.; MDText.com, Inc.: South Dartmouth, MA, USA, 2018. Available online: https://www.ncbi.nlm.nih.gov/books/NBK278973/ (accessed on 28 February 2019).

7. Pennington, K.L.; DeAngelis, M.M. Epidemiology of age-related macular degeneration (AMD): Associations with cardiovascular disease phenotypes and lipid factors. *Eye Vis. (Lond.)* **2016**, *3*, 34. [CrossRef]

8. Zhang, Q.Y.; Tie, L.J.; Wu, S.S.; Lv, P.L.; Huang, H.W.; Wang, W.Q.; Wang, H.; Ma, L. Overweight, Obesity, and Risk of Age-Related Macular Degeneration. *Investig. Ophthalmol. Vis. Sci.* **2016**, *57*, 1276–1283. [CrossRef]

9. Adams, M.K.; Simpson, J.A.; Aung, K.Z.; Makeyeva, G.A.; Giles, G.G.; English, D.R.; Hopper, J.; Guymer, R.H.; Baird, P.N.; Robman, L.D. Abdominal obesity and age-related macular degeneration. *Am. J. Epidemiol.* **2011**, *173*, 1246–1255. [CrossRef]

10. Chen, M.; Luo, C.; Zhao, J.; Devarajan, G.; Xu, H. Immune regulation in the aging retina. *Prog. Retin. Eye Res.* **2018**. [CrossRef]

11. Pinazo-Durán, M.D.; Gallego-Pinazo, R.; García-Medina, J.J.; Zanón-Moreno, V.; Nucci, C.; Dolz-Marco, R.; Martínez-Castillo, S.; Galbis-Estrada, C.; Marco-Ramírez, C.; López-Gálvez, M.I.; et al. Oxidative stress and its downstream signaling in aging eyes. *Clin. Interv. Aging* **2014**, *9*, 637–652. [CrossRef] [PubMed]

12. Chen, M.; Xu, H. Parainflammation, chronic inflammation, and age-related macular degeneration. *J. Leukoc. Biol.* **2015**, *98*, 713–725. [CrossRef]

13. Ehrlich, R.; Harris, A.; Kheradiya, N.S.; Winston, D.M.; Ciulla, T.A.; Wirostko, B. Age-related macular degeneration and the aging eye. *Clin. Interv. Aging* **2008**, *3*, 473–482. [PubMed]

14. Broadhead, G.K.; Grigg, J.R.; Chang, A.A.; McCluskey, P. Dietary modification and supplementation for the treatment of age-related macular degeneration. *Nutr. Rev.* **2015**, *73*, 448–462. [CrossRef] [PubMed]

15. Chew, E.Y. Nutrition effects on ocular diseases in the aging eye. *Investig. Ophthalmol. Vis. Sci.* **2013**, *54*, ORSF42–ORSF47. [CrossRef] [PubMed]

16. Huynh, T.P.; Mann, S.N.; Mandal, N.A. Botanical compounds: Effects on major eye diseases. *Evid.-Based Complement. Altern. Med.* **2013**, *2013*, 549174. [CrossRef] [PubMed]

17. José Bagur, M.; Alonso Salinas, G.L.; Jiménez-Monreal, A.M.; Chaouqi, S.; Llorens, S.; Martínez-Tomé, M.; Alonso, G.L. Saffron: An Old Medicinal Plant and a Potential Novel Functional Food. *Molecules* **2017**, *23*, 30. [CrossRef]

18. Christodoulou, E.; Kadoglou, N.P.; Kostomitsopoulos, N.; Valsami, G. Saffron: A natural product with potential pharmaceutical applications. *J. Pharm. Pharmacol.* **2015**, *67*, 1634–1649. [CrossRef] [PubMed]

19. Ríos, J.L.; Recio, M.C.; Giner, R.M.; Máñez, S. An Update Review of Saffron and its Active Constituents. *Phytother. Res.* **1996**, *10*, 189–193. [CrossRef]

20. Hosseini, A.; Razavi, B.M.; Hosseinzadeh, H. Saffron (Crocus sativus) petal as a new pharmacological target: A review. *Iran. J. Basic Med. Sci.* **2018**, *21*, 1091–1099. [PubMed]

21. WHO. *Monographs on Selected Medicinal Plants*; World Health Organization: Geneva, Switzerland, 2007; Volume 3. Available online: http://apps.who.int/medicinedocs/en/m/abstract/Js14213e/ (accessed on 28 February 2019).

22. Ghaffari, S.; Roshanravan, N. Saffron; An updated review on biological properties with special focus on cardiovascular effects. *Biomed. Pharmacother.* **2019**, *109*, 21–27. [CrossRef]

23. Poma, A.; Fontecchio, G.; Carlucci, G.; Chichiricò, G. Anti-inflammatory properties of drugs from saffron crocus. *Antiinflamm. Anti-Allergy Agents Med. Chem.* **2012**, *11*, 37–51. [CrossRef]

24. Pourmasoumi, M.; Hadi, A.; Najafgholizadeh, A.; Kafeshani, M.; Sahebkar, A. Clinical evidence on the effects of saffron (crocus sativus L.) on cardiovascular risk factors: A systematic review meta-analysis. *Pharmacol. Res.* **2019**, *139*, 348–359. [CrossRef] [PubMed]

25. Broadhead, G.K.; Chang, A.; Grigg, J.; McCluskey, P. Efficacy and Safety of Saffron Supplementation: Current Clinical Findings. *Crit. Rev. Food Sci. Nutr.* **2016**, *56*, 2767–2776. [CrossRef]

26. Rahiman, N.; Akaberi, M.; Sahebkar, A.; Emami, S.A.; Tayarani-Najaran, Z. Protective effects of saffron and its active components against oxidative stress and apoptosis in endothelial cells. *Microvasc. Res.* **2018**, *118*, 82–89. [CrossRef]

27. Pashirzad, M.; Shafiee, M.; Avan, A.; Ryzhikov, M.; Fiuji, H.; Bahreyni, A.; Khazaei, M.; Soleimanpour, S.; Hassanian, S.M. Therapeutic potency of crocin in the treatment of inflammatory diseases: Current status and perspective. *J. Cell. Physiol.* **2019**, 1–11. [CrossRef] [PubMed]

28. Abou-Hany, H.O.; Atef, H.; Said, E.; Elkashef, H.A.; Salem, H.A. Crocin mediated amelioration of oxidative burden and inflammatory cascade suppresses diabetic nephropathy progression in diabetic rats. *Chem. Biol. Interact.* **2018**, *284*, 90–100. [CrossRef] [PubMed]

29. Razavi, B.M.; Hosseinzadeh, H. Saffron: A promising natural medicine in the treatment of metabolic syndrome. *J. Sci. Food Agric.* **2017**, *97*, 1679–1685. [CrossRef]

30. Moher, D.; Liberati, A.; Tetzlaff, J.; Altman, D.G.; PRISMA Group. Preferred reporting items for systematic reviews and meta-analyses: The PRISMA statement. *PLoS Med.* **2009**, *6*, e1000097. [CrossRef] [PubMed]

31. Liberati, A.; Altman, D.G.; Tetzlaff, J.; Mulrow, C.; Gøtzsche, P.C.; Ioannidis, J.P.; Clarke, M.; Devereaux, P.J.; Kleijnen, J.; Moher, D. The PRISMA statement for reporting systematic reviews and meta-analyses of studies that evaluate health care interventions: Explanation and elaboration. *PLoS Med.* **2009**, *6*, e1000100. [CrossRef] [PubMed]

32. Falsini, B.; Piccardi, M.; Minnella, A.; Savastano, C.; Capoluongo, E.; Fadda, A.; Balestrazzi, E.; Maccarone, R.; Bisti, S. Influence of saffron supplementation on retinal flicker sensitivity in early age-related macular degeneration. *Investig. Ophthalmol. Vis. Sci.* **2010**, *51*, 6118–6124. [CrossRef]

33. Piccardi, M.; Marangoni, D.; Minnella, A.M.; Savastano, M.C.; Valentini, P.; Ambrosio, L.; Capoluongo, E.; Maccarone, R.; Bisti, S.; Falsini, B. A Longitudinal follow-up study of saffron supplementation in early age-related macular degeneration: Sustained benefits to central retinal function. *Evid.-Based Complement. Altern. Med.* **2012**, *2012*, 429124. [CrossRef] [PubMed]

34. Marangoni, D.; Falsini, B.; Piccardi, M.; Ambrosio, L.; Minnella, A.M.; Savastano, M.C.; Bisti, S.; Maccarone, R.; Fadda, A.; Mello, E.; et al. Functional effect of saffron supplementation and risk genotypes in early age-related macular degeneration: A preliminary report. *J. Transl. Med.* **2013**, *11*, 228. [CrossRef] [PubMed]

35. Lashay, A.; Sadough, G.; Ashrafi, E.; Lashay, M.; Movassat, M.; Akhondzadeh, S. Short-term outcomes of saffron supplementation in patients with age-related macular degeneration: A double-blind, Placebo-controlled, randomized trial. *Med. Hypothesis Discov. Innov. Ophthalmol.* **2016**, *5*, 32. [PubMed]

36. Riazi, A.; Panahi, Y.; Alishiri, A.A.; Hosseini, M.A.; Zarchi, A.A.K.; Sahebkar, A. The impact of saffron (crocus sativus) supplementation on visual function in patients with dry age-related macular degeneration. *Ital. J. Med.* **2017**, *11*, 758. [CrossRef]

37. Broadhead, G.K.; Grigg, J.R.; McCluskey, P.; Hong, T.; Schlub, T.E.; Chang, A.A. Saffron Therapy for the treatment of mild/ moderate age-related macular degeneration: A randomized clinical trial. *Graefe's Arch. Clin. Exp. Ophthalmol.* **2019**, *257*, 31–40. [CrossRef] [PubMed]

38. Bonyadi, M.H.J.; Yazdani, S.; Saadat, S. The ocular hypotensive effect of saffron extractin primary open angle glaucoma: A pilot study. *BMC Complement. Altern. Med.* **2014**, *14*, 399.

39. Sepahi, S.; Mohajeri, S.A.; Hosseini, S.M.; Khodaverdi, E.; Shoeibi, N.; Namdari, M.; Tabassi, S.A.S. Effects of crocin on diabetic maculopathy: A placebo-controlled randomized clinical trial. *Am. J. Ophthalmol.* **2018**, *190*, 89–98. [CrossRef]

40. Di Marco, F.; Romeo, S.; Nandasena, C.; Purushothuan, S.; Adams, C.; Bisti, S.; Stone, J. The time course of action of two neuroprotectants, dietary saffron and photobiomodulation, assessed in the rat retina. *Am. J. Neurodegener. Dis.* **2013**, *2*, 208–220.

41. Moshiri, M.; Vahabzadeh, M.; Hosseinzadeh, H. Clinical applications of saffron (crocus sativus) and its constituents: A review. *Drug Res.* **2015**, *65*, 287–295. [CrossRef]

42. Bisti, S.; Maccarone, R.; Falsini, B. Saffron and retina: Neuroprotection and pharmacokinetics. *Vis. Neurosci.* **2014**, *31*, 355–361. [CrossRef]

43. Alavizadeh, S.H.; Hosseinzadeh, H. Bioactivity assessment and toxicity of crocin: A comprehensive review. *Food Chem. Toxicol.* **2014**, *64*, 65–80. [CrossRef] [PubMed]

44. Schmidt, M.; Betti, G.; Hensel, A. Saffron in phytotherapy: Pharmacology and clinical uses. *Wien. Med. Wochenschr.* **2007**, *157*, 315–319. [CrossRef] [PubMed]

45. Mohamadpour, A.H.; Ayati, Z.; Parizadeh, M.R.; Rajbai, O.; Hosseinzadeh, H. Safety Evaluation of Crocin (a constituent of saffron) Tablets in Healthy Volunteers. *Iran. J. Basic Med. Sci.* **2013**, *16*, 39–46. [PubMed]

46. Modaghegh, M.H.; Shahabian, M.; Esmaeili, H.A.; Rajbai, O.; Hosseinzadeh, H. Safety evaluation of saffron (Crocus sativus) tablets in healthy volunteers. *Phytomedicine* **2008**, *15*, 1032–1037. [CrossRef] [PubMed]

47. Hausenblas, H.A.; Heekin, K.; Mutchie, H.L.; Anton, S. A systematic review of randomized controlled trials examining the effectiveness of saffron (Crocus sativus L.) on psychological and behavioral outcomes. *J. Integr. Med.* **2015**, *13*, 231–240. [CrossRef]

48. Ernst, E. Herbal medicinal products during pregnancy: Are they safe? *BJOG* **2002**, *109*, 227–235. [CrossRef] [PubMed]

49. Gherghel, D.; Griffiths, H.R.; Hilton, E.J.; Cunliffe, I.A.; Hosking, S.L. Systemic reduction in glutathione levels occurs in patients with primary open-angle glaucoma. *Investig. Ophthalmol. Vis. Sci.* **2005**, *46*, 877–883. [CrossRef] [PubMed]

50. Sung, Y.Y.; Kim, H.K. Crocin ameliorates atopic dermatitis symptoms by down regulation of Th2 response via blocking of NF-κB/STAT6 signaling pathways in mice. *Nutrients* **2018**, *10*, 1625. [CrossRef]

51. Lv, B.; Huo, F.; Zhu, Z.; Xu, Z.; Dang, X.; Chen, T.; Zhang, T.; Yang, X. Crocin Upregulates CX3CR1 expression by suppressing NF-κB/YY1 signaling and inhibiting lipopolysaccharide-induced microglial activation. *Neurochem. Res.* **2016**, *41*, 1949–1957. [CrossRef] [PubMed]

52. Li, Y.; Kakkar, R.; Wang, J. In vivo and in vitro approach to anti-arthritic and anti-inflammatory effect of crocetin by alteration of nuclear factor-E2-related factor 2/hem oxygenase (HO)-1 and NF-κB expression. *Front. Pharmacol.* **2018**, *9*, 1341. [CrossRef]

53. Giaccio, M. Crocetin from saffron: An active component of an ancient spice. *Crit. Rev. Food Sci. Nutr.* **2004**, *44*, 155–172. [CrossRef]

54. Xuan, B.; Zhou, Y.H.; Li, N.; Min, Z.D.; Chiou, G.C. Effects of crocin analogs on ocular blood flow and retinal function. *J. Ocul. Pharmacol. Ther.* **1999**, *15*, 143–152. [CrossRef]

55. Izzotti, A.; Saccà, S.C.; Longobardi, M.; Cartiglia, C. Sensitivity of ocular anterior chamber tissues to oxidative damage and its relevance to the pathogenesis of glaucoma. *Investig. Ophthalmol. Vis. Sci.* **2009**, *50*, 5251–5258. [CrossRef]

56. Saccà, S.C.; Gandolfi, S.; Bagnis, A.; Manni, G.; Damonte, G.; Traverso, C.E.; Izzotti, A. The Outflow Pathway: A Tissue with Morphological and Functional Unity. *J. Cell. Physiol.* **2016**, *231*, 1876–1893. [CrossRef]

57. Yaribeygi, H.; Zare, V.; Butler, A.E.; Barreto, G.E.; Sahebkar, A. Antidiabetic potential of saffron and its active constituents. *J. Cell. Physiol.* **2019**, *234*, 8610–8617. [CrossRef]

58. Corso, L.; Cavallero, A.; Baroni, D.; Garbati, P.; Prestipino, G.; Bisti, S.; Nobile, M.; Picco, C. Saffron reduces ATP-induced retinal cytotoxicity by targeting P2X7 receptors. *Purinergic Signal.* **2016**, *12*, 161–174. [CrossRef]

59. Maccarone, R.; Di Marco, S.; Bisti, S. Saffron supplement maintains morphology and function after exposure to damaging light in mammalian retina. *Investig. Ophthalmol. Vis. Sci.* **2008**, *49*, 1254–1261. [CrossRef]

60. Laabich, A.; Vissvesvaran, G.P.; Lieu, K.L.; Murata, K.; McGinn, T.E.; Manmoto, C.C.; Sinclair, J.R.; Karliga, I.; Leung, D.W.; Fawzi, A.; et al. Protective effect of crocin against lue light-and white light-mediated photoreceptor cell death in bovine and primate retinal primary cell culture. *Investig. Ophthalmol. Vis. Sci.* **2006**, *47*, 3156–3163. [CrossRef]

61. Waugh, N.; Loveman, E.; Colquitt, J.; Royle, P.; Yeong, J.L.; Hoad, G.; Lois, N. Treatments for dry age-related macular degeneration and Stargardt disease: A systematic review. *Health Technol. Assess.* **2018**, *22*, 1–167. [CrossRef]

62. Potnuri, A.G.; Allakonda, L.; Lahkar, M. Crocin attenuates cyclophosphamide induced testicular toxicity by preserving glutathione redox system. *Biomed. Pharmacother.* **2018**, *101*, 1740180. [CrossRef]

63. Xi, L.; Qian, Z.; Xu, G.; Zheng, S.; Sun, S.; Wen, N.; Sheng, L.; Shi, Y.; Zhang, Y. Beneficial impact of crocetin, a carotenoid from saffron, on insulin sensitivity in fructose-fed rats. *J. Nutr. Biochem.* **2007**, *18*, 64–72. [CrossRef]

64. Asai, A.; Nakano, T.; Takahashi, M.; Nagao, A. Orally administered crocetin and crocins are absorbed into blood plasma as crocetin and its glucuronide conjugates in mice. *J. Agric. Food Chem.* **2005**, *53*, 7302–7306. [CrossRef]

65. Umigai, N.; Murakami, K.; Ulit, M.V.; Antonio, L.S.; Shirotori, M.; Morikawa, H.; Nakano, T. The pharmacokinetic profile of crocetin in healthy adult human volunteers after a single oral administration. *Phytomedicine* **2011**, *18*, 575–578. [CrossRef]

66. Kell, G.; Rao, A.; Beccaria, G.; Clayton, P.; Inarejos-García, A.M.; Prodanov, M. affron® a novel saffron extract (*Crocus sativus* L.) improves mood in healthy adults over 4 weeks in a double-blind, parallel, randomized, placebo-controlled clinical trial. *Complement. Ther. Med.* **2017**, *33*, 58–64. [CrossRef]

67. Lopresti, A.L.; Drummond, P.D.; Inarejos-García, A.M.; Prodanov, M. affron®, a standardised extract from saffron (*Crocus sativus* L.) for the treatment of youth anxiety and depressive symptoms: A randomised, double-blind, placebo-controlled study. *J. Affect. Disord.* **2018**, *232*, 349–357. [CrossRef]

68. Tóth, B.; Hegyi, P.; Lantos, T.; Szakács, Z.; Kerémi, B.; Varga, G.; Tenk, J.; Pétervári, E.; Balaskó, M.; Rumbus, Z.; et al. The Efficacy of Saffron in the Treatment of Mild to Moderate Depression: A Meta-analysis. *Planta Med.* **2019**, *85*, 24–31. [CrossRef]

69. Chen, X.; Lu, L. Depression in Diabetic Retinopathy: A Review and Recommendation for Psychiatric Management. *Psychosomatics* **2016**, *57*, 465–471. [CrossRef]

70. Musch, D.C.; Niziol, L.M.; Janz, N.K.; Gillespie, B.W. Trends in and Predictors of Depression Among Participants in the Collaborative Initial Glaucoma Treatment Study (CIGTS). *Am. J. Ophthalmol.* **2019**, *197*, 128–135. [CrossRef]

71. European Food Safety Authority (EFSA). Food Supplements. 2019. Available online: https://www.efsa.europa.eu/en/topics/topic/food-supplements (accessed on 28 February 2019).

72. Office of Dietary Supplements (ODS). Dietary Supplements. Background Information. 2011. Available online: https://ods.od.nih.gov/factsheets/dietarysupplements-healthprofessional/#disc (accessed on 28 February 2019).

nutrients

MDPI

Review

The Role of Diet, Micronutrients and the Gut Microbiota in Age-Related Macular Degeneration: New Perspectives from the Gut–Retina Axis

Emanuele Rinninella [1,2], Maria Cristina Mele [1,2,*], Nicolò Merendino [3], Marco Cintoni [4], Gaia Anselmi [1], Aldo Caporossi [5,6], Antonio Gasbarrini [1,2] and Angelo Maria Minnella [5,6]

[1]   UOC di Nutrizione Clinica, Dipartimento di Scienze Gastroenterologiche, Endocrino-Metaboliche e Nefro-Urologiche, Fondazione Policlinico Universitario A. Gemelli IRCCS, Largo A. Gemelli 8, 00168 Rome, Italy; emanuele.rinninella@unicatt.it (E.R.); gaia.anselmi@gmail.com (G.A.); antonio.gasbarrini@unicatt.it (A.G.)

[2]   Istituto di Patologia Speciale Medica, Università Cattolica del Sacro Cuore, Largo F. Vito 1, 00168 Rome, Italy

[3]   Laboratorio di Nutrizione Cellulare e Molecolare, Dipartimento di Scienze Ecologiche e Biologiche (DEB), Università della Tuscia, Largo dell'Università snc, 01100 Viterbo, Italy; merendin@unitus.it

[4]   Scuola di Specializzazione in Scienza dell'Alimentazione, Università di Roma Tor Vergata, Via Montpellier 1, 00133 Rome, Italy; marco.cintoni@gmail.com

[5]   UOC di Oculistica, Dipartimento di Scienze dell'Invecchiamento, Neurologiche, Ortopediche e della Testa-Collo, Fondazione Policlinico Universitario A. Gemelli IRCCS, Largo A. Gemelli 8, 00168 Rome, Italy; aldo.caporossi@unicatt.it (A.C.); angelomaria.minnella@unicatt.it (A.M.M.)

[6]   Istituto di Oftalmologia, Università Cattolica del Sacro Cuore, Largo F. Vito 1, 00168 Rome, Italy

*   Correspondence: mariacristina.mele@unicatt.it; Tel.: +39-06-3015-6018

Received: 16 October 2018; Accepted: 31 October 2018; Published: 5 November 2018

Abstract: Age-related macular degeneration (AMD) is a complex multifactorial disease and the primary cause of legal and irreversible blindness among individuals aged ≥65 years in developed countries. Globally, it affects 30–50 million individuals, with an estimated increase of approximately 200 million by 2020 and approximately 300 million by 2040. Currently, the neovascular form may be able to be treated with the use of anti-VEGF drugs, while no effective treatments are available for the dry form. Many studies, such as the randomized controlled trials (RCTs) Age-Related Eye Disease Study (AREDS) and AREDS 2, have shown a potential role of micronutrient supplementation in lowering the risk of progression of the early stages of AMD. Recently, low-grade inflammation, sustained by dysbiosis and a leaky gut, has been shown to contribute to the development of AMD. Given the ascertained influence of the gut microbiota in systemic low-grade inflammation and its potential modulation by macro- and micro-nutrients, a potential role of diet in AMD has been proposed. This review discusses the role of the gut microbiota in the development of AMD. Using PubMed, Web of Science and Scopus, we searched for recent scientific evidence discussing the impact of dietary habits (high-fat and high-glucose or -fructose diets), micronutrients (vitamins C, E, and D, zinc, beta-carotene, lutein and zeaxanthin) and omega-3 fatty acids on the modulation of the gut microbiota and their relationship with AMD risk and progression.

Keywords: age-related macular degeneration; gut-retina axis; gut microbiota; dietary habits; micronutrients; fish oil; omega-3 polyunsaturated fatty acids; personalised medicine

1. Introduction

Age-related macular degeneration (AMD) is a complex multifactorial disease, and in developed countries, it represents the first cause of legal and irreversible blindness among individuals aged ≥65 years. Globally, it affects 30–50 million individuals, and despite the introduction of new therapies for prevention and treatment, it is expected to increase by tenfold (300 million) by 2040 [1]. Its prevalence is relatively low in young-adult individuals; however, it reaches nearly 12% in individuals aged >80 years [2]. The disease affects the quality of life and daily living activities of the patients, and it causes human and social burdens, as well as high economic costs for the entire healthcare system [3,4].

AMD, in the early stages, is characterized by the presence of hyaline deposits, referred to as "drusen", and hyper/hypopigmentations of the retinal epithelium in the retinal macular area, without visual impairment (age related maculopathy); however, it may evolve in the advanced dry form, referred to as "geographic atrophy" (GA), which is characterized by a significant loss of retinal pigment epithelium (RPE), or it may evolve in the wet (neovascular) form, sustained by abnormal choroidal neovascularization. These neovessels may undergo plasma or blood extravasation, leading to neurosensory or RPE detachment with fluid and/or blood accumulation. These changes, in turn, attract fibroblast migration and proliferation with further epithelium damage. At this stage, the GA or the neovascular form present moderate to severe visual loss.

It is unclear whether these forms are distinct or if they are different features of the same disease [5]. The neovascular form is treated with the use of anti-VEGF drugs; no effective treatments are available for the dry form, which relies on the prevention and control of risk factors, including specific nutritional intake and dietary supplements.

One potential approach to reduce the risk of AMD is the prescription of vitamins and other anti-oxidative micronutrients [6]. The main reason for this choice resides in their anti-inflammatory and anti-oxidant properties. Although the precise causes of AMD remain unknown, there is a clear role of inflammation in the pathophysiology of this disease. The RPE is fed by a dense vascular network, with high oxygen tension; moreover, a high rate of unsaturated fatty acid and photosensitizing compounds make the retina highly susceptible to reactive oxygen species (ROS) damage [7]. For example, one of the most recognized risk factors of AMD is light exposure: all light, even ambient natural light, can induce the formation of ROS in the RPE, which leads to the creation of lipid and protein peroxidation products [8]. Smoking is another well-known risk factor for AMD and is the most consistently identified modifiable risk factor, given its pro-oxidative and pro-inflammatory effect [5]. Many other factors, such as age, genetics, lifestyle, environmental factors and diet, may influence the risk and progression of AMD (Figure 1).

Obesity and overweight are associated with an increased risk of AMD in a dose-dependent fashion [9], while a reduction in the waist-to-hip ratio has been demonstrated to decrease the risk of AMD [10]. The impacts of high sugar and Western diets have also been shown in the risk and progression of AMD [11]. Furthermore, epidemiological studies strongly suggest the important contributions of several nutritional and non-nutritional compounds beyond the necessary energy intake to the risk of AMD. There has also been substantial progress in identifying the genetic variants that impact AMD risk.

Nutrients may act directly similar to antioxidant or anti-inflammatory compounds or indirectly by the gut microbiome; thus, there is a strong interest of the scientific community in the established efficacy of nutraceutics and functional foods rich in antioxidant and prebiotics in the prevention and as support to anti-AMD pharmacological treatments.

In the previous 10 years, an altered gut microbiota has been associated with many intestinal and extra-intestinal diseases, such as metabolic and inflammatory diseases, non-alcoholic fatty liver disease (NAFLD), cancer and obesity [12–15]. An increased intestinal permeability permits a higher translocation of bacterial products such as the endotoxin lipopolysaccharides (LPS) and pathogen-associated molecular pattern molecules (PAMPs), inducing low-grade inflammation in several tissues through the activation of pattern recognition receptors (PRRs). This biological crosstalk occurs also in dendritic cells, perivascular macrophages and RPE cells and may sustain ocular inflammation. Moreover, gut microbiota metabolites and products may modulate retina-specific immune cells [16]. Recently, an obesity-associated gut-microbiota has been shown to drive pathological angiogenesis toward aberrant choroidal neovascularization (CNV) in retinal tissue [16]. All this evidence supports the concept of a "gut–retina axis" in the pathogenesis of ocular diseases., It is known that the gut microbiota undergoes significant changes after the age of 65 years, reducing its richness and resilience to lifestyle changes, antibiotics, and diseases, particularly in frail patients [17] Also, the gut microbiota is a crucial player in the metabolism and absorption of several macro and micronutrients in the gut barrier, even those involved in AMD.

The identification of the relationship among diet, micronutrients, the gut microbiota, and host immunity is a new frontier in the treatment of many metabolic diseases. This review focuses on the interplay of the gut microbiota with diet and micronutrients in AMD, in the so-called "gut–retina axis".

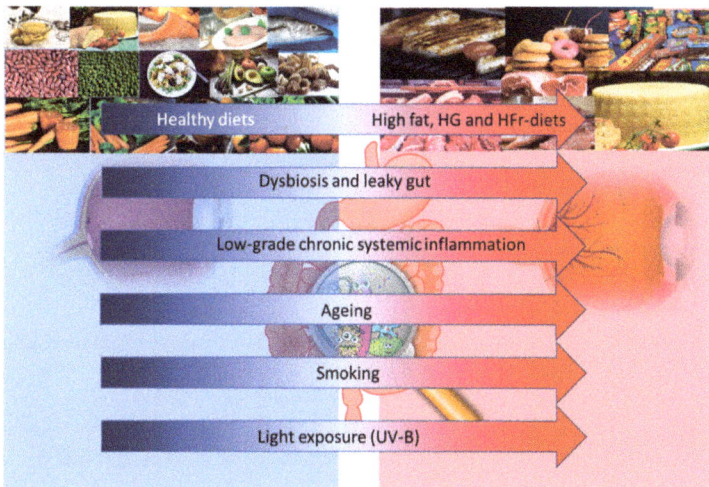

Figure 1. Altered dietary habits, dysbiosis and leaky gut, and low grade inflammation, together with aging, smoking and light exposure may influence the risk and progression of age-related macular degeneration (AMD). Abbreviations: HG: High Glucose; HFr: High Fructose; UV-B: Ultraviolet-B.

2. Literature Review Method

We searched and selected the most relevant publications between January 1998 and August 2018 on PubMed, Web of Science and Scopus with the search terms "age-related macular degeneration", "gut microbiota", "diet", "micronutrients", and "fish oil" and "omega-3 polyunsaturated fatty acids". We selected all experimental and epidemiologic studies, supporting or not, the link between dietary habits, micronutrients, the gut microbiota, and AMD.

3. Diet and Gut-Microbiota in Age-Related Macular Degeneration (AMD)

3.1. Gut Microbiota: A Diet-Driven Ecosystem

The gut microbiota is a complex ecosystem colonizing the gastrointestinal tract with a higher distribution in the colon (10^{14} cells per gram of feces). The gut microbiota is composed of a large number of bacteria (up to 1000 different species), archaea, fungi, and viruses. Overall, the number of resident microbial cells is 1.3 times greater than the number of eukaryotic cells of the whole human organism [18]. The gut microbiota has continuous cross-talk with its host, modulating the process of food absorption and maintaining the physiologic stimulation of the gut immunity system.

Although there are many known bacterial phyla in the gut microbiota, only a few phyla are predominantly represented and account for more than 160 species: among them, Firmicutes (mainly Gram-positive), Bacteroidetes (mainly Gram-negative), Actinobacteria and Proteobacteria are prevalent [19]. The former two (Firmicutes and Bacteroidetes) account for more than 90% of all phylogenetic types in both mice and humans; the less represented phyla are Verrucomicrobia and Fusobacteria [15,20].

The gut microbiota composition may vary according to dietary habits. For example, *ob/ob* mice (mutant mice that eat excessively due to mutations in the leptin gene) have a 50% reduction in the abundance of Bacteroidetes and a proportional increase in Firmicutes compared with those in lean mice [21]. The provision of a high-caloric, high-fat and simple sugar-based diet (Western diet) to wild-type mice leads to an overall decrease in the diversity of the gut microbiota, with a specific reduction in Bacteroidetes and a bloom of a single class of Firmicutes (Mollicutes) [22]. Turnbaugh et al. showed, in a mouse model, that this "obesity-associated gut microbiome" had an increased capacity to harvest energy from the diet, breaking down otherwise indigestible dietary polysaccharides. Moreover, this property is a transmissible trait: adult germ-free mice colonized (by gavage) with a microbiota of obese (*ob/ob*) donors exhibited a significantly higher increase in body fat over two weeks than mice colonized with lean donors' microbiota fed with the same quantity and caloric density of chow [15]. Moreover, a high-fat-diet (HFD) does not induce weight gain or hypercholesterolemia in the absence of gut microbiota, as shown in germ-free C57BL/6J mice compared to wild-type mice; germ-free mice also show enhanced insulin sensitivity with improved glucose tolerance compared to conventional mice on the same diet [23]. These findings confirm the pivotal role of the gut microbiota in driving the metabolic pathways of obesity and low-grade inflammation.

3.2. High-Fat Diets and the Gut Microbiota in AMD

It is well known that a HFD and unhealthy lifestyle may influence the progression of AMD in genetically predisposed individuals [24]. Chiu et al. specifically investigated the role of dietary patterns in the prevalence of AMD, collecting the nutritional habits of 4088 participants enrolled in the Age-Related Eye Disease Study (AREDS) (8103 eyes). The authors initially derived the dietary patterns by conducting principal component analysis (PCA) of food consumption data from the AREDS food frequency questionnaire (FFQ); they subsequently performed a logistic analysis and a qualitative comparative analysis (QCA) to evaluate the associations between dietary patterns and AMD, comparing the highest to lowest quintiles for each pattern. This cross-sectional study reported that a "Western diet", mainly composed of high-fat dairy products, butter or margarine, gravies, processed and red meats, eggs, sweets and desserts, energy drinks, refined grains and French fries, was strongly associated with a higher prevalence of advanced AMD (OR: 3.70; 95% CI, 2.31–5.92; $p < 0.0001$); in contrast, the so-called "oriental" dietary habit, mainly composed of vegetables, legumes, rice, whole grains, fruit, tomatoes, green leafy vegetables, low-fat dairy, fish, and seafood, was protective from advanced AMD (OR: 0.38; 95% CI, 0.27–0.54; $p < 0.0001$) [25].

A large epidemiological study, conducted in 21,287 participants from the Melbourne Collaborative Cohort Study, confirmed abdominal obesity as an independent risk factor for early (OR: 1.13; 95% CI, 1.01–1.26; $p = 0.03$) and late AMD (OR: 1.75; 95% CI, 1.11–2.76; $p = 0.02$) [26].

Andriessen et al. released the first evidence of the critical role of the gut microbiota in exacerbating CNV from a HFD. The authors randomized six-week-old C57BL/6J mice on a regular chow diet (RD; 16% kcal fat) or a high-fat diet (HFD; 60% kcal fat). The two groups were further randomized to receive (or not) neomycin (a non-gut permeable antibiotic) in their drinking water at the ninth week. At the eleventh week, all mice were subjected to a laser photocoagulation that induces choroidal neovascularization, mimicking neovascular AMD. Following sacrifice, HFD-fed mice showed a 60% increase in CNV compared to RD-fed controls. Surprisingly, HFD-fed mice treated with neomycin displayed a CNV level similar to RD-fed control mice, even if weight gain was consistent with the other HFD-diet mice. In the authors' opinion, this could explain the role of the gut microbiota in the aberrant choroidal neoangiogenesis, irrespective of the body weight.

Moreover, the type of diet significantly influenced the gut microbiota composition as shown by sequencing bacterial 16S rRNA extracted from the feces of mice: *Bacteroides* comprised 66% and Firmicutes comprised 33% of the total bacteria in RD-fed mice, whereas a sharp inversion of this ratio was found in HFD-fed mice (19% and 67%, respectively). Notably, neomycin-treated HFD-fed mice had a *Bacteroides* rate of 65% of the total bacteria (similar to RD-fed mice) and a Firmicutes rate of <10%. Of note, the HFD-diet harbored a modest presence of Actinobacteria and Spirochaetes. In turn, the different gut microbiota compositions in HFD-fed mice sustained an abnormal inflammatory response. Analysis via fluorescence-activated cell sorting (FACS) of retinas and sclera–choroid–RPE indicated a two-fold increase in mononuclear phagocytes and microglia of HFD-fed-mice compared to control RD-fed mice; in contrast, neomycin treatment abolished this effect. This gut dysbiosis-driven neovascular response was found to be linked to an increased intestinal permeability and chronic low-grade inflammation (elevated production of IL-6, IL-1beta, TNF-alpha and VEGF-A) [16]. It is not clear if the neomycin itself could act stabilizing Firmicutes rather than *Bacteroides* also in other animal contexts. However, the modulation of gut microbiota, even in HFD, appears a potential way to deal with AMD progression.

3.3. High-Glucose or -Fructose Diets and the Gut Microbiota in AMD

The excess of sugar in modern dietary habits has been linked to obesity and several metabolic diseases, including diabetes mellitus type II, NAFLD and cardiovascular diseases [27]. Inside these settings, a pivotal role is currently attributed to the gut microbiota, which is responsible for changes in intestinal permeability, leading to metabolic endotoxemia, insulin-resistance, lipid accumulation in tissues and organs and increasing plasmatic levels of inflammatory cytokines [28,29]. Moreover, many cross-sectional and prospective studies highlight a potential role of dietary carbohydrates in increasing the risk or progression of AMD, even if simple sugars play the most significant role in this context [30]. Their impact on the gut microbiota has recently been elucidated by Do et al. [31]: the authors assigned six-week-old C57BL/6J mice to receive four distinct dietary regimes: normal diet (ND), HFD, high-glucose diet (HGD), and high-fructose diet (HFrD). After 12 weeks, HFD- and HFrD-fed mice showed significantly higher plasmatic levels of blood glucose, total cholesterol, LDL and endotoxin than those of ND-fed mice. Moreover, HFD-, HGD- and HFrD-fed mice showed lower microbial diversity (fewer operational taxonomic units and lower Shannon indices) than ND-fed mice, with a lower abundance of Bacteroidetes and an increased abundance of Proteobacteria at the phylum level, as well as an increase of the *Desulfovibrio vulgaris* species. These metabolic and microbial differences were accompanied by a significant (2.5-fold greater) increase of gut permeability, demonstrated by a high plasma fluorescein isothiocyanate (FITC)-dextran concentration and a reduced expression of tight junction proteins, such as ZO-1 and Occludin. Consequently, the expression of inflammatory cytokines (TNF-alpha and IL-1beta) in the colon was higher in HFD-, HGD- and HFrD-fed mice than that in in ND-fed mice. This evidence suggests that a HGD and/or HFrD, as well as a HFD, can shape the gut microbiota, increasing the Firmicutes-to-Bacteroidetes ratio and the proportion of Proteobacteria, one of the best sources of lipopolysaccharides (LPS) [32]. Moreover, these dietary regimens significantly alter gut permeability, boosting metabolic endotoxemia and systemic inflammation through modulation of the gut microbiota. These changes were independent of body weight gain, which occurred in HFD, but not in HGD and HFrD.

A recent experimental study by Rowan et al. [33] explored the link among HGD, the gut microbiota and AMD in an animal model. To compare the effect on the RPE of the dietary glycemic load, the authors divided C57BL/6J wild-type mice to receive a high-glucose (HG) diet or low glucose (LG) diet for 12 months; half HG-fed mice were switched to the LG-diet after six months (HGxoLG mice). Histological features of dry AMD, such as photoreceptor cell damage, subretinal deposits, RPE vacuolation, hypopigmentation, thinning, and disorganization, were identified at the end of the study in the HG-fed diet but not in the LG-fed diet. Surprisingly, HGxoLG mice showed RPE features similar to LG-fed mice. Moreover, differences were noted in the gut microbial and metabolomic composition: LG-fed mice had similar amounts of Bacteroidetes and Firmicutes phyla, whereas HG-fed animals harbored bacteria of unknown classification.

Moreover, the microbiome of HG-fed mice was enriched in Firmicutes and Clostridia, which was related to a more advanced retinal damage score, whereas LG-fed mice showed an abundance of the Bacteroidales and Eysipelotrichi classes, which are both associated with protection against AMD features. Furthermore, the gut microbiota of HGxoLG-fed mice appears similar to that of LG-fed mice. The proof of an active role of the gut microbiota in AMD was also demonstrated by the assessment of microbial metabolites (serotonin, hippurate, tyrosine, and tryptophan), in which higher levels of serotonin were associated with protection against retinal damage in a diet dependent-manner. This study demonstrated a potential effect of an LG diet in reversing AMD features through modulation of the gut microbiota, paving the way for a novel nutritional approach to AMD.

4. Micronutrients

The crucial role of the gut microbiota in AMD is not only explained by the macronutrient quality, such as dietary habits; micronutrients also play an important role in shaping the gut microbiota, which, in turn, is an efficient player in mediating their protective effects. Table 1 reports the main human studies (case-control, cohort studies or randomized clinical trials) that have investigated the role of micronutrient intake in AMD.

Several nutritional supplements have recently been shown to reduce the risk of progression from early to late AMD [34]. The most important longitudinal studies to prove the effectiveness of vitamins and micronutrients in preventing the worsening of AMD remain the AREDS and AREDS2, which were both sponsored by the National Institutes of Health (NIH). The AREDS was an extensive, randomized, placebo-controlled, multicentric study, which enrolled 3640 patients aged 55–80 years with the diagnosis of AMD and followed the patients for a mean follow up of 6.3 years (1992–1998): the authors showed a 28% reduced risk (odds ratio (OR) = 0.72; 99% confidence interval (CI) = 0.52–0.98) of progression of AMD in patients who consumed a combination of high doses of vitamins C and E, beta-carotene and zinc [35]. In the second trial, AREDS2, published in May 2013, omega-3 fatty acids, as well as the antioxidants lutein and zeaxanthin (the major constituents of macular pigment, chemically similar to beta-carotene) were investigated for an additive effect in reducing the progression of AMD. The results did not show a further significant impact of these other micronutrients compared with the AREDS formulation alone. However, the AREDS2 design was quite intricate, and a real placebo group was not included. The authors concluded that lutein/zeaxanthin could be an appropriate beta-carotene substitute for safety reasons (an increased incidence of lung cancer was identified in former smokers in the beta-carotene group vs. no beta-carotene group), while the efficacy of omega-3 fatty acids remained unclear [36]. A secondary, exploratory subgroup analysis of the AREDS2 study performed a direct comparison of lutein/zeaxanthin vs. beta-carotene and showed a small yet significant risk reduction in the development of late AMD (hazard ratio (HR): 0.82; 95% CI: 0.69–0.96; $p = 0.02$) and neovascular AMD (HR: 0.78; 95% CI: 0.64–0.94; $p = 0.01$) in the lutein/zeaxanthin group. The authors concluded that lutein/zeaxanthin could be more appropriate than beta-carotene in the AREDS-type supplements [37].

Table 1. Main human studies investigating the role of micronutrients in AMD.

Year; Author [Ref.]	Study Design	Population	Micronutrients	Aims	Results
1994; Seddon J.M. et al. [38]	Case-control	356 subjects with advanced AMD and 520 control subjects	Vitamins A, C, E, Carotenoids	To assess association between dietary intake of carotenoids and vitamins A, C, and E and the risk of AMD.	Carotenoid intake reduces the risk of AMD (OR: 0.57; 95% CI, 0.35–0.92; $p = 0.02$); L and Z were most strongly associated with a reduced risk of AMD ($p = 0.001$). The intake of retinol was not appreciably related to AMD. Vitamin E or total vitamin C consumption or the intake of retinol were not associated with a reduced risk of AMD.
2004; Richer S. et al. (the LAST study) [39]	RCT	90 patients affected by atrophic AMD followed up in a period of 12 months: Group 1, L 10 mg; Group 2, L 10 mg + antioxidants/vitamins and minerals Group 3, placebo	L	To determine whether nutritional suppl. with L or L with antioxidants, vitamins, and minerals improves visual function in atrophic AMD.	Visual function and eye MPOD improved with L alone or L together with other nutrients. Patients who received the placebo had no significant changes in the measured findings.
2006; Delcourt C. et al. (the POLA Study) [40]	Cohort study	899 subjects ≥60 years, resident in Sète (Southern France)	L, Z	To assess association of plasma L/Z with the risk of AMD and cataract in the POLA Study.	The highest quintile of plasma Z was significantly associated with reduced risks of AMD (OR: 0.07; 95% CI, 0.01–0.58; $p = 0.005$), nuclear cataract (OR: 0.23; 95% CI, 0.08–0.68; $p = 0.003$) and any cataract (OR: 0.53; 95% CI, 0.31–0.89; $p = 0.01$). AMD was significantly associated with combined plasma L/Z (OR: 0.21; 95% CI, 0.05–0.79; $p = 0.01$) and tended to be associated with plasma L (OR: 0.31; 95% CI, 0.09–1.07; $p = 0.04$), whereas cataract showed no associations. Among other carotenoids, only beta-carotene showed a significant negative association with nuclear cataract, but not AMD.
2006; Seddon et al. (US Twin Study) [41]	Case-control	681 twins: 222 with AMD (intermediate or late stages) and 459 with no maculopathy	omega-3 PUFA	To evaluate modifiable risk and protective factors for AMD among elderly twins.	Dietary omega-3 PUFA intake inversely associated with AMD (OR: 0.55; 95% CI, 0.32–0.95). Cigarette smoking increases risk, while fish consumption and omega-3 fatty acid intake reduce risk of AMD.
2006; Chua B. et al. [42]	Cohort study	2335 subjects ≥49 years, participated in the Blue Mountains Eye Study (1992–1994) and re-examined after 5 years (1997–1999)	omega-3 PUFA	To assess association between dietary fat intake in the older cohort and incident AMD.	Participants with omega-3 PUFA intake had a lower risk of incident early AMD (OR: 0.41; 95% CI, 0.22–0.75). A 40% reduction of incident early AMD was associated with fish consumption at least once per week (OR: 0.58; 95% CI, 0.37–0.90), whereas fish consumption at least 3 times per week could reduce the incidence of late AMD (OR: 0.25; 95% CI, 0.06–1.00).
2007; SanGiovanni et al. (AREDS Study Report 20) [43]	Case-control	4519 AREDS Study participants: 4 AMD severity groups and a control group	omega-3 PUFA	To assess association of lipid intake with AMD in AREDS Study.	Dietary total omega-3 PUFA intake was inversely associated with neovascular AMD (OR, 0.61; 95% CI, 0.41–0.90). Higher intake of omega-3 PUFAs and fish was associated with decreased likelihood of having neovascular AMD.
2007; SanGiovanni et al. (AREDS Study Report 22) [44]	Case-control	4519 AREDS participants: 4 AMD severity groups and a control group	L/Z, Vitamins A and C, alpha-tocopherol	To evaluate the relationship among dietary carotenoids, vitamins A and C and alpha-tocopherol with prevalent AMD.	Dietary L/Z intake was inversely associated with neovascular AMD (OR: 0.65; 95% CI, 0.45–0.93), geographic atrophy (OR: 0.45; 95% CI, 0.24–0.86), and large or extensive intermediate drusen (OR: 0.7; 95% CI, 0.56–0.96), comparing the highest vs lowest quintiles of intake.
2008; SanGiovanni et al. (AREDS Study Report 23) [45]	Cohort study	2132 subjects from the clinical trial AREDS	omega-3 PUFA	To examine the association of neovascular AMD and CGA.	Decreased risk of progression from bilateral drusen to CGA among individuals who reported the highest levels of EPA (OR: 0.44, 95% CI, 0.23–0.87) and EPA+DHA (OR: 0.45, 95% CI, 0.23–0.90) consumption.

Table 1. Cont.

Year; Author [Ref.]	Study Design	Population	Micronutrients	Aims	Results
2008; Stringham, J.M. et al. [46]	Cohort study	40 healthy subjects	L, Z	To measure MPOD after L and Z supplementation for 6 months and evaluate visual improvement.	After 6 months, daily L/Z supplementation significantly increased MPOD and improved visual performance in glare for most subjects. At the 2-month time point, average MPOD had increased from 0.41 at baseline to 0.46. MPOD continued to increase at 4-month ($p = 0.032$) and 6-month ($p = 0.003$) time points, with increases from baseline of 0.10 and 0.16, respectively.
2008; Cho E. et al. [47]	Cohort study	71494 women and 41564 men with no diagnosis of AMD or cancer	L/Z	To assess association between L/Z intake and AMD risk by smoking status, vitamin C and E intakes, and body fatness.	L and Z intake was not associated with the risk of self-reported early AMD. There was a non-significant association between lutein and zeaxanthin intake and neovascular AMD risk.
2008; Tan J.S. et al. (the Blue Mountains Eye Study) [48]	Cohort study	3654 participants	Carotenoids, Vitamins A, C, E, Iron, Zinc	To assess incidence of early, late, and any AMD.	L/Z: participants in the top tertile of intake had a reduced risk of incident neovascular AMD (RR: 0.35; 95% CI, 0.13–0.92), and individuals with above median intakes had a reduced risk of indistinct soft or reticular drusen (RR: 0.66; 95% CI, 0.48–0.92). Beta-carotene: the highest compared with the lowest tertile of total beta-carotene intake predicted incident neovascular AMD (RR, 2.68; 95% CI, 1.03–6.96). Zinc: the RR that compared the top decile intake with the remaining population was 0.56 (95% CI, 0.32–0.97) for any AMD and 0.54 (95% CI, 0.30–0.97) for early AMD. Vitamin E: higher intakes predicted late AMD (RR compared with the lowest tertile, 2.83; 95% CI, 1.28–6.23; and RR, 2.55; 95% CI, 1.14–5.70 for the middle and highest tertiles, respectively).
2008; Newsome D.A. [49]	RCT	40 subjects randomly assigned to ZMC (25 mg) or placebo	Zinc (ZMC)	To assess association between use of ZMC and macular function in individuals with dry AMD.	After 6 months, the ZMC group showed improved visual acuity ($p < 0.0001$). ZMC (25 mg) twice daily was well tolerated and improved with macular function AMD.
2009; San Giovanni et al. (AREDS Study Report 30) [50]	Cohort study	1837 AREDS participants	omega-3 PUFA	To assess the association of dietary omega-3 PUFAs with progression to advanced AMD in subjects with a moderate risk of developing AMD.	Participants who reported highest omega-3 PUFA intake were 30% less likely than their peers to develop central geographic atrophy and neovascular AMD. Respective multivariate ORs are 0.65 (95% CI, 0.45–0.92; $p = 0.02$) and 0.68 (95% CI, 0.49–0.94; $p = 0.02$).
2011; Ho L. et al. (the Rotterdam Study) [51]	Case control	2167 individuals from the population-based Rotterdam Study	Zinc, beta-carotene, L/Z, EPA/DHA	To investigate the role of dietary nutrients in reducing the genetic risk of AMD conferred by the genetic variants CFH Y402H and LOC387715 A69S.	High dietary intake of nutrients with antioxidant properties reduces the risk of early AMD in individuals at high genetic risk. The results supported the possibility of biological interactions among LOC387715 A69S and CFH Y402H and zinc, beta-carotene, lutein/zeaxanthin, and EPA/DHA (all $p < 0.05$).
2012; Snellen E.L.M. et al. [52]	Case-control	72 cases and 66 controls	L	To assess the association between low antioxidant intake and neovascular AMD.	The prevalence rate of AMD in patients with low antioxidant intake and low L intake was approximately twice as high as that in patients with high intake: OR: 1.7; 95% CI, 0.8–3.7 and OR: 2.4; 95% CI, 1.1–5.1, respectively.
2012; Nolan J.M. et al. [53]	Cohort study	828 healthy subjects	L/Z	To investigate MPOD with respect to risk factors for AMD.	A statistically significant age-related reduction in MPOD was present in current and past smokers ($p < 0.01$), with a family history of AMD ($p < 0.01$). The enhanced risk that these variables represent for AMD may be attributable, at least in part, to a parallel deficiency of macular carotenoids.

Table 1. *Cont.*

Year; Author [Ref.]	Study Design	Population	Micronutrients	Aims	Results
2013; Souied E.H. et al. [54]	RCT	263 patients with early lesions of AMD received 840 mg/day of DHA and 270 mg/day of EPA or placebo	EPA, DHA	To evaluate the efficacy of DHA-enriched oral supplementation in preventing exudative AMD (time to occurrence of CNV, incidence of CNV developing in patients, changes in visual acuity, occurrence and progression of drusen, and changes in EPA plus DHA level in red blood cell membrane in RBCM).	Time to occurrence and incidence of CNV in the study eye were not significantly different between the DHA group (19.5 ± 10.9 months and 28.4%, respectively) and the placebo group (18.7 ± 10.6 months and 25.6%, respectively). In the DHA group, EPA plus DHA levels significantly increased in RBCM (+ 70%; $p < 0.001$). In the DHA- allocated group, patients steadily achieving the highest tertile of EPA plus DHA levels in RBCM had a significantly lower risk (−68%; $p = 0.047$; HR: 0.32; 95% CI, 0.10–0.99) of CNV developing over 3 years. No marked changes from baseline in best-corrected visual acuity, drusen progression, or geographic atrophy in the study eye were observed throughout the study in either group.
2014; Hammond, B.R. et al. [55]	RCT	115 young, healthy subjects (58 assigned to placebo and 57 assigned to L/Z (10/2 mg/day)	L, Z	To assess correlation between L and Z supplementation with MPOD, glare disability, photostress recovery, and chromatic contrast.	MPOD significantly increased in L/Z group versus placebo ($p < 0.001$ for both L and Z). Serum L and Z significantly increased by the first follow-up visit (at 3 months) and remained elevated throughout the intervention period of 1 year. There was a significant correlation between MPOD levels over time and visual performance.
2016; Aoki A. et al. [56]	Case-control	161 neovascular AMD cases and 369 control subjects	omega-3 PUFA, -tocopherol, Zinc, Vitamins D and C, beta-carotene	To assess micronutrient intake and neovascular AMD.	Low intakes of omega-3 PUFA, alpha-tocopherol, zinc, vitamin D, vitamin C, and beta-carotene were associated with neovascular AMD ($p < 0.0001$ for n-3 fatty acid, $p < 0.0001$ for alpha-tocopherol, $p < 0.0001$ for zinc, $p = 0.002$ for vitamin D, $p = 0.04$ for vitamin C, and $p = 0.0004$ for beta-carotene).
2017; Braakhuis A. et al. [57]	Case-control	149 controls; 42 cases with oxidative stress-related AMD	beta-carotene, Vitamin C	To assess association between the intake of dietary antioxidants and incidence of AMD.	Protective associations with higher consumption of vitamin C (OR: 0.63; 95% CI, 0.23–1.03; $p = 0.022$) and beta-carotene (OR: 0.56; 95% CI, 0.15–3.98; $p = 0.007$).

Abbreviations: AMD: age-related macular disease; CGA: central geographic atrophy; CNV: choroidal neovascularization; DHA: docosahexaenoic acid; EPA: eicosapentaenoic; L: Lutein; LAST study: Lutein Antioxidant Supplementation Trial study; MP: macular pigment; MPOD: macular pigment optical density; POLA study: Pathologies Oculaires Liées à l'Age study; RBCM: red blood cell membrane; RCT: randomized controlled trial; Z: Zeaxanthin; ZMC: zinc monocysteine; omega-3 PUFA: omega-3 polyunsaturated fatty acids.

4.1. Vitamins C and E

Vitamin C is a potent antioxidant and a cofactor for many enzymatic reactions. Moreover, vitamin E is considered the most effective scavenger of free radicals [58]. Vitamins C and E are non-enzymic antioxidants that protect against oxidative stress, a significant contributory factor in the progression of neovascular AMD [56]. Vitamin C is the most important water-soluble antioxidant in the human body and the primary antioxidant in the eye. Vitamin C is present at high concentrations in the cornea, central corneal epithelium, lachrymal film, vitreous humor and aqueous humor [59]. Vitamin E comprises a group of essential eight lipid-soluble compounds, including tocopherols and tocotrienols, capable of penetrating into cellular membranes through a hydrophobic side and donating a phenolic hydrogen to reduce free radicals, thus preventing the propagation of ROS and the peroxidation of cellular and subcellular membrane phospholipids. One of the primary therapeutic applications of vitamin E is NASH (non-alcoholic steatohepatitis), a hepatic inflammatory syndrome sustained by low-grade systemic inflammation and lipid peroxidation often associated with metabolic syndrome and insulin-resistance [60]. Alpha-tocopherol is the most studied form of Vitamin E and has been shown to reduce the biomarkers of total body oxidative stress and inflammation [61,62]. In contrast to other water-soluble vitamins (i.e., thiamin, biotin, or folate) that, although they cannot be synthesized de novo by our body, can be produced by the normal microflora of the large intestine, vitamin C cannot be synthesized de novo in humans; however, it is obtained from dietary sources (fruits and vegetables) via intestinal absorption with a Na+ dependent carrier-mediated process [63]. The main vegetal sources of vitamin E include wheat germ oil, extra virgin olive oil, hazelnuts, and peanuts; animal sources include fish, oysters, eggs, and butter.

An animal study on mice published in 1985 showed that ascorbic acid might play a role in protecting the retina from oxidative insults by light. The retinas of rats that received ascorbate supplement showed significantly lesser damage than the retinas of unsupplemented rats [64]. The idea that vitamins C and E could have a protective effect for age-related retinal disease results from several epidemiological and animal studies. In these studies, contradictory results are reported correlating the intake of vitamin E or C and the prevention or reduction of the risk of AMD.

A recent case-control study conducted by Braakhuis et al. indicated that a higher intake of vitamin C was associated with a reduced risk of oxidative stress-related eye diseases (OR: 0.63; 95% CI, 0.23–1.03; $p = 0.022$) [57].

Another case-control study that enrolled 161 neovascular AMD cases and 369 population-based control subjects demonstrated that dietary intake of alpha-tocopherol and vitamin C was associated with a reduced risk of neovascular AMD [56].

The Beaver Dam Eye study indicated a correlation between low vitamin E intake and developing large drusen [65], whereas the Eye Disease Case-control Study Group did not identify a significant interaction between the risk of AMD and vitamin E [38].

The reason for these contradictory results could be because four different tocopherols are available, each with different biological activities and absorptions.

The Eye Disease Control Study reported that lower Vitamin C plasma levels were related to an increased AMD risk. Moreover, high plasma concentrations were not protective, i.e., vitamin or total vitamin C consumption was not associated with a statistically significantly reduced risk of AMD, although a potentially lower risk of AMD was suggested [38]. Furthermore, the population cohort Blue Mountains Eye Study showed a positive correlation between vitamin E intake and AMD with an RR of incident late AMD for participants in the middle and highest tertiles of total vitamin E intake of 2.83 (95% CI, 1.28–6.23; $p = 0.0099$) and 2.55 (95% CI, 1.14–5.70; $p = 0.022$), respectively, compared with the lowest tertile.

Therefore, the link between neovascular AMD and the consumption of vitamin C and vitamin E (alpha -tocopherol) remains controversial, with some studies showing a significant relationship and other studies indicating no relationship.

The redox state strongly modulates the gut microbiota: the oxidative stress generated in a HFD mouse model was found to alter the gut microbiota composition, increasing *E. coli* and *Enterococcus* and decreasing *Lactobacilli* compared to those in the control group [66]. Altered expression of anti-inflammatory pathways, such as superoxide dismutase (SOD), has been reported in inflammatory bowel diseases (IBD) [67]. In IBD, there is an imbalance of the gut microbiota, with a decline in the diversity of Firmicutes (a specific decrease in the *Clostridium leptum* groups, particularly *Faecalibacterium prausnitzii*) and an increase of Proteobacteria (such as Enterobacteriaceae and specifically *E. coli*) [68]. Moreover, the relative abundance of Bacteroidetes is increased in Crohn's Disease (CD) compared with healthy controls [69].

Previous studies have indicated that both vitamins C and E are protective following mucosal tissue damage in chemical-induced colitis models [70,71]. It has also been demonstrated that natural antioxidants may regulate the gut microbiota composition by scavenging excessive free radicals and supporting the cellular and humoral immune responses [72].

Recent findings in an animal model of ileal pouchitis suggest that an antioxidant diet, enriched in vitamins C and E, selenium, and retinoic acid, may reshape the gut microbial community toward an anti-inflammatory profile, mitigating mucosal inflammation. This capacity appears to be mediated by an increase in the relative percent of Bacteroidetes and a decrease in Firmicutes at the phylum level, with an overall increase in alpha-diversity (Shannon diversity index) [73]. This evidence is confirmed by a human study conducted in pregnant women in the second trimester, which showed that higher intakes of vitamin E were associated with a decrease in Proteobacteria and Firmicutes and an increase in Bacteroidetes [74]. The association between micronutrient intakes, including Vitamins C and E, and gut microbiota variations was further assessed in a small group of free-living adults with stable cystic fibrosis (CF): the authors found that vitamin C and E intakes were positively correlated with Firmicutes and its lower taxa (i.e., Clostridium) and negatively associated with Bacteroidetes [75]. Clearly, the issue must be further elucidated, particularly at lower taxonomic levels; moreover, a recent study conducted on iron-deficient infants and toddlers showed an increase of the relative abundance of the genus *Roseburia* (phylum Firmicutes), a butyrate producer, in the group supplemented with iron + Vitamin E compared to that in the only iron-supplemented group [76].

A recent animal study conducted on early-weaned piglets explored the effects of an antioxidant blend, including vitamins C and E, on the oxidative stress generated by weaning stress. The study confirmed the antioxidant capacity of these micronutrients in scavenging free radicals and restoring the gut microbiota microenvironment, increasing *Lactobacillus* and *Bifidobacterium* counts, and decreasing *E. coli* counts in the gut environment [77]. However, the antioxidant property of these compounds appears to be empowered by their synergistic effect, according to the so-called "antioxidant network theory" [78].

4.2. Zinc

Zinc is a trace mineral responsible for the metabolism of nucleic acids, signal transduction, protein folding, and gene expression. Zinc is involved as a cofactor in more than 300 enzymatic reactions in vivo [79,80].

In the eye, zinc plays an important anti-oxidant role, and it is a cofactor of many active ocular enzymes, including superoxide dismutase and catalase. Zinc can be found at high concentrations in the human retina, RPE, and choroid; moreover, it is involved in the formation of electrical signals of photoreceptors. Low levels of zinc are associated with poor night vision and the degradation of RPE and photoreceptors [81].

Recent evidence indicates that zinc also has a close interaction with the complement system, which may further represent an important factor in determining the beneficial effects in AMD [82]. Moreover, observational studies have suggested that individuals who eat a diet rich in antioxidant minerals (selenium and zinc) may be less likely to develop AMD [83].

In vitro studies have suggested that AREDS vitamins and zinc supplementation attenuate angiogenesis and endothelial-macrophage interactions, thereby reducing the progression of AMD [84]. Finally, data from an animal model of light-induced retinal degeneration suggest that integration with zinc induces changes in gene expression, as well as enhances the antioxidative power in the retina and reduces the oxidative damage that arises from intense light exposure. [85].

A randomized, prospective trial [49] showed, after six months of treatment, an improvement of visual acuity and contrast sensitivity and a shortening of the macular light flash recovery times in patients who received zinc supplementation.

Several clinical studies have been designed to define the exact dose of zinc supplementation and its beneficial effects. Experimental data [86] suggest that low amounts of zinc protect RPE cells in culture from stress-induced effects, whereas higher amounts of zinc have the opposite influence. These effects are dependent, in part, on the "health status" of the cells. It also appears that zinc-induced death of RPE cells can be attenuated by compounds such as antioxidants (alpha-tocopherol, Trolox, and metipranolol) or cellular energy substrates (pyruvate and oxaloacetate). Therefore, a combination of zinc and antioxidants or energy substrates instead of zinc alone should represent a safer and more effective treatment for diseases, such as AMD.

AREDS indicated that 80 mg of zinc oxide, alone or in combination with antioxidants, significantly reduced the risk of progression to advanced AMD. AREDS2 showed that a low dose of zinc (25 mg) instead of a high dose (80 mg) displayed no difference in the primary outcome of progression to late AMD. The Blue Mountains Study, a population-cohort study, indicated a protective effect of dietary zinc intake for early or any AMD with a potential threshold effect (>15.8 mg/day) [48].

Zinc is absorbed mainly in the small intestine as well as in the stomach and large intestine via a non-specific, unsaturable diffusion-mediated mechanism and saturable carrier-mediated component [79]. Zinc is indispensable for the growth of most organisms. The amount of zinc inside cells is highly regulated, as too little zinc does not support growth, while too much zinc is toxic. Moreover, numerous bacterial cells require zinc uptake systems for growth and virulence [87]. Considering that the gut houses the majority of an individual's microbes, in recent years, numerous animal studies have been conducted with the aim of elucidating the impact of dietary zinc on the gut microbiota. One study, which collected feces throughout a five-week time course of dietary zinc manipulation, demonstrated that excess nutritional zinc alters the diversity and structure of the gut microbiota in mice [88].

Another study, conducted on chickens, demonstrated that quantities of zinc in the gastrointestinal tract are reduced in conventional chicks compared to limited-flora chicks, which suggests that the microbiota affect the availability of this trace element [87].

A dramatic compositional and functional remodeling of the gut microbiota of *Gallus gallus* was assessed under chronic zinc deficient conditions [89]. Another study on pigeons [90] showed different effects of zinc-methionine supplementation on intestinal bacterial growth depending on the dosage: an increase of Bacillaceae, *Lactobacillus*, *Enterococcus* and *Bifidobacterium* populations and a decrease of *Escherichia coli* were identified at the dosage of 2 mg, whereas an overall lower count of *Lactobacillus*, *Enterococcus* and *Bifidobacterium* populations occurred at the dosage of 10 mg. Further studies confirm these data; Engberg et al. [91] showed that supplementation with zinc bacitracin significantly reduced the number of coliform bacteria. Ren et al. [92] identified a significant increase in the amounts of *Lactobacillus* and *Bifidobacterium* and a decrease in the amounts of *E. coli*, *Staphylococcus* and *Enterococcus* in the feces of dogs supplemented with zinc- enriched probiotics compared to those in controls.

4.3. Carotenoids

Carotenoids are pigments responsible for the yellow, orange, and red colors of many fruits and vegetables and are divided into two classes: xanthophylls (lutein, zeaxanthin and the isomer meso-zeaxanthin) and carotenes (alpha-carotene, beta-carotene, and lycopene). Apart from the carotenoids present in major foodstuffs (e.g., melon, carrots, eggs, shrimps, lobsters, and salmons), the human diet includes carotenoids from spices such as saffron, paprika, and annatto. Due to their intense orange to red colors, carotenoids are also widely used as colorants in the food-processing industry [93].

The bioavailability of carotenoids can be influenced by dietary and phytological factors, according to the different species of carotenoids, and depending on the molecular linkage, amount of carotenoids consumed in a meal, effectors of absorption and bioconversion, nutrient status of the host, genetic factors, host-related factors, and interactions [94].

The mechanism of carotenoid absorption starts with the mechanical and enzymatic disruption of the food matrix followed by their emulsification and micellization in the intestinal lumen. The mixed micelles are absorbed by the small intestinal epithelium (enterocytes) through simple diffusion. However, studies have suggested the existence of receptor-mediated transport of beta-carotene and lutein in the apical membrane of enterocytes [95–97].

4.3.1. Beta-Carotene

Carotenoid supplementation can prevent and reduce the risk of AMD and other ocular diseases [98]. Carotenoids are potent antioxidants that able to reduce the systemic oxidative stress that influences the macula. Alpha-carotene, beta-carotene, and lycopene have been found in human RPE and choroid to protect these tissues against light-induced oxidative damage and locally produced free radicals.

Several studies have shown that carotenoids are widely used to treat oxidative stress-induced ocular diseases, such as AMD and cataract [40,99–101].

The retinal content of macular carotenoids is inversely associated with the incidence of AMD [102,103]. Moreover, clinical studies have shown that carotenoid supplementation can improve visual performance in some subjects [46,55,104].

The original AREDS formula containing beta-carotene has been demonstrated to reduce the progression to advanced AMD. However, when lutein/zeaxanthin replaced beta-carotene in the original AREDS formulation, an increase in efficacy was observed [105].

The initial enthusiasm for supplementation with beta-carotene was tempered by the emergence of evidence that high doses of beta-carotene (30 mg daily) could be harmful in smokers, causing an increase in the incidence of lung cancer. This side effect does not occur in non-smokers or when the carotenoid is administered in lower doses. Therefore, it is recommended to avoid supplementation with beta-carotene in smokers with AMD.

The question of whether beta-carotene has protective or harmful effects on AMD remains to be clarified. A recent case-control study indicated a protective association between dietary intake of beta-carotene and oxidative stress-related eye diseases (OR: 0.56; 95% CI, 0.15–0.98; $p = 0.007$) [57]. However, contradictory results originated from the Blue Mountains Study, which showed a significantly positive correlation between beta-carotene dietary intake and AMD (R: 1.36; 95% CI, 1.02–1.81 per 1-SD increase; $p = 0.039$), with a significant trend across increasing tertiles of dietary beta-carotene intake. [48].

Nevertheless, it remains clear that beta-carotene represents a protective agent against a variety of other chronic diseases or cardiovascular diseases.

The previously described study on patients affected by CF showed that high intakes of beta-carotene were negatively correlated with Bacteroidetes and positively correlated with Firmicutes and their lower taxa (e.g., *Clostridium*). It is plausible that antioxidant vitamin (such as beta-carotene) supplement may counteract the impact of increased oxidative stress on gut bacterial members [75].

A study conducted to identify functional alterations of the gut metagenome related to symptomatic atherosclerosis suggests that high levels of beta-carotene in the serum of healthy controls could be due to the potential production of this anti-oxidant by the gut microbiota [106]. Even if poor data are available on this topic, it is plausible to suggest that the anti-inflammatory effects of beta-carotene are also mediated by the gut microbiota or its transformation to Vitamin A.

4.3.2. Lutein and Zeaxanthin

Lutein, zeaxanthin and the isomer meso-zeaxanthin are the predominant carotenoids and belong to the class of xanthophylls which accumulate in the retina and are constituents of the retinal macular pigment (MP) [107]. Lutein and zeaxanthin play a pivotal role in maintaining the morphology and function of the macula, displaying their antioxidant activity through the absorption of damaging blue light, neutralization of photosensitizers and active oxygen species, and scavenging of free radicals [108].

In the retina, meso-zeaxanthin can be found in the center, zeaxanthin in the mid-periphery and lutein in the periphery of the macula [109].

Lutein is a potent antioxidant: high dietary intakes enhance immune function and also reduce the risk of developing chronic diseases, such as cancer and cardiovascular diseases [110]. Moreover, studies have shown that a supplement containing lutein, zeaxanthin and blackcurrant extract has beneficial effects on visual functioning [39], and high MP provides protection against the development of many retinal diseases, particularly AMD; in contrast, low MP increases the risk of these diseases [53,111].

Most vegetables (spinach, kale, lettuce, asparagus, and broccoli) and pistachio nuts contain only lutein, whereas corn and eggs contain both lutein and zeaxanthin. Fat consumption (i.e., a salad dressing, extra virgin olive oil or whole eggs) together with carotenoid intake have been shown to increase the absorption and bioavailability of some carotenoids, such as lutein [112].

Xanthophylls (lutein, zeaxanthin, and meso-zeaxanthin) seem to be more easily released from the food matrix and more efficiently micellized than carotenes, such as beta-carotene, and then absorbed by intestinal cells. [113].

The roles of lutein and zeaxanthin have been investigated in cohort studies, case-control studies and clinical trials. Overall, evidence [44,47,52] suggests a significant association between lutein/zeaxanthin intake and risk reduction for advanced AMD, both neovascular AMD and geographic atrophy, while the correlation with early AMD is unclear.

Evidence also suggests that lutein/zeaxanthin intake may have a protective effect against AMD in patients with a high genetic predisposition [114]. The Carotenoids in Age-Related Eye Disease Study (CAREDS) trial indicates that specific genes related to xanthophylls are associated with the development of AMD: in particular, nine genes were found to be associated with AMD, seven of which were associated with either lutein/zeaxanthin levels in the serum and macula [115].

In the Rotterdam Study [51] and the Blue Mountains Eye Study [48], high dietary intakes of lutein/zeaxanthin reduced the risk of early AMD among participants at a high genetic risk; no similar association was identified for individuals with a low genetic risk.

The macular pigment optical density (MPOD) is a measure of the attenuation of blue light by the macular pigment; therefore, it is correlated with the amount of lutein/zeaxanthin in the macula.

An increased intake of lutein and zeaxanthin through diet or supplementation has been demonstrated to increase MPOD levels, improve visual function, and reduce the risk of age-related eye diseases [116].

An important clinical trial evaluating the impact of lutein on MPOD is the Lutein Antioxidant Supplementation Trial (LAST) [39]. The results of this study show that MPOD can be modulated in patients with AMD. Moreover, the visual acuity, visual function, photo-stress recovery time, and contrast sensitivity were significantly improved in the group that received 10 mg of lutein plus antioxidants compared to placebo or lutein alone. A meta-analysis published in 2016, which analyzed 20 randomized controlled trials, confirmed the significant benefits of lutein, zeaxanthin and meso-zeaxanthin supplementation on MPOD in AMD patients and healthy subjects with a dose-response relationship [117].

The Blue Mountains Study, based on a cohort of 3654 Australian patients, indicated that participants within the top tertile of dietary intake of lutein and zeaxanthin ($\geq$942 g/day) compared with the remaining population were significantly less likely to develop neovascular AMD (RR: 0.35; 95% CI, 0.13–0.92; p = 0.033), and individuals above the median (743 g) were also less likely to develop indistinct soft or reticular drusen (RR: 0.66; 95% CI, 0.48–0.92; p = 0.013) [48].

The AREDS2 [36] study and its previously described secondary analysis [37] confirmed the role of lutein/zeaxanthin in reducing the risk of late AMD and neovascular AMD.

A dominant population of *Bifidobacteria* and *Lactobacilli* in the gut microbiota has been associated with many benefits on human health, such as the inhibition of gut pathogens [118], prevention of colon cancer [119], synthesis of vitamins and enhancement of the immune system [120].

In 2013, a phytotherapy study on humans [121] identified two products, composed of blackcurrant extract powder, lactoferrin, and lutein, as good prebiotics that significantly promoted the growth of *Bifidobacteria* and *Lactobacilli* and reduced other bacteria populations, such as *Bacteroides* spp and *Clostridium* spp. Moreover, these compounds were also demonstrated to decrease the activity of beta-glucuronidase, an enzyme involved in colorectal carcinogenesis [122].

4.4. Vitamin D

Vitamin D exists in two primary forms: vitamin D2 (ergocalciferol), derived from plants, and vitamin D_3 (cholecalciferol), derived from animal sources (animal fats, eggs, and fish oil). An essential source of vitamin D_3 in humans derives from skin exposure to sunlight (ultraviolet radiation, UV-B), by photolysis of 7-dehydrocholesterol. Once assumed or produced by the skin, vitamin D must be activated by two necessary hydroxylation processes in the liver (25(OH) hydroxylation) and kidneys (1,25(OH)$_2$ hydroxylation).

Vitamin D is a secosteroid, which acts as a steroid hormone by binding to the vitamin D receptor (VDR) and activating the transcription of genes involved in mineral and bone homeostasis, cell proliferation, differentiation, and apoptosis. Another mechanism of action is the non-genomic pathway, which involves secondary messengers and cytosolic kinase systems in target cells expressing specific receptors on the cellular membrane. Vitamin D is well known for its role in bone mineralization via phosphorus and calcium homeostasis. However, the VDR is ubiquitously expressed in other body tissues, such as intestinal, immune, endothelial and smooth vascular cells [123,124]. Moreover, vitamin D has also been investigated for its role in the regulation of immune function, inflammation, control of cell proliferation and apoptosis; potential inverse associations between vitamin D and chronic diseases (cardiovascular, autoimmune and cancer diseases) have thus been suggested [125,126].

More recently, a role of vitamin D in AMD has been proposed: VDR and the renal hydroxylases involved in Vitamin D metabolism (CYP27B1 and CYP24B1) have been found in the retina, RPE, and choroid [127–129]. In these tissues, vitamin D may act as a paracrine/autocrine hormone, thereby regulating oxidation, inflammation and angiogenesis. A recent review by Layana et al. [130] exposed several plausible mechanisms of action of vitamin D on AMD pathophysiology: protection against oxidative stress due to the generation of free ROS and lysosomal enzymes; possible inhibition of amyloid beta protein deposits, considered a primary activator of the complement cascade and inflammation; suppression of pro-inflammatory cytokines secreted by macrophages and microglia; and inhibition of angiogenesis, mediated by the inhibition of the transcription of hypoxia-inducible factor (HIF-1), induction of endothelial cell apoptosis and inhibition of the production of metalloproteinase, MMP-9.

One potential role of vitamin D deficiency in AMD is supported by many observational, cross-sectional and case-control studies, which correlated serum vitamin D levels or dietary vitamin intake with the risk of early or late AMD, although the present studies are heterogeneous and prospective studies are currently lacking [130]. However, although there is no robust evidence of a causal role of vitamin D deficiency in AMD pathogenesis, a protective role of vitamin D against chronic retinal inflammation should be considered in dietary programs and future studies.

A pivotal role in low-grade systemic inflammation, a constant of all chronic inflammatory diseases, is exerted by the gut permeability. Prolonged inflammatory dietary habits and dysbiosis may lead to a leakage of the gut barrier, potential disruption of the blood–brain barrier, and neuroinflammation [131]. Vitamin D promotes a gut barrier function that protects the integrity of the intestinal barrier. Its actions range from the regulation of tight junction proteins to the suppression of gut epithelial cell apoptosis and the stimulation of the expression of antimicrobial peptides, such as defensins and cathelicidin, by epithelial cells and monocytes. Furthermore, it regulates gut immunity toward an anti-inflammatory pattern, inhibiting pro-inflammatory Th-1 and Th-17 cells and stimulating T regulatory cells in ulcerative colitis (UC). The VDR is highly expressed in the intestine, and a low VDR expression or dysfunction is frequently found in patients with IBD. Moreover, a vitamin D deficiency has been correlated with disease activity, inflammation and clinical relapse in IBD [132]. Vitamin D was recently found to regulate gut permeability and the gut microbiota composition from the embryonal age [133,134].

Recent human studies support the evidence of a gut microbiota modulation by Vitamin D. A cross-sectional study, which investigated the association between vitamin D intake and the gut microbiota composition in healthy subjects, found that individuals in the highest vitamin D intake group had more abundant *Prevotella* and less abundant *Haemophilus* and *Veillonella* species [135]. An interventional human study indicated that high dose vitamin D supplementation promotes a shift in the gut microbiota on the upper intestinal tract, increasing the bacterial richness and decreasing Gammaproteobacteria [136]. The effect of vitamin D intake on modulation of the gut microbiota is mainly found in inflammatory diseases, such as UC and CD, where its administration seems to reduce intestinal inflammation and increase the abundance of several beneficial bacterial strains [137,138]. In another model of inflammatory disease, CF, vitamin D-insufficient CF patients were found to harbor numerous, potentially pathogenic species compared with vitamin D-sufficient CF patients in the gut and airway microbiota. The same study randomized vitamin D-insufficient patients to receive vitamin D_3 or placebo. After 12 weeks of treatment, the *Lactococcus* species were enriched in patients who received vitamin D_3, whereas *Veillonella* and Erysipelotrichaceae were significantly enriched in patients treated with placebo [139]. Taken together, these preliminary results show a potential beneficial effect of vitamin D on the systemic inflammatory status through enforcement of the gut barrier and positive modulation of the gut microbiota. This evidence could potentially be found to support the AMD model given the recent findings of Andriessen et al. [16]. However, this hypothesis should be confirmed by further studies.

4.5. Omega-3 Fatty Acids

Several studies have shown that the supplementation of omega-3 polyunsaturated fatty acids (PUFAs) provides multiple health benefits against different chronic degenerative diseases, such as cardiovascular diseases [140], rheumatoid arthritis [141], IBD [142], depression [143] and cancer [144].

The omega-3 PUFAs eicosapentaenoic acid (EPA; C20:5 ω3) and docosahexaenoic acid (DHA; C22:6 ω3) are the two main bioactive forms in humans. These fatty acids can be synthesized from the dietary precursor, an essential fatty acid, linolenic acid (ALA, C18:3), even if the synthesis pathway is quite complex, and dietary uptake of EPA and DHA-rich foods (seafood sources, such as sardines or salmon, nuts and seeds) is recommended. However, EPA and DHA are widely used as nutritional supplements, often as nutraceuticals.

Aging is characterized by an increase in the concentration of some pro-inflammatory molecules in the circulation, a phenomenon that has been termed "inflammaging" [145,146]. Low-grade inflammation (LGI) is associated with the age-related decline of many functional systems and increased risks of ill-health, poor well-being and mortality.

Controlling LGI could prevent or reduce the age-related functional decline associated with mental health and wellbeing as well as retinal health. The human retina contains lipid profiles enriched in long-chain polyunsaturated fatty acids (LC-PUFAs) and very long-chain polyunsaturated fatty acids (VLC-PUFAs) that are essential for regular retinal structure and function [147].

Epidemiological, clinical and experimental studies show that dietary omega-3 PUFAs (DHA and EPA) are associated with a reduced incidence of AMD and have a protective role in AMD progression. A randomized prospective study, the Nutritional AMD Treatment-2 (NAT-2) trial, demonstrated that patients who achieved red blood cell membrane EPA/DHA levels were significantly protected against AMD compared with the placebo group, having low levels of EPA/DHA [43,54]. The Eye Disease Case-Control Study in the US demonstrated that a higher intake of omega-3 fatty acids is inversely associated with AMD (OR: 0.55; 95% CI, 0.32–0.95). Moreover, the reduction in the risk of AMD with a higher intake of omega-3 fatty acids was primarily identified among subjects with a linoleic acid intake (an omega-6 fatty acid) below the median ($p < 0.001$) [41]. The Blue Mountains Eye Study also demonstrated a protective effect of high fish consumption in reducing the risk of incidence of late AMD in individuals in the most top quintile of intake [42]. Other studies report a lower risk of developing central geographic atrophy and neovascular AMD in individuals who consume higher levels of EPA/DHA [45,50]. These findings are in contrast with those of the AREDS2 study, concluding that the addition of DHA + EPA to the original AREDS formulation (vitamin C, vitamin E, beta-carotene, zinc, and copper) did not further reduce the risk of progression of AMD. However, one potential explanation is that the design, setting intake or subjects of AREDS2 did not permit an adequate expression of the prophylactic potential of omega-3 PUFA.

Furthermore, many studies have demonstrated that increased levels of omega-3 fatty acids in the diet do not prevent or slow the progression of AMD in accordance with the AREDS 2 trial [148].

Summing up the current evidence, beneficial effects on macular physiology and protection against degeneration can be assumed considering the extraordinarily high concentration of long-chain-PUFAs in retinal cell membranes. In general, epidemiological studies support the recommendation that a dietary fatty fish intake (e.g., salmon, tuna sardine, mackerel, and trout) or fish oil supplements are associated with a lower risk of AMD; however, it is prudent to advise patients of the actual scientific uncertainty regarding their real effectiveness.

Different bacterial taxa modulate immune functionality toward a *pro* or *anti*-inflammatory pattern. Thus, the composition of the microbiota community determines, in part, the level of resistance to infection and susceptibility to inflammatory diseases. A reduction of healthy bacteria, such as *Lactobacilli* and *Bifidobacteria*, has been associated with many metabolic disorders, such as cardiovascular diseases, diabetes, and obesity, all associated with metabolic endotoxemia, due to lipopolysaccharides (LPS) translocation across the intestinal epithelium and systemic inflammation. Many of these conditions are elicited by a high-fat diet or Western dietary patterns, decreasing Bacteroidetes and increasing both Firmicutes and Proteobacteria [149].

High (saturated) fat diets also promote the blooming of endotoxin-producing and sulfate-reducing bacteria, such as *Bilophila wadsworthia* (Proteobacteria phylum), associated with mucus layer degradation, low-grade inflammation and insulin-resistance [150,151]. In contrast, a fish oil-based diet has been demonstrated, in a murine model, to promote *Lactobacillus* and *Akkermansia muciniphila* blooming, associated with reduced inflammation and gut barrier improvement [152]. Several human studies have shown that EPA- and DHA-enriched diets may restore the Firmicutes/Bacteroidetes ratio, boosting Lachospiraceae taxa, resulting in an increased SCFA synthesis [153,154]. The improvement of the gut intestinal barrier is also reported by a human study showing a direct correlation between a higher total intake of omega-3 PUFA and lower serum zonulin concentration, a protein of intestinal tight junctions, for which serum levels are the hallmark of chronic low-grade inflammation and leaky gut [155]. The impact of PUFA on the gut microbiota composition appears to be crucial from the time of gestation and early life; moreover, a reduction of omega-3 PUFA in the maternal diet has been associated with a significant depletion of Epsilonproteobacteria, Bacteroides, and *Akkermansia* and a higher relative abundance of *Clostridia* in offspring [156]. For further investigations on the impact of omega-3 PUFA on the gut microbiota, we refer the reader to a comprehensive review of the literature by Costantini et al. [157]. Recently, a model of transgenic mice that exhibited an endogenously high omega-3 PUFA tissue content and a lower omega-6 to omega-3 PUFA ratio (fat-1 mice) has been demonstrated to maintain a lean phenotype compared with the wild-type counterpart (WT-mice), even when fed a high fat/high sucrose diet. At the basis of these results, there is a lower gut permeability, as shown by the ZO-1 immunostaining (higher in transgenic mice) and the plasma LPS concentration (assessed by liquid chromatography-tandem mass spectrometry), which is lower in fat-1 mice that had low plasma LPS concentrations. The authors also identified several differences in the gut microbiota composition between the two models, with a significantly higher diversity level and a more abundant population of Verrucomicrobia phylum (*Akkermansia* genus) in fat-1 mice than in WT mice. More interestingly, WT mice colonized with fat-1 fecal microbiota exhibited a significant improvement in glucose tolerance, a lower total weight gain than their WT counterparts, and a reduction of intestinal permeability, comparable to fat-1 mice, after transplantation with fat-1 fecal microbiota [158]. This evidence highlights the role of -3 PUFA in modulation of the gut microbiota, thus opening new horizons in the treatment of metabolic and inflammatory diseases linked to leaky gut, such as inflammatory bowel disease, diabetes, obesity, cancer, neuropathologies and, likely, AMD.

5. Conclusions and Future Perspectives

AMD is an invalidating disease with an increasing incidence due to the higher prevalence of elderly individuals in Western countries. Currently, there are few treatments available to manage its course.

Many observational studies have shown the potential role of micronutrient supplementation in lowering the risk of progression of the early stages of AMD. Moreover, an animal model highlighted the role of high-fat and high simple-sugar diets on the development of AMD through a derangement of the gut microbiota that leads to systemic low-grade inflammation. Furthermore, recent evidence indicates a strict interaction between the gut microbiota and retina that is referred to as the "gut-retina axis". A better understanding of the mechanisms that underlie this connection may help clinicians to prompt new models of personalized care of AMD based on the promotion of healthy nutritional habits and adequate micronutrient intake. These practices could modulate the gut microbiota toward a reduction of dysbiosis, leaky gut and LGI and, consequently, retinal damage. Further studies are required to elucidate whether modulation of the gut microbiota through dietary interventions can delay the course of this frequent disease in the clinical setting.

Author Contributions: M.C.M. conceived the topic and revised the final version, E.R. prepared the manuscript, N.M. edited the nutritional section, G.A. generated the schematic diagrams, M.C. selected bibliographic sources, A.C. edited the ophthalmologic section, A.G. revised the final version, A.M. edited the ophthalmologic section and improved English language. All the authors made significant contributions to this review article.

Funding: This research received no external funding.

Conflicts of Interest: The authors declare no conflicts of interest.

References

1. Wong, W.L.; Su, X.; Li, X.; Cheung, C.M.; Klein, R.; Cheng, C.Y.; Wong, T.Y. Global prevalence of age-related macular degeneration and disease burden projection for 2020 and 2040: A systematic review and meta-analysis. *Lancet Glob. Health.* **2014**, *2*, e106–e116. [CrossRef]
2. Friedman, D.S.; O'Colmain, B.J.; Muñoz, B.; Tomany, S.C.; McCarty, C.; de Jong, P.T.; Nemesure, B.; Mitchell, P.; Kempen, J. Prevalence of age-related macular degeneration in the United States. *Arch. Ophthalmol.* **2004**, *122*, 564–572. [PubMed]
3. Klein, R. Overview of progress in the epidemiology of age-related macular degeneration. *Ophthalmic Epidemiol.* **2007**, *14*, 184–187. [CrossRef] [PubMed]
4. The Global Economic Cost of Visual Impairment, March 2010. Available online: http://www.icoph.org/dynamic/attachments/resources/globalcostofvi_finalreport.pdf. (accessed on 18 June 2018).
5. de Jong, P.T. Age-related macular degeneration. *N. Engl. J. Med.* **2008**, *358*, 2606–2617. [CrossRef] [PubMed]
6. Evans, J.R.; Lawrenson, J.G. Antioxidant vitamin and mineral supplements for slowing the progression of age-related macular degeneration. *Cochrane Database Syst. Rev.* **2012**. [CrossRef]
7. Carneiro, Â.; Andrade, J.P. Nutritional and Lifestyle Interventions for Age-Related Macular Degeneration: A Review. *Oxid. Med. Cell Longev.* **2017**. [CrossRef]
8. Margrain, T.H.; Boulton, M.; Marshall, J.; Sliney, D.H. Do blue light filters confer protection against age-related macular degeneration? *Prog. Retin. Eye Res.* **2004**, *23*, 523–531. [CrossRef] [PubMed]

9. Zhang, Q.Y.; Tie, L.J.; Wu, S.S.; Lv, P.L.; Huang, H.W.; Wang, W.Q.; Wang, H.; Ma, L. Overweight, obesity, and risk of age-related macular degeneration. *Investig. Ophthalmol. Vis. Sci.* **2016**, *57*, 1276–1283. [CrossRef] [PubMed]

10. Peeters, A.; Magliano, D.J.; Stevens, J.; Duncan, B.B.; Klein, R.; Wong, T.Y. Changes in abdominal obesity and age-related macular degeneration: The Atherosclerosis Risk in Communities Study. *Arch. Ophthalmol.* **2008**. [CrossRef] [PubMed]

11. Chiu, C.J.; Taylor, A. Dietary hyperglycemia, glycemic index and metabolic retinal diseases. *Prog. Retin. Eye Res.* **2011**, *30*, 18–53. [CrossRef] [PubMed]

12. Lau, K.; Srivatsav, V.; Rizwan, A.; Nashed, A.; Liu, R.; Shen, R.; Akhtar, M. Bridging the gap between gut microbial dysbiosis and cardiovascular diseases. *Nutrients* **2017**, *9*, 859. [CrossRef] [PubMed]

13. Ma, J.; Zhou, Q.; Li, H. Gut microbiota and nonalcoholic fatty liver disease: insights on mechanisms and therapy. *Nutrients* **2017**, *9*, 1124. [CrossRef] [PubMed]

14. Ponziani, F.R.; Bhoori, S.; Castelli, C.; Putignani, L.; Rivoltini, L.; Del Chierico, F.; Sanguinetti, M.; Morelli, D.; Paroni Sterbini, F.; Petito, V.; et al. Hepatocellular carcinoma is associated with gut microbiota profile and inflammation in nonalcoholic fatty liver disease. *Hepatology* **2018**. [CrossRef] [PubMed]

15. Turnbaugh, P.J.; Ley, R.E.; Mahowald, M.A.; Magrini, V.; Mardis, E.R.; Gordon, J.I. An obesity-associated gut microbiome with increased capacity for energy harvest. *Nature* **2006**, *444*, 1027–1031. [CrossRef] [PubMed]

16. Andriessen, E.M.; Wilson, A.M.; Mawambo, G.; Dejda, A.; Miloudi, K.; Sennlaub, F.; Sapieha, P. Gut microbiota influences pathological angiogenesis in obesity-driven choroidal neovascularization. *EMBO Mol. Med.* **2016**, *8*, 1366–1379. [CrossRef] [PubMed]

17. Ticinesi, A.; Lauretani, F.; Milani, C.; Nouvenne, A.; Tana, C.; Del Rio, D.; Maggio, M.; Ventura, M.; Meschi, T. Aging gut microbiota at the cross-road between nutrition, physical frailty, and sarcopenia: is there a gut-muscle axis? *Nutrients* **2017**, *9*, 1303. [CrossRef] [PubMed]

18. Sender, R.; Fuchs, S.; Milo, R. Are we really vastly outnumbered? Revisiting the ratio of bacterial to host cells in humans. *Cell* **2016**, *164*, 337–340. [CrossRef] [PubMed]

19. Laterza, L.; Rizzatti, G.; Gaetani, E.; Chiusolo, P.; Gasbarrini, A. The gut microbiota and immune system relationship in human graft-versus-host disease. *Mediterr. J. Hematol. Infect Dis.* **2016**, *8*, e2016025. [CrossRef] [PubMed]

20. Eckburg, P.B.; Bik, E.M.; Bernstein, C.N.; Purdom, E.; Dethlefsen, L.; Sargent, M.; Gill, S.R.; Nelson, K.E.; Relman, D.A. Diversity of the human intestinal microbial flora. *Science* **2005**, *308*, 1635–1638. [CrossRef] [PubMed]

21. Ley, R.E.; Bäckhed, F.; Turnbaugh, P.; Lozupone, C.A.; Knight, R.D.; Gordon, J.I. Obesity alters gut microbial ecology. *Proc. Natl. Acad. Sci. USA* **2005**, *102*, 11070–11075. [CrossRef] [PubMed]

22. Clarke, S.F.; Murphy, E.F.; Nilaweera, K.; Ross, P.R.; Shanahan, F.; O'Toole, P.W.; Cotter, P.D. The gut microbiota and its relationship to diet and obesity: New insights. *Gut Microbes* **2012**, *3*, 186–202. [CrossRef] [PubMed]

23. Rabot, S.; Membrez, M.; Bruneau, A.; Gérard, P.; Harach, T.; Moser, M.; Raymond, F.; Mansourian, R.; Chou, C.J. Germ-free C57BL/6J mice are resistant to high-fat-diet-induced insulin resistance and have altered cholesterol metabolism. *FASEB J.* **2010**, *24*, 4948–4959. [CrossRef] [PubMed]

24. Lim, L.S.; Mitchell, P.; Seddon, J.M.; Holz, F.G.; Wong, T.Y. Age-related macular degeneration. *Lancet* **2012**, *379*, 1728–1738. [CrossRef]

25. Chiu, C.J.; Chang, M.L.; Zhang, F.F.; Li, T.; Gensler, G.; Schleicher, M.; Taylor, A. The relationship of major American dietary patterns to age-related macular degeneration. *Am. J. Ophthalmol.* **2014**, *158*, 118–127. [CrossRef] [PubMed]

26. Adams, M.K.; Simpson, J.A.; Aung, K.Z.; Makeyeva, G.A.; Giles, G.G.; English, D.R.; Hopper, J.; Guymer, R.H.; Baird, P.N.; Robman, L.D. Abdominal obesity and age-related macular degeneration. *Am. J. Epidemiol.* **2011**, *173*, 1246–1255. [CrossRef] [PubMed]

27. Stanhope, K.L. Sugar consumption, metabolic disease and obesity: The state of the controversy. *Crit. Rev. Clin. Lab. Sci.* **2016**, *53*, 52–67. [CrossRef] [PubMed]

28. Williams, L.M.; Campbell, F.M.; Drew, J.E.; Koch, C.; Hoggard, N.; Rees, W.D.; Kamolrat, T.; Ngo, H.T.; Steffensen, I.-L.; Gray, S.R. The development of diet-induced obesity and glucose intolerance in c57bl/6 mice on a high-fat diet consists of distinct phases. *PLoS ONE* **2014**, *9*, e106159. [CrossRef] [PubMed]

29. Lim, S.M.; Jeong, J.J.; Woo, K.H.; Han, M.J.; Kim, D.H. Lactobacillus sakei ok67 ameliorates high-fat diet–induced blood glucose intolerance and obesity in mice by inhibiting gut microbiota lipopolysaccharide production and inducing colon tight junction protein expression. *Nutr. Res.* **2016**, *36*, 337–348. [CrossRef] [PubMed]

30. Schleicher, M.; Weikel, K.; Garber, C.; Taylor, A. Diminishing risk for age-related macular degeneration with nutrition: A current view. *Nutrients* **2013**, *5*, 2405–2456. [CrossRef] [PubMed]

31. Do, M.H.; Lee, E.; Oh, M.J.; Kim, Y.; Park, H.Y. High-Glucose or -Fructose Diet Cause Changes of the Gut Microbiota and Metabolic Disorders in Mice without Body Weight Change. *Nutrients* **2018**, *10*, 761. [CrossRef] [PubMed]

32. Rizzatti, G.; Lopetuso, L.R.; Gibino, G.; Binda, C.; Gasbarrini, A. Proteobacteria: A common factor in human diseases. *Biomed. Res. Int.* **2017**. [CrossRef] [PubMed]

33. Rowan, S.; Jiang, S.; Korem, T.; Szymanski, J.; Chang, M.L.; Szelog, J.; Cassalman, C.; Dasuri, K.; McGuire, C.; Nagai, R.; et al. Involvement of a gut-retina axis in protection against dietary glycemia-induced age-related macular degeneration. *Proc. Natl. Acad. Sci. USA* **2017**, *114*, E4472–E4481. [CrossRef] [PubMed]

34. Schmidl, D.; Garhöfer, G.; Schmetterer, L. Nutritional supplements in age-related macular degeneration. *Acta Ophthalmol.* **2015**, *93*, 105–121. [CrossRef] [PubMed]

35. Age-Related Eye Disease Study Research Group. A randomized, placebo-controlled, clinical trial of high-dose supplementation with vitamins C and E, beta carotene, and zinc for age-related macular degeneration and vision loss. *Arch. Ophthalmol.* **2001**, *119*, 1417–1436. [CrossRef]

36. Chew, E.Y.; Clemons, T.E.; SanGiovanni, J.P.; Danis, R.; Ferris, F.L.; Elman, M.; Antoszyk, A.; Ruby, A.; Orth, D.; Bressler, S.; et al. Lutein + zeaxanthin and omega-3 fatty acids for age-related macular degeneration: The Age-Related Eye Disease Study 2 (AREDS2) randomized clinical trial. *JAMA* **2013**, *309*, 2005–2015.

37. Age-Related Eye Disease Study 2 (AREDS2) Research Group. Secondary analyses of the effects of lutein/zeaxanthin on age-related macular degeneration progression: AREDS2 report No. 3. *JAMA Ophthalmol.* **2014**, *132*, 142–149. [CrossRef] [PubMed]

38. Seddon, J.M.; Ajani, U.A.; Sperduto, R.D.; Hiller, R.; Blair, N.; Burton, T.C.; Farber, M.D.; Gragoudas, E.S.; Haller, J.; Miller, D.T. Dietary carotenoids, vitamins A, C, and E, and advanced age-related macular degeneration. *JAMA* **1994**, *272*, 1413–1420. [CrossRef] [PubMed]

39. Richer, S.; Stiles, W.; Statkute, L.; Pulido, J.; Frankowski, J.; Rudy, D.; Pei, K.; Tsipursky, M.; Nyland, J. Double-masked, placebo-controlled, randomized trial of lutein and antioxidant supplementation in the intervention of atrophic age-related macular degeneration: The veterans LAST study (Lutein Antioxidant Supplementation Trial). *Optometry* **2004**, *75*, 216–230. [CrossRef]

40. Delcourt, C.; Carrière, I.; Delage, M.; Barberger-Gateau, P.; Schalch, W. Plasma Lutein and Zeaxanthin and Other Carotenoids as Modifiable Risk Factors for Age-Related Maculopathy and Cataract: The POLA Study. *Investig. Ophthalmol. Vis. Sci.* **2006**, *47*, 2329–2335. [CrossRef] [PubMed]

41. Seddon, J.M.; George, S.; Rosner, B. Cigarette, smoking, fish consumption, omega 3 fatty acid intake and association with age-related macular degeneration: The US Twin Study of age-related macular degeneration. *Arch. Ophthlmol.* **2006**, *124*, 995–1001. [CrossRef] [PubMed]

42. Chua, B.; Flood, V.; Rochtchina, E.; Wang, J.J.; Smith, W.; Mitchell, P. Dietary fatty acids and the 5-year incidence of age-related maculopathy. *Arch. Ophthalmol.* **2006**, *124*, 981–986. [CrossRef] [PubMed]

43. SanGiovanni, J.P.; Chew, E.Y.; Clemons, T.E.; Davis, M.D.; Ferris, F.L. 3rd; Gensler, G.R.; Kurinij, N.; Lindblad, A.S.; Milton, R.C.; Seddon, J.M.; et al. The relationship of dietary lipid intake and age-related macular degeneration in a case-control study: AREDS Report No. 20. *Arch. Ophthalmol.* **2007**, *125*, 671–679. [PubMed]

44. Age-Related Eye Disease Study Research Group. The relationship of dietary carotenoid and vitamin A, E, and C intake with age-related macular degeneration in a case-control study: AREDS Report No. 22. *Arch. Ophthalmol.* **2007**, *125*, 1225–1232. [CrossRef] [PubMed]

45. SanGiovanni, J.P.; Chew, E.Y.; Agrón, E.; Clemons, T.E.; Ferris, F.L., 3rd; Gensler, G.; Lindblad, A.S.; Milton, R.C.; Seddon, J.M.; Klein, R.; et al. The relationship of dietary omega-3 long-chain polyunsaturated fatty acid intake with incident age-related macular degeneration: AREDS report No. 23. *Arch. Ophthalmol.* **2008**, *126*, 1274–1279. [PubMed]

46. Stringham, J.M.; Hammond, B.R. Macular pigment and visual performance under glare conditions. *Optom. Vis. Sci.* **2008**, *85*, 82–88. [CrossRef] [PubMed]

47. Cho, E.; Hankinson, S.E.; Rosner, B.; Willett, W.C.; Colditz, G.A. Prospective study of lutein/zeaxanthin intake and risk of age-related macular degeneration. *Am. J. Clin. Nutr.* **2008**, *87*, 1837–1843. [CrossRef] [PubMed]

48. Tan, J.S.; Wang, J.J.; Flood, V.; Rochtchina, E.; Smith, W.; Mitchell, P. Dietary antioxidants and the long-term incidence of age-related macular degeneration: The Blue Mountains eye study. *Ophthalmology* **2008**, *115*, 334–341. [CrossRef] [PubMed]

49. Newsome, D.A. A randomized, prospective, placebo-controlled clinical trial of a novel zinc-monocysteine compound in age-related macular degeneration. *Curr. Eye Res.* **2008**, *33*, 591–598. [CrossRef] [PubMed]

50. SanGiovanni, J.P.; Agron, E.; Clemons, T.E.; Chew, E.Y. Omega-3 long-chain polyunsaturated fatty acid intake inversely associated with 12-year progression to advanced age-related macular degeneration. *Arch. Ophthalmol.* **2009**, *127*, 110–112. [CrossRef] [PubMed]

51. Ho, L.; van Leeuwen, R.; Witteman, J.C.; van Duijn, C.M.; Uitterlinden, A.G.; Hofman, A.; de Jong, P.T.; Vingerling, J.R.; Klaver, C.C. Reducing the genetic risk of age-related macular degeneration with dietary antioxidants, zinc, and omega-3 fatty acids: The Rotterdam study. *Arch. Ophthalmol.* **2011**, *129*, 758–766. [CrossRef] [PubMed]

52. Snellen, E.L.M.; Verbeek, A.L.M.; van den Hoogen, G.W.P.; Cruysberg, J.R.M.; Hoyng, C.B. Neovascular age-related macular degeneration and its relationship to antioxidant intake. *Acta Ophthalmol. Scand.* **2002**, *80*, 368–371. [CrossRef] [PubMed]

53. Nolan, J.M.; Akkali, M.C.; Loughman, J.; Howard, A.N.; Beatty, S. Macular carotenoid supplementation in subjects with atypical spatial profiles of macular pigment. *Exp. Eye Res.* **2012**, *101*, 9–15. [CrossRef] [PubMed]

54. Souied, E.H.; Delcourt, C.; Querques, G.; Bassols, A.; Merle, B.; Zourdani, A.; Smith, T.; Benlian, P. Nutritional AMD Treatment 2 Study Group. Oral docosahexaenoic acid in the prevention of exudative age-related macular degeneration: The Nutritional AMD Treatment 2 study. *Ophthalmology* **2013**, *120*, 1619–1631. [CrossRef] [PubMed]

55. Hammond, B.R.; Fletcher, L.M.; Roos, F. A double-blind, placebo- controlled study on the effects of lutein and zeaxanthin on photostress recovery, glare disability, and chromatic contrast. *Investig. Ophthalmol. Vis. Sci.* **2014**, *55*, 8583–8589. [CrossRef] [PubMed]

56. Aoki, A.; Inoue, M.; Nguyen, E. Dietary *n*-3 Fatty Acid, α-Tocopherol, Zinc, vitamin D, vitamin C, and β-carotene are Associated with Age-Related Macular Degeneration in Japan. *Sci. Rep.* **2016**, *6*, 20723. [CrossRef] [PubMed]

57. Braakhuis, A.; Raman, R.; Vaghefi, E. The Association between Dietary Intake of Antioxidants and Ocular Disease. *Diseases* **2017**, *5*, 3. [CrossRef] [PubMed]

58. Rinninella, E.; Pizzoferrato, M.; Cintoni, M.; Servidei, S.; Mele, M.C. Nutritional support in mitochondrial diseases: The state of the art. *Eur. Rev. Med. Pharmacol. Sci.* **2018**, *22*, 4288–4298. [PubMed]

59. Ambati, J.; Fowler, B.J. Mechanisms of age-related macular degeneration. *Neuron* **2012**, *75*, 26–39. [CrossRef] [PubMed]

60. Schwenger, K.J.; Allard, J.P. Clinical approaches to non-alcoholic fatty liver disease. *World J. Gastroenterol.* **2014**, *20*, 1712–1723. [CrossRef] [PubMed]

61. Singh, U.; Devaraj, S.; Jialal, I. Vitamin E, oxidative stress, and inflammation. *Annu Rev. Nutr.* **2005**, *25*, 151–174. [CrossRef] [PubMed]

62. Westergren, T.; Kalikstad, B. Dosage and formulation issues: Oral vitamin E therapy in children. *Eur. J. Clin. Pharmacol.* **2010**, *66*, 109–118. [CrossRef] [PubMed]

63. Subramanian, V.S.; Srinivasan, P.; Wildman, A.; Marchant, J.S.; Said, H.M. Molecular mechanism(s) involved in differential expression of vitamin C transporters along the intestinal tract. *Am. J. Physiol. Gastrointest. Liver Physiol.* **2017**, *312*, G340–G347. [CrossRef] [PubMed]

64. Li, Z.Y.; Tso, M.O.; Wang, H.M.; Organisciak, D.T. Amelioration of photic injury in rat retina by ascorbic acid: A histopathologic study. *Investig. Ophthalmol. Vis. Sci.* **1985**, *26*, 1589–1598.

65. VandenLangerberg, G.M.; Mares-Perlman, J.A.; Klein, R. Associations between antioxidant and zinc intake and the 5-years incidence of early age-related maculopathy in the Beaver Dam Eye Study. *Am. J. Epidemiol.* **1998**, *148*, 204–214. [CrossRef]

66. Qiao, Y.; Sun, J.; Ding, Y.; Le, G.; Shi, Y. Alterations of the gut microbiota in high-fat diet mice is strongly linked to oxidative stress. *Appl. Microbiol. Biotechnol.* **2013**, *97*, 1689–1697. [CrossRef] [PubMed]

67. Kruidenier, L.; Kuiper, I.; van Duijn, W.; Marklund, S.L.; van Hogezand, R.A.; Lamers, C.B.; Verspaget, H.W. Differential mucosal expression of three superoxide dismutase isoforms in inflammatory bowel disease. *J. Pathol.* **2003**, *201*, 7–16. [CrossRef] [PubMed]

68. Matsuoka, K.; Kanai, T. The gut microbiota and inflammatory bowel disease. *Semin. Immunopathol.* **2015**, *37*, 47–55. [CrossRef] [PubMed]

69. Wright, E.K.; Kamm, M.A.; Teo, S.M.; Inouye, M.; Wagner, J.; Kirkwood, C.D. Recent advances in characterizing the gastrointestinal microbiome in Crohn's disease: A systematic review. *Inflamm. Bowel Dis.* **2015**, *21*, 1219–1228. [PubMed]

70. Yan, H.; Wang, H.; Zhang, X.; Li, X.; Yu, J. Ascorbic acid ameliorates oxidative stress and inflammation in dextran sulfate sodium-induced ulcerative colitis in mice. *Int. J. Clin. Exp. Med.* **2015**, *8*, 20245–20253. [PubMed]

71. Tahan, G.; Aytac, E.; Aytekin, H.; Gunduz, F.; Dogusoy, G.; Aydin, S.; Tahan, V.; Uzun, H. Vitamin E has a dual effect of anti-inflammatory and antioxidant activities in acetic acid-induced ulcerative colitis in rats. *Can. J. Surg.* **2011**, *54*, 333–338. [CrossRef] [PubMed]

72. Maggini, M.; Eva, S.W.; Beveridge, S.; Dietrich, H.H. Selected vitamins and trace elements support immune function by strengthening epithelial barriers and cellular and humoral immune responses. *Br. J. Nutr.* **2007**, *98*, S29–S35. [CrossRef] [PubMed]

73. Pierre, J.F.; Hinterleitner, R.; Bouziat, R.; Hubert, N.A.; Leone, V.; Miyoshi, J.; Jabri, B.; Chang, E.B. Dietary antioxidant micronutrients alter mucosal inflammatory risk in a murine model of genetic and microbial susceptibility. *J. Nutr. Biochem.* **2018**, *54*, 95–104. [CrossRef] [PubMed]

74. Mandal, S.; Godfrey, K.M.; McDonald, D.; Treuren, W.V.; Bjørnholt, J.V.; Midtvedt, T.; Moen, B.; Rudi, K.; Knight, R.; Brantsæter, A.L.; et al. Fat and vitamin intakes during pregnancy have stronger relations with a pro-inflammatory maternal microbiota than does carbohydrate intake. *Microbiome* **2016**, *4*, 55. [CrossRef] [PubMed]

75. Li, L.; Krause, L.; Somerset, S. Associations between micronutrient intakes and gut microbiota in a group of adults with cystic fibrosis. *Clin. Nutr.* **2017**, *36*, 1097–1104. [CrossRef] [PubMed]

76. Tang, M.; Frank, D.N.; Sherlock, L.; Ir, D.; Robertson, C.E.; Krebs, N.F. Effect of vitamin E with therapeutic iron supplementation on iron repletion and gut microbiome in US iron deficient infants and toddlers. *J. Pediatr. Gastroenterol. Nutr.* **2016**, *63*, 379–385. [CrossRef] [PubMed]

77. Xu, J.; Xu, C.; Chen, X. Regulation of an antioxidant blend on intestinal redox status and major microbiota in early weaned piglets. *Nutrition* **2014**, *30*, 584–589. [CrossRef] [PubMed]

78. Packer, L.; Colman, C. *The Antioxidant Miracle*; Wiley & Sons: New York, NY, USA, 1999.

79. Skrypnik, K.; Suliburska, J. Association between the gut microbiota and mineral metabolism. *J. Sci. Food Agric.* **2017**, *98*, 2449–2460. [CrossRef] [PubMed]

80. Wang, Y.; Yi, L.; Zhao, M.L. Effects of zinc–methionine on growth performance, intestinal flora and immune function in pigeon squabs. *Br. Poult. Sci.* **2014**, *3*, 403–408. [CrossRef] [PubMed]

81. Leure-duPree, A.E.; McClain, C.J. The Effect of severe zinc deficiency on the morphology of the rat retinal pigment epithelium. *Investig. Ophthalmol. Vis. Sci.* **1982**, *23*, 425–434.

82. Smailhodzic, D.; van Asten, F.; Blom, A.M.; Mohlin, F.C.; den Hollander, A.I.; van de Ven, J.P.; van Huet, R.A.; Groenewoud, J.M.; Tian, Y.; Berendschot, T.T.; et al. Zinc supplementation inhibits complement activation in age-related macular degeneration. *PLoS ONE* **2014**, *9*, e112682. [CrossRef] [PubMed]

83. Zampatti, S.; Ricci, F.; Cusumano, A.; Marsella, L.T.; Novelli, G.; Giardina, E. Review of nutrient actions on age-related macular degeneration. *Nutr. Res.* **2014**, *34*, 95–105. [CrossRef] [PubMed]

84. Zeng, S.; Hernandez, J.; Mullins, R.F. Effects of antioxidant components of AREDS vitamins and zinc ions on endothelial cell activation: Implications for macular degeneration. *Investig. Ophthalmol. Vis. Sci.* **2012**, *53*, 1041–1047. [CrossRef] [PubMed]

85. Organisciak, D.; Wong, P.; Rapp, C.; Darrow, R.; Ziesel, A.; Rangarajan, R.; Lang, J. Light-induced retinal degeneration is prevented by zinc, a component in the agerelated eye disease study formulation. *Photochem. Photobiol.* **2012**. [CrossRef] [PubMed]

86. Wood, J.P.; Osborne, N.N. Zinc and energy requirements in induction of oxidative stress to retinal pigmented epithelial cells. *Neurochem. Res.* **2003**, *28*, 1525–1533. [CrossRef] [PubMed]

87. Gielda, L.M.; DiRita, V.J. Zinc competition among the intestinal microbiota. *MBio* **2012**. [CrossRef] [PubMed]

88. Zackular, J.; Moore, A.J. Dietary Zinc Alters the Microbiota and Decreases Resistance to Clostridium difficile Infection. *Nat. Med.* **2016**, *22*, 1330–1334. [CrossRef] [PubMed]

89. Reed, S.; Neuman, H.; Moscovich, S. Chronic Zinc Deficiency Alters Chick Gut Microbiota Composition and Function. *Nutrients* **2015**, *7*, 9768–9784. [CrossRef] [PubMed]

90. Watson, H.; Mitra, S.; Croden, F.C.; Taylor, M.; Wood, H.M.; Perry, S.L.; Spencer, J.A.; Quirke, P.; Toogood, G.J.; Lawton, C.L.; et al. A randomised trial of the effect of omega-3 polyunsaturated fatty acid supplements on the human intestinal microbiota. *Gut* **2017**. [CrossRef] [PubMed]

91. Engberg, R.M.; Hedemann, M.S.; Leser, T.D.; Jensen, B.B. Effect of zinc bacitracin and salinomycin on intestinal microflora and performance of broilers. *Poult. Sci.* **2000**, *79*, 1311–1319. [CrossRef] [PubMed]

92. Ren, Z.; Zhao, Z.; Wang, Y.; Huang, K. Preparation of selenium/zinc-enriched probiotics and their effect on blood selenium and zinc concentrations, antioxidant capacities, and intestinal microflora in canine. *Biol. Trace Elem. Res.* **2011**, *141*, 170–183. [CrossRef] [PubMed]

93. Chhikara, N.; Kushwaha, K.; Sharma, P.; Gat, Y.; Panghal, A. Bioactive compounds of beetroot and utilization in food processing industry: A critical review. *Food Chem.* **2019**, *272*, 192–200. [CrossRef] [PubMed]

94. Castenmiller, J.J.; West, C.E. Bioavailability and bioconversion of carotenoids. *Annu. Rev. Nutr.* **1998**, *18*, 19–38. [CrossRef] [PubMed]

95. Reboul, E.; Abou, L.; Mikail, C.; Ghiringhelli, O.; André, M.; Portugal, H.; Jourdheuil-Rahmani, D.; Amiot, M.J.; Lairon, D.; Borel, P. Lutein transport by Caco-2 TC-7 cells occurs partly by a facilitated process involving the scavenger receptor class B type I (SR-BI). *Biochem. J.* **2005**, *387*, 455–461. [CrossRef] [PubMed]

96. Van Bennekum, A.; Werder, M.; Thuahnai, S.T.; Han, C.H.; Duong, P.; Williams, D.L.; Wettstein, P.; Schulthess, G.; Philips, M.C.; Hauser, H. Class B scavenger receptor-mediated intestinal absorption of dietary beta-carotene and cholesterol. *Biochemistry* **2005**, *44*, 4517–4525. [CrossRef] [PubMed]

97. During, A.; Dawson, H.D.; Harrison, E.H. Carotenoid transport is decreased and expression of the lipid transporters SR-BI, NPC1L1, and ABCA1 is downregulated in Caco-2 cells treated with ezetimibe. *J. Nutr.* **2005**, *135*, 2305–2312. [CrossRef] [PubMed]

98. Binxing, L.; Preejith, P.V.; Zhengqing, S. Retinal Accumulation of Zeaxanthin, Lutein, and β-Carotene in Mice Deficient in Carotenoid Cleavage Enzymes. *Exp. Eye Res.* **2017**, *159*, 123–131.

99. Leung, I.Y.; Sandstrom, M.M.; Zucker, C.L.; Neuringer, M.; Max Snodderly, D. Nutritional manipulation of primate retinas. IV Effects of n–3 fatty acids, lutein, and zeaxanthin on S-cones and rods in the foveal region. *Exp. Eye Res.* **2005**, *81*, 513–529. [CrossRef] [PubMed]

100. Trevithick-Sutton, C.C.; Foote, C.S.; Collins, M.; Trevithick, J.R. The retinal carotenoids, zeaxanthin and lutein scavenge superoxide and hydroxyl radicals: A chemiluminescence and ESR study. *Mol. Vis.* **2006**, *12*, 1127–1135. [PubMed]

101. Vu, H.T.; Robman, L.; McCarty, C.A.; Taylor, H.R.; Hodge, A. Does dietary lutein and zeaxanthin increase the risk of age related macular degeneration? The Melbourne Visual Impairment Project. *Br. J. Ophthalmol.* **2006**, *90*, 389–393. [CrossRef] [PubMed]

102. Landrum, J.T.; Bone, R.A.; Joa, H.; Kilburn, M.D.; Moore, L.L.; Sprague, K.E. A one year study of the macular pigment: The effect of 140 days of a lutein supplement. *Exp. Eye Res.* **1997**, *65*, 57–62. [CrossRef] [PubMed]

103. LaRowe, T.L.; Mares, J.A.; Snodderly, D.M.; Klein, M.L.; Wooten, B.R.; Chappell, R. Macular pigment density and age-related maculopathy in the Carotenoids in Age-Related Eye Disease Study. An ancillary study of the women's health initiative. *Ophthalmology* **2008**, *115*, 876–883. [CrossRef] [PubMed]

104. Nolan, J.M.; Power, R.; Stringham, J.; Dennison, J.; Stack, J.; Kelly, D.; Moran, R.; Akuffo, K.O.; Corcoran, L.; Beatty, S. Enrichment of Macular Pigment Enhances Contrast Sensitivity in Subjects Free of Retinal Disease: Central Retinal Enrichment Supplementation Trials-Report 1. *Investig. Ophthalmol. Vis. Sci.* **2016**, *57*, 3429–3439. [CrossRef] [PubMed]

105. Chew, E.Y. Nutrition Effects on Ocular Diseases in the Aging Eye. *Investig. Ophthalmol. Vis. Sci.* **2013**, *54*, ORSF42–ORSF47. [CrossRef] [PubMed]

106. Karlsson, F.H.; Fåk, F.; Nookaew, I.; Tremaroli, V.; Fagerberg, B.; Petranovic, D.; Bäckhed, F.; Nielsen, J. Symptomatic atherosclerosis is associated with an altered gut metagenome. *Nat. Commun.* **2012**, *3*, 1245. [CrossRef] [PubMed]

107. Arteni, A.A.; Fradot, M.; Galzerano, D.; Mendes-Pinto, M.M.; Sahel, J.A.; Picaud, S.; Robert, B.; Pascal, A.A. Structure and conformation of the carotenoids in human retinal macular pigment. *PLoS ONE* **2015**, *10*, e0135779. [CrossRef] [PubMed]

108. Bartlett, H.; Howells, O.; Eperjesi, F. The role of macular pigment assessment in clinical practice: A review. *Clin. Exp. Optom.* **2010**, *93*, 300–308. [CrossRef] [PubMed]

109. Nolan, J.M.; Meagher, K.; Kashani, S.; Beatty, S. What is meso-zeaxanthin, and where does it come from? *Eye* **2013**, *27*, 899–905. [CrossRef] [PubMed]

110. Hadden, W.L.; Watkins, R.H.; Levy, L.W.; Regalado, E.; Rivadeneira, D.M.; van Breemen, R.B.; Schwartz, S.J. Carotenoids composition of marigold (Tagetes erecta) flower extract used as nutritional supplement. *J. Agric. Food Chem.* **1999**, *47*, 4189–4194. [CrossRef] [PubMed]

111. Nolan, J.M.; Stack, J.; O' Donovan, O.; Loane, E.; Beatty, S. Risk factors for age-related maculopathy are associated with a relative lack of macular pigment. *Exp. Eye Res.* **2007**, *84*, 61–74. [CrossRef] [PubMed]

112. Eisenhauer, B.; Natoli, S.; Liew, G.; Flood, V.M. Lutein and Zeaxanthin food sources, bioavailability and dietary variety in age related macular degeneration protection. *Nutrients* **2017**, *9*, 120. [CrossRef] [PubMed]

113. Yonekura, L.; Nagao, A. Intestinal absorption of dietary carotenoids. *Mol. Nutr. Food Res.* **2007**, *51*, 107–115. [CrossRef] [PubMed]

114. Wang, J.J.; Buitendijk, G.H.; Rochtchina, E.; Lee, K.E.; Klein, B.E.; van Duijn, C.M.; Flood, V.M.; Meuer, S.M.; Attia, J.; Myers, C.; et al. Genetic susceptibility, dietary antioxidants, and long-term incidence of age-related macular degeneration in two populations. *Ophthalmology* **2014**, *121*, 667–675. [CrossRef] [PubMed]

115. Meyers, K.J.; Mares, J.A.; Igo, R.P.Jr.; Truitt, B.; Liu, Z.; Millen, A.E.; Klein, M.; Johnson, E.J.; Engelman, C.D.; Karki, C.K.; et al. Genetic evidence for role of carotenoids in age-related macular degeneration in the Carotenoids in Age-related Eye Disease study (CAREDS). *Investig. Ophthalmol. Vis. Sci.* **2014**, *55*, 587–599. [CrossRef] [PubMed]

116. Bernstein, P.S.; Delori, F.C.; Richer, S.; van Kuijk, F.J.; Wenzel, A.J. The Value of Measurement of Macular Carotenoid Pigment Optical Densities and Distributions in Age-Related Macular Degeneration and Other Retinal Disorders. *Vis. Res.* **2010**, *50*, 716–728. [CrossRef] [PubMed]

117. Ma, L.; Liu, R.; Du, J.H.; Liu, T.; Wu, S.S.; Liu, X.H. Lutein, Zeaxanthin and Meso-zeaxanthin Supplementation Associated with Macular Pigment Optical Density. *Nutrients.* **2016**, *8*, 426. [CrossRef] [PubMed]

118. Gibson, G.R.; Wang, X. Enhancement of bifidobacteria from human gut contents by oligofructose using continuous culture. *FEMS Microbiol. Lett.* **1994**, *118*, 121–127. [CrossRef] [PubMed]

119. Burns, A.J.; Rowland, I.R. Anti-carcinogenicity of probiotics and prebiotics. *Curr. Issues Intest. Microbiol.* **2000**, *1*, 13–24. [PubMed]

120. Gill, H.S. Stimulation of the immune system by lactic cultures. *Int. Dairy J.* **1998**, *8*, 535–544. [CrossRef]

121. Molan, A.L.; Liu, Z.; Plimmer, G. Evaluation of the Effect of Blackcurrant Products on Gut Microbiota and on Markers of Risk for Colon Cancer in Humans. *Phytother. Res.* **2014**, *28*, 416–422. [CrossRef] [PubMed]

122. Gill, C.I.; Rowland, I.R. Diet and cancer: Assessing the risk. *Br. J. Nutr.* **2002**, *88*, S73–S87. [CrossRef] [PubMed]

123. Bakke, D.; Sun, J. Ancient nuclear receptor vdr with new functions: Microbiome and inflammation. *Inflamm. Bowel Dis.* **2018**, *24*, 1149–1154. [CrossRef] [PubMed]

124. Kassi, E.; Adamopoulos, C.; Basdra, E.K.; Papavassiliou, A.G. Role of vitamin D in atherosclerosis. *Circulation* **2013**, *128*, 2517–2531. [CrossRef] [PubMed]

125. Holick, M.F. Vitamin D deficiency. *N. Engl. J. Med.* **2007**, *357*, 266–281. [CrossRef] [PubMed]

126. Bikle, D.D. Vitamin D metabolism, mechanism of action, and clinical applications. *Chem. Biol.* **2014**, *21*, 319–329. [CrossRef] [PubMed]

127. Johnson, J.A.; Grande, J.P.; Roche, P.C.; Campbell, R.J.; Kumar, R. Immuno-localization of the calcitriol receptor, calbindin-D28k and the plasma membrane calcium pump in the human eye. *Curr. Eye Res.* **1995**, *14*, 101–108. [CrossRef] [PubMed]

128. Morrison, M.A.; Silveira, A.C.; Huynh, N.; Jun, G.; Smith, S.E.; Zacharaki, F.; Sato, H.; Loomis, S.; Andreoli, M.T.; Adams, S.M.; et al. Systems biology-based analysis implicates a novel role for vitamin D metabolism in the pathogenesis of age-related macular degeneration. *Hum. Genom.* **2011**, *5*, 538–568. [CrossRef]

129. Choi, D.; Appukuttan, B.; Binek, S.J.; Planck, S.R.; Stout, J.T.; Rosenbaum, J.T.; Smith, J.R. Prediction of cis-regulatory elements controlling genes differentially expressed by retinal and choroidal vascular endothelial cells. *J. Ocul. Biol. Dis. Inform.* **2008**, *1*, 37–45. [CrossRef] [PubMed]

130. Layana, A.G.; Minnella, A.M.; Garhöfer, G.; Aslam, T.; Holz, F.G.; Leys, A.; Silva, R.; Delcourt, C.; Souied, E.; Seddon, J.M. Vitamin D and Age-Related Macular Degeneration. *Nutrients* **2017**, *9*, E1120. [CrossRef] [PubMed]

131. Riccio, P.; Rossano, R. Diet, Gut Microbiota, and Vitamins D + A in Multiple Sclerosis. *Neurotherapeutics* **2018**, *15*, 75–91. [CrossRef] [PubMed]

132. Gubatan, J.; Moss, A.C. Vitamin D in inflammatory bowel disease: More than just a supplement. *Curr. Opin. Gastroenterol.* **2018**, *34*, 217–225. [CrossRef] [PubMed]

133. Villa, C.R.; Taibi, A.; Chen, J.; Ward, W.E.; Comelli, E.M. Colonic Bacteroides are positively associated with trabecular bone structure and programmed by maternal vitamin D in male but not female offspring in an obesogenic environment. *Int. J. Obes.* **2018**, *42*, 696–703. [CrossRef] [PubMed]

134. Talsness, C.E.; Penders, J.; Jansen, E.H.J.M.; Damoiseaux, J.; Thijs, C.; Mommers, M. Influence of vitamin D on key bacterial taxa in infant microbiota in the KOALA Birth Cohort Study. *PLoS ONE* **2017**, *12*, e0188011. [CrossRef] [PubMed]

135. Luthold, R.V.; Fernandes, G.R.; Franco-de-Moraes, A.C.; Folchetti, L.G.; Ferreira, S.R. Gut microbiota interactions with the immunomodulatory role of vitamin D in normal individuals. *Metabolism* **2017**, *69*, 76–86. [CrossRef] [PubMed]

136. Bashir, M.; Prietl, B.; Tauschmann, M.; Mautner, S.I.; Kump, P.K.; Treiber, G.; Wurm, P.; Gorkiewicz, G.; Högenauer, C.; Pieber, T.R. Effects of high doses of vitamin D_3 on mucosa-associated gut microbiome vary between regions of the human gastrointestinal tract. *Eur. J. Nutr.* **2016**, *55*, 1479–1489. [CrossRef] [PubMed]

137. Garg, M.; Hendy, P.; Ding, J.N.; Shaw, S.; Hold, G.; Hart, A. The effect of vitamin D on intestinal inflammation and faecal microbiota in patients with ulcerative colitis. *J. Crohn's Colitis* **2018**, *12*, 963–972. [CrossRef] [PubMed]

138. Schäffler, H.; Herlemann, D.P.; Klinitzke, P.; Berlin, P.; Kreikemeyer, B.; Jaster, R.; Lamprecht, G. Vitamin D administration leads to a shift of the intestinal bacterial composition in Crohn's disease patients, but not in healthy controls. *J. Dig. Dis.* **2018**, *19*, 225–234. [CrossRef] [PubMed]

139. Kanhere, M.; He, J.; Chassaing, B.; Ziegler, T.R.; Alvarez, J.A.; Ivie, E.A.; Hao, L.; Hanfelt, J.; Gewirtz, A.T.; Tangpricha, V. Bolus Weekly Vitamin D_3 Supplementation Impacts Gut and Airway Microbiota in Adults with Cystic Fibrosis: A Double-Blind, Randomized, Placebo-Controlled Clinical Trial. *J. Clin. Endocrinol. MeTable* **2018**, *103*, 564–574. [CrossRef] [PubMed]

140. Watanabe, Y.; Tatsuno, I. Omega-3 polyunsaturated fatty acids for cardiovascular diseases: Present, past and future. *Expert Rev. Clin. Pharmacol.* **2017**, *10*, 865–873. [CrossRef] [PubMed]

141. Miles, E.A.; Calder, P.C. Influence of marine n-3 polyunsaturated fatty acids on immune function and a systematic review of their effects on clinical outcomes in rheumatoid arthritis. *Br. J. Nutr.* **2012**, *107*, S171–S184. [CrossRef] [PubMed]

142. Calder, P.C. Fatty acids and immune function: Relevance to inflammatory bowel diseases. *Int. Rev. Immunol.* **2009**, *28*, 506–534. [CrossRef] [PubMed]

143. Arnold, L.E.; Young, A.S.; Belury, M.A.; Cole, R.M.; Gracious, B.; Seidenfeld, A.M.; Wolfson, H.; Fristad, M.A. Omega-3 fatty acids plasma levels before and after supplementation: Correlation with mood and clinical outcomes in the omega-3 and therapy studies. *J. Child Adolesc. Psychopharmacol.* **2017**, *27*, 223–233. [CrossRef] [PubMed]

144. Merendino, N.; Costantini, L.; Manzi, L.; Molinari, R.; D'Eliseo, D.; Velotti, F. Dietary omega-3 polyunsaturated fatty acid DHA: A potential adjuvant in the treatment of cancer. *Biomed. Res. Int.* **2013**. [CrossRef] [PubMed]

145. Franceschi, M.; Bonafè, S.; Valensin, F.; Olivieri, M.; de Luca, E.; Ottaviani, G.; de Benedictis, G. Inflammaging: An evolutionary perspective on immunosenescence. *Ann. N.Y. Acad. Sci.* **2000**, *908*, 244–254. [CrossRef] [PubMed]

146. Franceschi, M.; Capri, D.; Monti, S.; Giunta, S.; Olivieri, F.; Sevini, F.; Panourgia, M.P.; Invidia, L.; Celani, L.; Scurti, M.; et al. Inflammaging and anti-inflammaging: A systemic perspective on aging and longevity emerged from studies in humans. *Mech. Ageing Dev.* **2007**, *128*, 92–105. [CrossRef] [PubMed]

147. Gong, Y.; Fu, Z.; Liegl, R.; Chen, J.; Hellström, A.; Smith, L.E. ω-3 and ω-6 long-chain PUFAs and their enzymatic metabolites in neovascular eye diseases. *Am. J. Clin. Nutr.* **2017**, *106*, 16–26. [CrossRef] [PubMed]

148. Lawrenson, J.G.; Evans, J.R. Omega 3 fatty acids for preventing or slowing the progression of age-related macular degeneration. *Cochrane Database Syst. Rev.* **2015**, *9*, CD010015. [CrossRef] [PubMed]

149. Hildebrandt, M.A.; Hoffmann, C.; Sherrill-Mix, S.A.; Keilbaugh, S.A.; Hamady, M.; Chen, Y.Y.; Knight, R.; Ahima, R.S.; Bushman, F.; Wu, G.D. High-fat diet determines the composition of the murine gut microbiome independently of obesity. *Gastroenterology* **2009**, *137*, 1716–1724. [CrossRef] [PubMed]

150. Zhang, C.; Zhang, M.; Wang, S.; Han, R.; Cao, Y.; Hua, W.; Mao, Y.; Zhang, X.; Pang, X.; Wei, C.; et al. Interaction between gut microbiota, host genetics and diet relevant to development of metabolic syndrome in mice. *ISME J.* **2010**, *4*, 232–241. [CrossRef] [PubMed]

151. Devkota, S.; Wang, Y.; Musch, M.W.; Leone, V.; Fehlner-Peach, H.; Nadimpalli, A.; Antonopoulos, D.A.; Jabri, B.; Chang, E.B. Dietary-fat-induced taurocholic acid promotes pathobiont expansion and colitis in Il10−/− mice. *Nature* **2012**, *487*, 104–108. [CrossRef] [PubMed]

152. David, L.A.; Maurice, C.F.; Carmody, R.N.; Gootenberg, D.B.; Button, J.E.; Wolfe, B.E.; Ling, A.V.; Devlin, A.S.; Varma, Y.; Fischbach, M.A.; et al. Diet rapidly and reproducibly alters the human gut microbiome. *Nature* **2014**, *505*, 559–563. [CrossRef] [PubMed]

153. Noriega, B.S.; Sanchez-Gonzalez, M.A.; Salyakina, D.; Coffman, J. Understanding the impact of omega-3 rich diet on the gut microbiota. *Case Rep. Med.* **2016**. [CrossRef] [PubMed]

154. Menni, C.; Zierer, J.; Pallister, T.; Jackson, M.A.; Long, T.; Mohney, R.P.; Steves, C.J.; Spector, T.D.; Valdes, A.M. Omega-3 fatty acids correlate with gut microbiome diversity and production of N-carbamylglutamate in middle aged and elderly women. *Sci. Rep.* **2017**, *7*, 11079. [CrossRef] [PubMed]

155. Mokkala, K.; Roytio, H.; Munukka, E.; Pietila, S.; Ekblad, U.; Ronnemaa, T.; Eerola, E.; Laiho, A.; Laitinen, K. Gut Microbiota Richness and Composition and Dietary Intake of Overweight Pregnant Women Are Related to Serum Zonulin Concentration, a Marker for Intestinal Permeability. *J. Nutr.* **2016**, *146*, 1694–1700. [CrossRef] [PubMed]

156. Robertson, R.C.; Kaliannan, K.; Strain, C.R.; Ross, R.P.; Stanton, C.; Kang, J.X. Maternal omega-3 fatty acids regulate offspring obesity through persistent modulation of gut microbiota. *Microbiome* **2018**, *6*, 95. [CrossRef] [PubMed]

157. Costantini, L.; Molinari, R.; Farinon, B.; Merendino, N. Impact of Omega-3 Fatty Acids on the Gut Microbiota. *Int. J. Mol. Sci.* **2017**, *18*, 2645. [CrossRef] [PubMed]

158. Bidu, C.; Escoula, Q.; Bellenger, S.; Spor, A.; Galan, M.; Geissler, A.; Bouchot, A.; Dardevet, D.; Morio-Liondor, B.; Cani, P.D.; et al. The transplantation of ω3 pufa-altered gut microbiota of fat-1 mice to wild-type littermates prevents obesity and associated metabolic disorders. *Diabetes* **2018**, *67*, 1512–1523. [CrossRef] [PubMed]

nutrients

MDPI

Review

Cataract Preventive Role of Isolated Phytoconstituents: Findings from a Decade of Research

Vuanghao Lim [1,2], Edward Schneider [2], Hongli Wu [3,4,*] and Iok-Hou Pang [3,4,*]

1 Integrative Medicine Cluster, Advanced Medical and Dental Institute, Universiti Sains Malaysia, Bertam, Kepala Batas, Penang 13200, Malaysia; vlim@usm.my
2 Botanical Research Institute of Texas (BRIT), 1700 University Drive, Fort Worth, TX 76107-3400, USA; eschneider@brit.org
3 Department of Pharmaceutical Sciences, System College of Pharmacy, University of North Texas Health Science Center, Fort Worth, TX 76107, USA
4 North Texas Eye Research Institute, University of North Texas Health Science Center, Fort Worth, TX 76107, USA
* Correspondence: Hongli.Wu@unthsc.edu (H.W.); iok-hou.pang@unthsc.edu (I.-H.P.); Tel.: +1-817-735-7617 (H.W.); +1-817-735-2960 (J.-H.P.)

Received: 6 September 2018; Accepted: 15 October 2018; Published: 26 October 2018

Abstract: Cataract is an eye disease with clouding of the eye lens leading to disrupted vision, which often develops slowly and causes blurriness of the eyesight. Although the restoration of the vision in people with cataract is conducted through surgery, the costs and risks remain an issue. Botanical drugs have been evaluated for their potential efficacies in reducing cataract formation decades ago and major active phytoconstituents were isolated from the plant extracts. The aim of this review is to find effective phytoconstituents in cataract treatments in vitro, ex vivo, and in vivo. A literature search was synthesized from the databases of Pubmed, Science Direct, Google Scholar, Web of Science, and Scopus using different combinations of keywords. Selection of all manuscripts were based on inclusion and exclusion criteria together with analysis of publication year, plant species, isolated phytoconstituents, and evaluated cataract activities. Scientists have focused their attention not only for anti-cataract activity in vitro, but also in ex vivo and in vivo from the review of active phytoconstituents in medicinal plants. In our present review, we identified 58 active phytoconstituents with strong anti-cataract effects at in vitro and ex vivo with lack of in vivo studies. Considering the benefits of anti-cataract activities require critical evaluation, more in vivo and clinical trials need to be conducted to increase our understanding on the possible mechanisms of action and the therapeutic effects.

Keywords: cataract; phytoconstituents; lens; preclinical models; drug discovery

1. Introduction

The ocular lens is located at the anterior segment of the eye that, together with the cornea, provides the refractive power of the eye. The mature lens is composed of a core of primary lens fiber cells, layers of secondary lens fiber cells, and one layer of anterior lens epithelial cells, which covers the anterior surface of the lens [1]. The major function of the lens is to maintain transparency so that the light can be properly focused on the retina. Unfortunately, the delicate balance required for lens transparency can be easily disturbed by oxidative stress, aging, and UV radiation, and cataracts develop as a result [1].

Cataracts are the most common cause of vision loss in people over the age of 40 and are the leading cause of blindness in the world [2]. Cataracts are defined as lens opacification that prevents a sharply defined image from reaching the retina. As a result, cataract patients have clouded, blurred, or

Nutrients **2018**, *10*, 1580

dim visions, which significantly affect their daily life. According to a report from the World Health Organization, nearly 40 million people are blind worldwide, almost half of them are due to cataract [3]. Although cataract-related vision loss can be corrected by replacement with synthetic lenses, cataract surgery is a costly procedure and may develop complications like infectious endophthalmitis, posterior capsule rupture during surgery, post-operative macular edema, and posterior capsule opacity (also called posterior capsule opacification). In developing countries, many cataract patients cannot have their vision restored due to financial concerns or lack of medical resources. Therefore, identifying a safe compound that can reduce the incidence or delay the onset of cataract is an important step in finding new treatments for cataract.

There have been many compounds evaluated for their potential efficacies in reducing cataract formation. In this article, we focus on active ingredients derived from plants. Phytoconstituents are a trove of often structurally complicated compounds with interesting biological functions. They themselves or their derivatives have always been important sources of pharmacologically active agents. To provide a comprehensive review of potentially useful anti-cataract phytoconstituents, we searched, selected, and extracted the appropriate information from published literature according to the following procedures. We feel that, by listing the comprehensive collection of phytocontituents in one place, this manuscript serves as an overview and perhaps an inspiration to prompt additional studies in this important research area. Collaborative efforts between phytochemists and cataract researchers are promisingly fruitful.

2. Materials and Methods

2.1. Literature Search

Literature search of articles published from January 2008 to December 2017 was performed. We searched the databases of Pubmed, Science direct, Google Scholar, Web of Science and Scopus using different combinations of keywords: lens epithelial cells, sodium selenite-, ultraviolet radiation-, steroid induced, oxygen-, H_2O_2-induced opacity/cataract, congenital/juvenile cataract, transgenic/knockout mice with cataract, diabetic cataract, spontaneous cataract, isolated phytoconstituents, medicinal plants.

2.2. Study Selection

The selection of the manuscripts was based on the following inclusion criteria: isolation of phytoconstituents from plants, and selection of phytoconstituent(s) with the most potent anti-cataractogenesis activities. Exclusion of manuscripts from this review involves synthesized/commercialized compounds, phytoconstituents screening without isolated compounds, activities with extracts only and isolated phytoconstituents without names or articles that did not meet the inclusion criteria. The selection process is summarized in Figure 1.

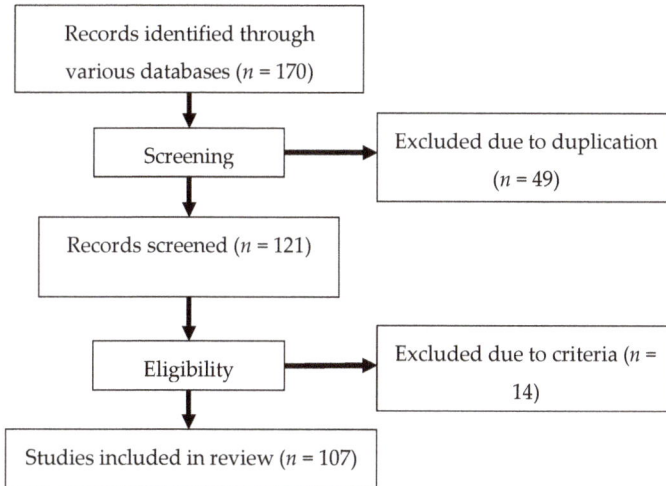

Figure 1. Flow chart of review process in article selection.

2.3. Data Extraction

All the selected manuscripts were analyzed for year of publication, plant species, family, part of plant, solvent extraction, isolation method, isolated phytoconstituents, anti-cataract activities (in vitro, ex vivo or in vivo), route of administration (in vivo), dose or concentration for IC_{50}, treatment duration (in vivo) and isolated phytoconstituent(s) with the strongest activity(s), as well as their structural formula. The extracted data are presented in Tables 1 and 2 throughout this article.

3. Results

3.1. Experimental Cataract Models

There are a large number of in vitro and in vivo models that mimic certain aspects of the pathophysiological features of human cataracts. They have been used to demonstrate the potential therapeutic effects of phytochemicals. In this section, we describe the most commonly used models in order to aid the understanding and appraisal of results. During our literature search, most of the phytochemicals were tested in in vitro or ex vivo models only. Only a dozen or so were assessed in in vivo cataract models. Nevertheless, for completion's sake, we list both in vitro and in vivo models. In vitro models discussed are hydrogen peroxide (H_2O_2)-, xylose-, galactose-induced lens opacity, aldose reductase (AR) activity assay, and advanced glycation end products (AGE) formation. In vivo models include sodium selenite-, ultraviolet (UV) radiation-, and steroid-induced cataracts. These models have been widely used to study the mechanisms of cataract and serve as the screening platform of anti-cataract therapies with the long-term goal to treat cataract in humans.

3.2. In Vitro Models

3.2.1. Oxidative Stress Model

H_2O_2-Induced Cataract

It is widely accepted that oxidative stress is the major factor for the development of cataracts. Hydrogen peroxide (H_2O_2) is the major reactive oxygen species (ROS). H_2O_2 is mainly generated in vivo by the detoxification of superoxide (O_2^-) radical by superoxide dismutase (SOD) through the dismutation reaction [4,5]. Alternatively, H_2O_2 can be produced by a number of oxidase enzymes including monoamine oxidases and peroxisomal pathway for β-oxidation of fatty acids. In the lens,

H$_2$O$_2$ can also be generated by the photochemical reaction [4]. Most human tissues, including the lens, are exposed to some level of H$_2$O$_2$, with the mitochondria being the major site for production. Previous studies have shown the strong association between H$_2$O$_2$ overproduction and cataract development. Cataract patients had elevated H$_2$O$_2$ in both the aqueous body and lens ranging from seven- to 30-fold higher than normal [5]. Lens organ ex vivo culture with H$_2$O$_2$ in the medium is a common experimental model of cataract. This type of cataract is characterized by loss of GSH and increased protein oxidation. To establish the model, rat or porcine lenses are dissected and cultured in TC-199 medium containing 200 to 1000 μM H$_2$O$_2$ with final osmolarity of 298 ± 2 mOsm/L. The lens are usually harvested after 24 to 96 h to induce cataracts [6,7].

3.2.2. Diabetic Cataract

Cataract is a major cause of visual impairment in patient with diabetes mellitus. Both clinical and basic research studies have indicated the strong association between diabetes and cataract formation [8,9]. The molecular mechanisms that may be involved in diabetic cataract include polyol pathway flux, increased formation of AGEs, osmotic stress, and elevated oxidative stress.

Aldose Reductase (AR) Activity

The polyol pathway, also known as sorbitol-AR pathway, is a two-step process that converts glucose to fructose. AR, the first and rate-limiting enzyme in the pathway, reduces glucose to sorbitol using nicotinamide adenine dinucleotide phosphate (NADPH) as a cofactor. Sorbitol is then converted to fructose by sorbitol dehydrogenase (SDH) [10]. As a sugar alcohol, sorbitol does not diffuse across cell membranes readily. When accumulating intracellularly, it produces osmotic stress on cells by driving water into the lens that may eventually cause diabetic cataract. Therefore, one of the possibilities to prevent the onset of diabetic cataract is to use AR inhibitor (ARI) [11]. The in vitro ARI assay was used to evaluate if the compound can inhibit the polyol pathway. Briefly, the reaction mixture contains 50 μM potassium phosphate buffer pH 6.2, 0.4 mM lithium sulfate, 5 μM 2-mercaptoethanol, 10 μM DL-glyceraldehyde, 0.1 μM NADPH, and freshly-prepared AR enzyme. The reaction is initiated by the addition of NADPH at 37 °C. The AR activity is determined indirectly by a spectrophotometer that measures NADPH absorption [10].

Xylose-Induced Lens Opacity

Another in vitro diabetic cataract model is xylose-induced lens opacity. Glucose, galactose, and xylose are all known to induce cataract. Among these three sugars, xylose is the most effective molecule in producing cataracts due to the fact that it is the preferred substrate of AR in the lens. Kinoshita and colleagues first established xylose-induced lens in 1974 [12]. They cultured rat lenses in 4 mL of medium containing 30 mM xylose for six days. They observed a progressive development of lens opacity accompanied by increased osmotic stress and lens swelling [12].

Galactose-Induced Lens Opacity

Compared with glucose, galactose has higher affinity with AR and its reduction product galactitol is more difficult to be metabolized by sorbitol dehydrogenase than sorbitol. Therefore, high galactose is more likely to induce sugar cataract than high glucose itself [13]. There are several methods available to establish galactosemic cataract. For example, rat galactosemic cataract can be induced by 30% or 50% galactose diet. Glactose-induced lens opacity can also be achieved by daily intraperitoneal injection of 30–50% galactose solution or daily retrobulbar injection of 20% galactose solution. Another cost efficient way to induce galactosemic cataract is to feed rat 10% galactose solution for 18 days. For in vitro lens culture, 30 mM galactose is added in the culture medium for 72 h incubation [13].

Formation of Advanced Glycation End (AGE) Products

Another important factor that is involved with the pathogenesis of diabetic cataract is the formation of AGEs. In diabetic patients with cataract, the elevated glucose starts forming covalent adducts with the lens proteins through a non-enzymatic process called glycation [14,15]. This process is known as one of the most important forms of post-translational modification of proteins under hyperglycemic conditions. Many studies have shown that protein glycation-induced AGEs play a pivotal role in diabetic cataract formation. Therefore, AGE formation assay is used to examine the potential anti-cataract potential of tested compounds. To determine the amount of AGEs, a reaction mixture containing 10 mg/mL of bovine serum albumin and 0.5 M fructose and glucose are mixed with tested compounds. After 15 days of incubation, the fluorescent intensity is measured using a spectrofluorometric detector with an excitation wavelength of 350 nm and an emission wavelength of 450 nm [16,17].

3.3. In Vivo Models

Commonly used in vivo models represent specific pathogenesis aspects of human cataract. For example, the diabetic cataract rodent model focuses on mechanisms involved in diabetes-related cataract; selenite-induced cataract addresses oxidative damage-induced cataract; the UV- and steroid-induced models represent their respective associated pathological changes. In various studies, drug effects in these in vivo models correlate well with the pharmacodynamics properties shown in appropriate in vitro models.

3.3.1. Diabetic Cataract

The in vivo diabetic cataract model can be established by using streptozotocin (STZ). After intraperitoneal (i.p.) or intravenous (i.v.) injection, STZ enters the pancreatic β-cell through the glucose transporter 2 transporter (Glut-2) resulting in hyperglycemia [16]. Moreover, STZ is also a source of free radicals that may lead to DNA oxidative damage and subsequent β-cell death. STZ can be administered as a single high dose (e.g., 160 to 240 mg/kg) or as multiple low doses (e.g., 40 mg/kg for 5 days) [18].

Another commonly used diabetic cataract model is AR transgenic mice. The ubiquitous transgenic and lens-specific AR transgenic mice were developed to further prove that polyol accumulation is responsible for diabetic cataract. In both models, sorbitol accumulates in the lens, causing osmotic swelling, and eventually leading to accelerated diabetic cataract formation [10,19].

3.3.2. Selenite-Induced Cataract

Selenite-induced cataract is an effective, rapid, and reproducible model of nuclear cataracts. Selenite cataract is usually produced either by a single dose (19–30 µM/kg body weight) or repeated smaller dosage of sodium selenite (40–50 nmol/g body weight) subcutaneous injection to suckling rat of 10–14 days of age [20]. It has been proposed that selenite treatment leads to altered metabolism in lens epithelium, including loss of small antioxidant molecules such as glutathione (GSH), decreased rate of epithelial cell differentiation, and increased DNA oxidation damage. Such extensive alterations to the epithelium leads to disrupted calcium homeostasis and calcium accumulation in the nucleus of the lens. Increased calcium activates calcium dependent protease m-calpain (calpain II) which results in rapid proteolysis, precipitation of crystallins, and eventually cataract development in rodent lenses [21,22].

3.3.3. UV-Induced Cataract

UV radiation is a major contributor to the pathogenesis of cataract. The strong energy in the UV light can directly cause a DNA lesion in the lens by inducing thymine dimer formation. More importantly, UV can induce cataract formation by the generation of ROS that indirectly induce oxidative damage to DNA by disturbing cell proliferation in the lens epithelium, altering kinetic properties of enzymes in the energy metabolism, increasing insoluble and decreasing soluble protein, and disturbing the sodium potassium balance, leading to aberrant water balance in the lens [23]. It has been widely accepted that cataract formation is related to oxidative stress induced by continued intraocular penetration of UV light and consequent photochemical generation of ROS such as superoxide and singlet oxygen and their oxidant derivatives such as hydrogen peroxide and hydroxyl radical [24]. Sprague-Dawley rats or mice are exposed to 8 kJ/m^2 UV-B radiation for 15 min to induce cataracts [25].

3.3.4. Steroid-Induced Cataract

As the steroid hormones, glucocorticoids (GCs) have strong anti-inflammatory effects. By binding with the glucocorticoid receptor (GR), GCs have the ability to inhibit all stages of the inflammatory response [26]. Due to its strong anti-inflammatory effects, GCs are widely used in the management of many clinical conditions, including autoimmune disorders, allergies, and asthma, and they also play important roles in chemotherapy and preventing the rejection after solid organ transplantation. However, prolonged use of GCs is associated with the development of posterior subcapsular and nuclear cataracts [26]. The chick embryo has been used to establish an experimental model to study the response of the lens to GCs. When dexamethasone (0.02 μmol/egg) is administered, the lenses of chicken embryos become cataract within 48 h. More recently, the mammalian lens has also been used to establish the steroid-induced cataract models. For example, Brown-Norway rats given a daily 1% prednisolone acetate instillation of a total volume of 1.0 mg/kg or a daily intramuscular injection of 0.8–1.0 mg/kg prednisolone acetate for 10 months successfully induced morphological changes similar to those found in human steroid-induced cataracts [27,28].

3.4. Anti-Cataract Phytoconstituents

Based on our literature strategy listed above, the following phytoconstituents are listed in alphabetical order. They have been shown to possess potential anti-cataract efficacy according to the described study models.

3.4.1. 1-*O*-Galloyl-β-D-glucose (β-Glucogallin)

Molecular formula: C$_{13}$H$_{16}$O$_{10}$ (332.262 g/mol), Melting point: 214–216 °C.

β-Glucogallin isolated from the aqueous fruit extract of *Emblica officinalis Gaertn.* (emblic, Indian gooseberry) or *Phyllanthus emblica* Linn. (Euphorbeaceae) (gooseberry) [29] shows potent activity against human AR in vitro with an IC$_{50}$ of 17 μM [30]. Treatment with this compound prevented the sorbitol accumulation by 73% (30 μM) in transgenic human AR expressing lenses ex vivo [30]. This result substantiated the in vitro assay using shared substrate glyceraldehyde at IC$_{50}$ of 58 μM. Treatment with β-Glucogallin produced a significant decrease of sorbitol levels in macrophages [31]. Computational molecular docking studies exhibited favorable binding to the active site of between human AR and β-glucogallin. This corroborates the inhibition result of sorbitol production under hyperglycemic conditions in earlier experiments [30].

3.4.2. 1,3-Di-*O*-caffeoylquinic Acid

Molecular formula: C$_{25}$H$_{24}$O$_{12}$ (516.45 g/mol).

1,3-di-*O*-caffeoylquinic acid has been isolated from *Artemisia iwayomogi* (haninjin) and *Xanthium strumarium* (rough cocklebur) as inhibitor for rat lens AR (RLAR), recombinant human AR (RHAR) and advanced glycation end-product (AGE) inhibitory activities. The compound inhibited RLAR with IC$_{50}$ values of 0.22–1.90 μM [32,33]. This result was supported by inhibition of RHAR at IC$_{50}$ of 0.81 μM. In AGE inhibitory activity, 1,3-di-*O*-caffeoylquinic acid suppressed at IC$_{50}$ of 24.85 μM [33].

3.4.3. 1,5-Di-hydroxy-1,5-di-[(*E*)-3-(4-hydroxyphenyl)-2-propenoic]-3-pentanonyl Ester (DHDP)

Molecular formula: $C_{23}H_{22}O_9$ (442.41 g/mol).

A novel polyphenolic inhibitor of AR, DHDP was isolated from *Lysimachia christinae* (gold coin grass, jinqiancao) using AR affinity-based ultrafiltration-HPLC profiling method. The reversible inhibitory activity of RHAR was recorded at IC_{50} value of 194.7 μM with sorbitol content of 1002.3 μg/g of lens weight. The effect of DHDP was further investigated in in silico using computer simulation of binding by molecular docking. DHDP was predicted to block the AR active site by binding and preventing the formation of product [34].

3.4.4. 1,5-Di-*O*-caffeoylquinic Acid

Molecular formula: $C_{24}H_{24}O_{11}$ (488.44 g/mol).

A new method of enzyme assay-guided high-performance liquid chromatography microfractionation and elution-extrusion counter-current chromatography of roots ethanolic extract of *Nardostachys chinensis* (spikenard) afforded six secondary metabolites with 1,5-di-*O*-caffeoylquinic acid as the most potent inhibitor against RLAR activity (IC_{50} = 2.98 μM). The compound was reported as the first time isolated from the plant [35].

3.4.5. 1,3,6-Trihydroxy-2-methoxymethylanthraquinone

Molecular formula: $C_{16}H_{12}O_6$ (300.26 g/mol).

Bioassay-guided fractionation of *Knoxia valerianoides* (hongdaji) methanolic root extract afforded eight secondary metabolites with 1,3,6-trihydroxy-2-methoxymethylanthraquinone showing the highest inhibition against AGE formation at IC_{50} value of 52.7 μM. The same phytoconstituent also exhibited strong inhibitory activity against RLAR with IC_{50} value of 3.0 μM [36].

3.4.6. 1,2,3,6-Tetra-*O*-galloyl-β-D-glucose

Molecular formula: $C_{34}H_{28}O_{22}$ (788.57 g/mol).

1,2,3,6-tetra-*O*-galloyl-β-D-glucose was isolated from the methanolic seeds extract of *Cornus officinalis* (cornus tree, shan zhu yu) after repeated Sephadex column chromatography. Appeared as an off-white amorphous powder [37], 1,2,3,6-tetra-*O*-galloyl-β-D-glucose showed the most potent inhibitory activity (IC_{50} = 0.70 μM) compared to other secondary metabolites. In addition, AGE formation was also reduced to IC_{50} value of 1.99 μM. This compound was further evaluated for its inhibitory effect on ex vivo cataractogenesis activity using rat lenses induced with xylose 20 mM. Treatment with 1,2,3,6-tetra-*O*-galloyl-β-D-glucose significantly reduced the opacities of the lenses after two days at the concentration of 80 μM [38].

3.4.7. 1,3,5,8-Tetrahydroxyxanthone

Molecular formula: $C_{13}H_8O_6$ (260.19 g/mol).

Several xanthones have been isolated from the ethanolic extract of *Swertia mussotii* Franch (yinchen) as inhibitors for RLAR activity. The most potent inhibition was shown by 1,3,5,8-tetrahydroxyxanthone with IC_{50} of 0.0886 μM [39]. The compound appeared in slight yellow powder with 98.6% purity.

3.4.8. 2″,4″-*O*-Diacetylquercitrin

Molecular formula: $C_{25}H_{24}O_{13}$ (532.11 g/mol), Melting point: 187 °C.

2″,4″-*O*-Diacetylquercitrin was isolated from *Melastoma sanguineum* (red melastome, fox-tongued melatsome) as a yellow amorphous powder. This compound exhibited the strongest inhibition against RLAR and AGE activities among all the isolated phytoconstituents. IC_{50} inhibitory activities for RLAR and AGE were recorded at 0.077 μM and 11.46 μM, respectively. Compared to the positive standards, aminoguanidine (IC_{50} = 965.9 μM, AGE) and 3,3-tetramethyleneglutaric acid (IC_{50} = 28.8 μM, RLAR), 2″,4″-*O*-Diacetylquercitrin inhibited 87 (AGE) and 374 (RLAR) times more efficaciously [40].

3.4.9. 3-Isomangostin

Molecular formula: $C_{24}H_{26}O_6$ (410.46 g/mol), Melting point: 182–183 °C.

Three main constituents of dichloromethane extract from root bark of *Garcinia mangostana* Linn (mangosteen) were isolated from the hexane/methanol fraction. The result of the study indicated that 3-isomangostin possessed the highest RLAR inhibitory activity with at IC_{50} value of 3.28 μM. The presence of cyclization of the prenyl group at the position-two carbon with xanthone derivative enhanced the structure-activity relationship [41].

3.4.10. 3′,4-Dihydroxy-3,5′-dimethoxy-bibenzyl (Gigantol)

Molecular formula: $C_{16}H_{18}O_4$ (274.316 g/mol), Melting point: 135 °C.

Gigantol is a bibenzyl-type phenolic compound presents in most herbs of Orchidaceae family [42]. It has been isolated from the stems of various *Dendrobium* genus such as *Dendrobium aurantiacum* var. *denneanum* (die qiao shi hu) [43,44] and *Dendrobium chrysotoxum* Lindl (fried-egg orchid) [45,46] for anti-cataract activities. As a white solid, gigantol suppresses the damage of rat lenses both in vitro and in vivo in galactose-induced cataractogenesis. The delay in lens turbidity was caused by the inhibition of AR and inducible nitric oxide synthase mRNA expression at an IC_{50} of 239.4 μM (65.7 μg/mL) and 32.0 μM (8.8 μg/mL), respectively [43]. Gigantol isolated from *Dendrobium chrysotoxum* Lindl interpolated into the DNA base pairs in AR gene with a binding constant of 1.85×10^3 L/mol, thus, suppressed the gene expression [46].

3.4.11. 3′,5′-Dimethoxy-(1,1′-biphenyl)-3,4-diol 3-*O*-β-D-glucopyranoside

Molecular formula: $C_{20}H_{24}O_9$ (406.42 g/mol).

The leaves and twigs of *Osteomeles schwerinae* C. K. Schneid. (hu xi xiao shi ji) were examined for their possible inhibitory activity on RLAR. Four secondary metabolites have been isolated from the $CHCl_3$-MeOH fraction of the EtOH extract and found that 3′,5′-dimethoxy-(1,1′-biphenyl)-3,4-diol 3-*O*-β-D-glucopyranoside to be the most potent inhibitor against RLAR activity at IC_{50} value of 3.8 μM. The phytoconstituent was obtained as a brownish powder [47].

3.4.12. 3,5-Di-*O*-caffeoylquinic Acid

Molecular formula: $C_{25}H_{24}O_{12}$ (516.45 g/mol), Melting point: 184–187 °C.

Methanolic extract of the stems and leaves of *Erigeron annuus* (annual fleabane, daisy fleabane) afforded 16 secondary metabolites. 3,5-di-*O*-caffeoylquinic acid appeared as pale-yellow powder and isolated from the ethyl acetate-soluble fraction after repeated column chromatography [48–50]. The same constituent was also isolated from *Aster koraiensis* (Korean starwart) [51], *Xanthium strumarium* (clotbur, common cocklebur) [33], *Artemisia iwayomogi* (haninjin) [32] and *Artemisia montana* [52]. 3,5-di-*O*-caffeoylquinic acid was reported as the most significant inhibitory activities against AGEs, RLAR and ex vivo xylose-induced lens opacity assays from all isolated constituents. It attenuates AGE formation with IC_{50} values ranging from 6 μM to 32 μM, and inhibits RLAR with IC_{50} values of 0.2 to 5 μM. These findings are further substantiated by its ability in inhibition of galactitol accumulation at an IC_{50} of 153 μM [33] and prevention of xylose-induced opacity of lenses at a concentration of 10 μM [48].

3.4.13. 4-*O*-Butylpaeoniflorin and Palbinone

Molecular formula of 4-*O*-butylpaeoniflorin: $C_{27}H_{36}O_{11}$ (536.22 g/mol), Melting point: 173–175 °C.
Molecular formula of Palbinone: $C_{22}H_{30}O_4$ (359.47 g/mol), Melting point: 254–255 °C.

Both 4-*O*-butylpaeoniflorin and Palbinone were isolated from methanolic extract of the cortex of *Paeonia suffruticosa* (tree peony, mudan, moutan) with highest inhibitory activities of RLAR (palbinone) and AGE (4-*O*-butylpaeoniflorin) compared to other isolated phytoconstituents [53]. Palbinone appeared as red needles with $[\alpha]_D$ −223.8° ($CHCl_3$) and absorbed UV at 237 (log ε: 3.2) and 387 nm (log ε: 3.0) [54]. Isolated from the butanol fraction. 4-*O*-butylpaeoniflorin was found as an optically active white foam, $[\alpha]_D^{25}$–7.8 (*c* 0.14, MeOH) and later confirmed as an extraction artifact after HPLC analysis. Palbinone inhibits RLAR at an IC_{50} value of 11.4 μM. It was suggested that the absence of ring E, side chain of ring D together with double bonds and a conjugated carbonyl group on the ring D played the inhibitory properties. Unlike palbinone, 4-*O*-butylpaeoniflorin inhibited (IC_{50} = 10.8 μM) for AGE activity. The chemical moiety of hydroxy groups in the benzoyl connected to the sugar unit complement the activity [53].

3.4.14. 4,5-Di-*O*-trans-caffeoyl-D-quinic Acid

Molecular formula: $C_{25}H_{24}O_{12}$ (516.45 g/mol).

Caffeoylquinic acid analog, 4,5-Di-*O*-trans-caffeoyl-D-quinic acid isolated from *Hydrangea macrophylla* var. *thunbergii* (bigleaf hydrangea) and *Ilex paraguariensis* (Yerba mate) showed the strongest inhibitory activity against RLAR at IC_{50} value of 0.29 μM [55]. Inhibitory effect of quinic acid with two caffeoyl groups assisted the potency.

3.4.15. 5-*O*-Feruloly Quinic Acid

Molecular formula: $C_{17}H_{20}O_9$ (368.33 g/mol)

Bioassay-guided isolation of root methanolic extract of *Aralia continentalis* Kitag. (dong bei tu dang gui) produced 18 secondary metabolites. 5-*O*-Feruloly quinic acid was isolated from the ethyl acetate fraction as an amorphous white powder. It had a highest inhibitory activity of RLAR at IC_{50} value of 14.2 μM among all other phytoconstituents [56].

3.4.16. 5,7,4′ Trihydroxyisoflavone (Genistein)

Molecular formula: $C_{15}H_{10}O_5$ (270.24 g/mol), Melting point: 297–298 °C.

Genistein appears as colorless plates and isolated from the roots of *Pueraria lobata* (kudzu, Japanese arrowroot) [57,58] and stem bark of *Maackia amurensis* (Amur maackia) [59]. Both plants are native to Eastern Asia and used as traditional medicine in China, Korea, and Japan. Genistein shows a significant dose-dependent inhibition on RLAR activity (IC_{50} = 9.48 μM) compared to the positive control, TMG (3,3-tetramethyleneglutaric acid) (IC_{50} = 28.70 μM). Nevertheless, IC_{50} was recorded higher at 57.1 μM for the same activity compared to quercetin IC_{50} = 10.1 μM [59]. In an ex vivo lens opacity study genistein suppressed xylose-induced lens opacity at 5 μg/mL (18.5 μM). Further analysis with human lens epithelia cells (LECs; HLE-B3 cells) found that the expression of TGF-β2, αβ-crystallin, and fibronectin mRNAs were reduced, suggesting genistein is protective against lens opacity with antioxidative effects [60]. It is proposed that the chemical moiety with free hydroxyl group at C-7 of genistein attributes to the inhibitory of AR [59].

3.4.17. 20(*S*)-Ginsenoside Rh2

Molecular formula: $C_{36}H_{62}O_8$ (622.87 g/mol).

20(*S*)-Ginsenoside Rh2 is classified under triterpene glycosides and isolated from the root of *Panax ginseng* C. A. Meyer, (ginseng). It has been used traditionally in East Asia for many years ago with many main active constituents, ginsenosides have been isolated. In RHAR inhibitory activity, 20(*S*)-Ginsenoside Rh2 showed the most potent inhibitor with an IC_{50} of 147.4 μM among all other isolated ginsenosides. It was suggested that the moiety of hydroxyl group at the carbon-20 enhanced the AR activity relationship [61].

3.4.18. Acteoside

Molecular formula: $C_{29}H_{36}O_{15}$ (624.58 g/mol), Melting point: 143–146 °C.

Isolated as yellowish amorphous powder from methanolic extract of *Abeliophyllum distichum* (forsythia) and leaves and stem ethanolic extracts of *Brandisia hancei* (laijiangteng), acteoside showed the highest RLAR inhibitory activities at IC_{50} values ranging 0.83 μM and 1.39 μM compared to four other isolated phenolic glycosides from each plant, respectively. The isolation was conducted by high-speed counter current chromatography using a solvent system of ethyl acetate:n-butanol:water [62]. Isolated acteoside from *Brandisia hancei* showed potent AGE inhibitory activity with an IC_{50} value of 5.11 μM [63].

3.4.19. Basilicumin [7-(3-hydroxypropyl)-3-methyl-8-$β$-*O*-D-glucoside-2H-chromen-2-one]

Molecular formula: $C_{19}H_{24}O_9$ (396.38 g/mol).

Basilicumin was isolated from *Ocimum basilicum* (basil). It exhibits potent inhibitory activity against AR (AKR1B1) and aldehyde reductase (AKR1A1) compared to the second phytoconstituent isolated, ocimunone. Basilicumin inhibited AKR1A1 at IC_{50} value of 0.78 μM and 2.1 μM for AKR1B1 activity. It was suggested that coumarin and glucose scaffold in basilicumin moiety enhance the activity [64].

3.4.20. Caffeic Acid

Molecular formula: $C_9H_8O_4$ (180.16 g/mol), Melting point: 223–225 °C.

Caffeic acid appears as white amorphous powder and classified as a hydroxycinnamic acid [65]. In an attempt to find potential cataractogenesis inhibitors from plants, caffeic acid has been isolated from a few plants with potential activity. Isolation of caffeic acid for RLAR activity has been shown from methanolic extract of *Dipsacus asper* (xuduan), *Erigeron annuus* (L.) Pers., and *Phellinus linteus* (black hoof mushroom, meshima, song gen, meshimakobu, sanghwang) with the highest activity among all isolated secondary metabolites at IC_{50} values of 16.7 μM to 55 μM. Comparable results were observed for RHAR (IC_{50} = 55 to 210 μM) and AGE activities (IC_{50} = 7.6 μM) [66–68]. Interestingly, no inhibitory effects were observed for caffeic acid isolated from *Perilla frutescens* L. [69] and *Prunella vulgaris* L. [70].

3.4.21. Canangafruiticoside E

Molecular formula: $C_{25}H_{32}O_{12}$ (524.51 g/mol).

The repeated column chromatography of methanolic flower bud extract of *Cananga odorata* Hook. F. and Thomson generated 25 secondary metabolites and they were tested for RLAR inhibitory activity. The result of the study indicated that among the isolated constituents, canangafruiticoside E possessed the highest activity (IC_{50} = 0.8 μM) [71].

3.4.22. Capsofulvesin A [((2S)-l-*O*-(6Z,9Z,12Z,15Zoctadecatetraenoyl)-2-*O*-(4Z,10Z,13Zhexadecatetraenoyl) -3-*O*-$β$-D-galactopyranosyl Glycerol)]

Molecular formula: $C_{45}H_{72}O_{10}$ (773.04 g/mol).

The isolation of Capsofulvesin A from ethanolic extract of *Capsosiphon fulvescens* (one of the green algae) showed the strongest RLAR inhibitory activity among all other secondary metabolites, albeit moderate activity at IC_{50} value of 52.5 μM. However, the constituent did not show any inhibition against AGE activity [72].

3.4.23. Caryatin-3′ methyl ether-7-*O*-$β$-D-glucoside

Molecular formula: $C_{24}H_{26}O_{12}$ (506.14 g/mol).

The bark of the pecan tree (*Carya illinoinensis* (Wangenh) K. Koch) has shown good inhibition of AR activity with few compounds have been isolated. Among them, caryatin-3′ methyl ether-7-*O*-$β$-D-glucoside exhibits the most powerful activity in suppressing the lens AR levels in diabetic cataract rats [73]. The catechol moiety on the B ring of caryatin-3′ methyl ether-7-*O*-$β$-D-glucoside was suggested to inhibit AR in comparison to the activity of other compounds isolated. In addition, the potent AR activity was also supported by the presence of neighboring *O*-methyl group in phenolics and an OH group at C-4′ [73–75]. Caryatin-3′ methyl ether-7-*O*-β-D-glucoside is physically yellow amorphous powder with UV (MeOH) λ_{max} absorption at 350, 330, and 260 nm [73].

3.4.24. C-Phycocyanin (C-PC)

Molecular formula: $C_{33}H_{38}N_4O_6$ (586.67 g/mol).

C-Phycocyanin (C-PC), a prominent phytoconstituent found in the stromal surface of thylakoid membranes of *Spirulina platensis* (a blue-green algae) is a biliprotein that functions to capture light energy to chlorophyll A [76–78]. As C-PC is miscible in water but not alcohol and esters, most of the isolations of C-PC use water extraction method [79]. C-PC attenuates selenite-induced cataractogenesis both in vitro and in vivo rat model [78,80]. In vitro study showed C-PC recorded low degree of opacification at 200 µg C-PC with 100 µM sodium selenite [78]. The purified C-PC was active toward the in vivo selenite mediated cataractogenesis showing only slight opacification at 200 mg/kg [78]. Same concentration was observed for naphthalene- and galactose-induced cataract rat models [81]. The protective effect of C-PC in these models were proven from the increment of glutathione, soluble proteins, and water content levels of the lens [79]. Histology study indicated the protection of the lens from oxidative damage. Restoration of lenticular micro-architecture was found with C-PC treated group [77]. C-PC maintains the lens transparency by transcriptional regulation of crystallin, redox genes, and apoptotic cascade mRNA expression [80]. Furthermore, C-PC was suggested to possess protective effects on human LEC by abrogating D-galactose-induced apoptosis through the mitochondrial pathway (p53 and Bcl-2 family protein expression) and unfolded protein response pathway (GRP78 and CHOP expression) [82].

3.4.25. Davallialactone

Molecular formula: $C_{25}H_{20}O_9$ (464.43 g/mol).

Davallialactone was isolated as yellow amorphous powder from the active ethyl acetate fraction of fruiting body of *Phellinus linteus*. Davallialactone possessed the most potent inhibitory against RLAR and RHAR among all the isolated compounds with IC_{50} values of 0.33 µM and 0.56 µM, respectively. The inhibitory activities were nine times (RLAR) and 11 times (RHAR) compared to that of quercetin (IC_{50} = 2.91 µM; RLAR and IC_{50} = 6.27 µM; RHAR) [67].

3.4.26. Delphinidin 3-*O*-β-galactopyranoside-3′-*O*-β-glucopyranoside

Molecular formula: $C_{27}H_{31}O_{17}$ (627.52 g/mol).

Anthocyanin delphinidin 3-*O*-β-galactopyranoside-3′-*O*-β-glucopyranoside was isolated from the methanolic extract of the air-dried fruit pericarp of *Litchi chinensis* Sonn (lychee). This fruit is a tropical and subtropical edible fruit native to Southeast Asia. Delphinidin 3-*O*-β-galactopyranoside-3′-*O*-β-glucopyranoside exhibits the most significant inhibitory activity in RLAR assay with an IC_{50} value of 0.23 µg/mL (0.37 µM) compared to the positive control, tetramethylene glutaric acid (IC_{50} = 0.48 µg/mL) [83].

3.4.27. Desmethylanhydroicaritin

Molecular formula: $C_{20}H_{18}O_6$ (356.36 g/mol), Melting point: 220–222 °C.

The isolation of repeated chromatography of the CH_2Cl_2 fraction over a silica-gel column and Sephadex LH20 from root methanolic extract of *Sophora flavescens* (kushen) afforded desmethylanhydroicaritin. Desmethylanhydroicaritin exerted remarkable inhibitory activity of RLAR with IC_{50} value of 0.95 µM. Comparable results were observed in RHAR and AGE inhibitions where IC_{50} values were observed at 0.45 µM and 294.6 µM, respectively. The presence of prenyl and lavandulyl groups enhanced the RLAR and RHAR inhibitory activities. The 3-hydroxyl group at prenylated flavonoids was suggested for the structural contribution for inhibition of AGE formation [84].

3.4.28. Ellagic Acid

Molecular formula: $C_{14}H_6O_8$ (302.19 g/mol), Melting point: $\geq$350 °C.

During a search for possible cataractogenesis activities for isolated ellagic acid, three plants from Korea were found with most potent inhibitions. Ellagic acid isolated from *Phellinus linteus*, *Geranium thunbergii* (Thunberg's geranium), and *Syzygium cumini* (L.) Skeels (jambolan, Java plum, black plum) inhibits RLAR activity with IC_{50} value ranging from 0.12 μM to 6.9 μM [67,85–87]. The compound was also effective in the inhibition of AGE formation (IC_{50} = 26.0 μM). In RHAR assay, the activity of ellagic acid (IC_{50} = 1.37 μM) from *Phellinus linteus* was more potent than that of quercetin (IC_{50} = 6.27 μM) [67]. This was substantiated by its inhibition (42.5%) of galactitol accumulation in rat lenses incubated in high glucose with 485.6 μg/lens wet weight [85].

3.4.29. Epiberberine

Molecular formula: $C_{20}H_{18}NO_4^+$ (336.36 g/mol), Melting point: 187 °C.

The bioassay-guided isolation of the rhizome of *Coptis chinensis* Franch (Chinese goldthread) afforded seven secondary metabolites with epiberberine exhibited the highest inhibitory of RLAR activity. The IC_{50} of the reported value was 100 μM. Conversely, epiberberine showed a comparable result against RHAR with IC_{50} value of 168.1 μM. The chemical moiety of dioxymethylene (ring D) and its oxidized form (ring A) was suggested to enhance the AR inhibitory activities, albeit in moderate effects [88].

3.4.30. Geraniin

Molecular formula: $C_{41}H_{28}O_{27}$ (952.64 g/mol), Melting point: 360 °C.

The anti-cataract activities of *Nephelium lappaceum* (rambutan) [89] and *Geranium thunbergii* [85] lead to the isolation of geraniin with good yield. Geraniin was isolated from the ethanolic rind extract of *Nephelium lappaceum* as the major bioactive compound. This compound exhibits better AR activity with an IC_{50} of 0.15 μM at approximately 40% higher compared to quercetin IC_{50} = 5.76 μM [89]. Geraniin isolated from *Geranium thunbergii* shows slightly higher concentration of IC_{50} (8.54 μM) in the same activity, however, using rat lens as the source of enzyme [85]. In AGE assay, the activity of geraniin was 96% of inhibition after incubation time of seven days at the concentration of 20 μg/mL (21 μM). Galactitol accumulation in rat lenses incubated with high galactose was inhibited at 39.9% by geraniin with 507.5 μg/lens wet weight (g). It was concluded that geraniin isolated from both plants is a promising agent in the prevention or treatment of diabetic complications.

3.4.31. Hipolon

Molecular formula: $C_{12}H_{12}O_4$ (220.22 g/mol), Melting point: 237.5–238.5 °C.

Three inhibitors have been isolated from ethanolic extract of *Phellinus merrillii* (willow) fruiting body and identified as hispidin, hispolon, and inotilone. Hipolon showed highest inhibition against RLAR activity (IC_{50} = 9.47 μM) among the three suggesting that phenolic chemical moiety enhanced the activity [90].

3.4.32. Hirsutrin

Molecular formula: $C_{21}H_{20}O_{12}$ (462.40 g/mol), Melting point: 156–157 °C.

Hirsutrin was isolated together with six nonanthocyanin and five anthocyanin compounds from *Zea mays* L. (corn) for anti-cataractogenesis activity. Isolation of hirsutrin was conducted through bioassay-guided fractionation of ethanolic extract from the kernel of *Zea mays* L. using repeated column chromatography from ethyl acetate fraction. Hirsutrin showed the highest inhibitory activity in RLAR with an IC_{50} value of 4.78 μM and inhibitory constant (K_i) at 7.21×10^{-7} M from secondary plots of Lineweaver-Burk plots for RHAR assay. Further inhibition by hirsutrin on galactitol formation in rat lens (33.8% inhibition) and erythrocytes (15.7 µM, 32.5% inhibition) supported the efficacy of hirsutrin as the most effective AR inhibitors compared to all isolated compounds [91].

3.4.33. Hopeafuran

Molecular formula: $C_{28}H_{18}O_7$ (466.43 g/mol), Melting point: 131–134 °C.

Hopeafuran, classified under oligostilbenoids was isolated from the bark of *Shorea roxburghii* (white meranti) and exhibits the highest RLAR inhibitory activity compared to other isolated secondary metabolites from the same plant. This phytoconstituent inhibits the AR enzyme at an IC_{50} value of 6.9 µg/mL (14.8 µM) [92].

3.4.34. Hypolaetin 7-*O*-[6‴-*O*-acetyl-β-D-allopyranosyl-(1→2)]-6″-*O*-acetyl-β-D-glucopyranoside

Molecular formula: $C_{31}H_{34}O_{19}$ (710.59 g/mol).

Hypolaetin 7-*O*-[6‴-*O*-acetyl-β-D-allopyranosyl-(1→2)]-6″-*O*-acetyl-β-D-glucopyranoside was isolated from *Sideritis brevibracteata* (Dağ çayı) [93] and appeared as yellow powder [94,95]. This plant is native to Turkey and widely used as an herbal tea in folk medicine. Isolated hypolaetin has shown the most potent inhibitory activity of AR with IC_{50} value of 0.66 µM [93].

3.4.35. Isocampneoside II

Molecular formula: $C_{29}H_{36}O_{16}$ (640.58 g/mol).

Isocampneoside II is an active phenylethanoid glycoside isolated from acetone-H_2O (7:3, *v/v*) seeds extract of *Paulownia coreana* (kiri, paotong) at room temperature for 72 h. *Paulownia coreana* is long cultivated in Eastern Asia, particularly Korea and has been used traditionally in medicines for certain ailments [66]. A total of nine potential inhibitors have been isolated from this plant, however Isocampneoside II is the most potent inhibitor in anti-cataract activities. This compound significantly and uncompetitively inhibited RHAR activity with an IC_{50} value of 9.72 µM [66].

3.4.36. Isorhamnetin-3-glucoside

Molecular formula: $C_{22}H_{22}O_{12}$ (478.406 g/mol), Melting point: 168–172 °C.

Cochlospermum religiosum (silk-cotton tree, buttercup tree) has been reported to possess anticataract activity [96]. Purification of hot 95% ethanolic leaves extract of *C. religiosum* yielded isorhamnetin-3-glucoside. This bioactive compound was obtained as yellow needles and identified as flavonoids with yellowish orange color in alkali, pink in Mg-HCl and reaction with Fe^{3+} gives olive green color. Isorhamnetin-3-glucoside at the concentration of 25 µg/mL (52 µM) inhibited further formation of vacuoles and opacity on sodium selenite-induced lens opacity of rat pups. The antioxidant property of isorhamnetin-3-glucoside was suggested to complement its anticataract activity [97].

3.4.37. Kaempferol

Molecular formula: $C_{15}H_{10}O_6$ (286.23 g/mol), Melting point: 276–278 °C.

Kaempferol isolated from *Litsea japonica* (Thunb.) Juss. (hamabiwa) showed the most potent against RLAR inhibitory activity with IC_{50} value of 1.10 µM among all phytoconstituents isolated [98]. The same constituent was also isolated from *Agrimonia pilosa* Ledeb (hairy agrimony, hangul), *Allium victorialis* (victory onion), and *Paulownia coreana* with IC_{50} values of 15.2 µM (RLAR) [99], 1.10 µM (RLAR) [100], and 45.58 µM (RHAR) [66], respectively. The cataract prevention was further supported by inhibition of AGE activity at IC_{50} of 36.01 µM from *Allium victorialis* [100].

3.4.38. Kakkalide

Molecular formula: $C_{28}H_{32}O_{15}$ (608.549 g/mol), Melting point: 251–253 °C.

Kakkalide was isolated from *Viola hondoensis* W. Becker et H Boss (ri ben qiu guo jin cai) for its AR inhibitory activity. This plant is widely distributed in southern Korea and has been used to as traditional medicine in the form of expectorant. AR activity-guided isolation using column chromatography on a silica gel and gel filtration column afforded kakkalide. Kakkalide significant inhibited AR from Sprague-Dawley rat lenses at an IC_{50} of 0.34 µg/mL (0.56 µM), more potent than that of the positive control, tetramethylene glutaric acid (IC_{50} = 0.48 µg/mL) [101].

3.4.39. Lucidumol A [(24*S*)-24,25-Dihydroxylanost-8-ene-3,7-dione]

Molecular formula: $C_{30}H_{48}O_4$ (472.69 g/mol).

Lucidumol A is a new triterpenoid isolated from the ethanolic extract of the fruiting body of *Ganoderma lucidum* (lingzhi mushroom, reishi mushroom) from a thorough fractionation process [102]. Obtained as a white amorphous powder [103], lucidumol A suppressed the strongest AR activity with an IC_{50} of 19.1 µM compared to all other reported isolated secondary metabolites including ganoderic acid Df (IC_{50} = 22.8 µM) [104], ganoderic acid C2 (IC_{50} = 43.8 µM) [105], ganoderol B (IC_{50} = 110.1 µM) [106], and others.

3.4.40. Lupeol

Molecular formula: $C_{30}H_{50}O$ (426.72 g/mol), Melting point: 120–122 °C.

Lupeol, a pentacyclic triterpenoid has been isolated from the ethanolic flower extract of *Musa* sp. var. Nanjangud rasa bale (banana) [107] and methanolic leaf extract of *Vernonia cinereal* (purple fleabane) [108] with anti-cataractogenesis activities. Repeated silica gel chromatography after fractionation from both plants yielded lupeol as white needles. It inhibits human recombinant AR activity at IC_{50} of 1.53 µg/mL (3.6 µM) [52]. Similar inhibition was observed for AGE with inhibition in the range of 79–82% [107]. The potent activity of lupeol was substantiated with in vivo study using selenite-induced cataract formation in Sprague-Dawley rat pups. Lupeol attenuated the formation of vacuoles and opacity of rat pups lenses at the concentration of 25 µg/g in a dose-dependent manner in selenite-induced cataractogenesis [108].

3.4.41. Luteolin (2-(3,4-dihydroxyphenyl)-5,7-dihydroxy-4-chromomenone)

Molecular formula: $C_{15}H_{10}O_6$ (286.24 g/mol), Melting point: >320 °C.

The potential anti-cataract effect of luteolin is well known [109,110]. As yellow crystalline, luteolin has been isolated from various plants including *Platycodon grandiflorum* (balloon flower, Chinese bellflower) [111], *Vitex negundo* (Chinese chastetree, horseshoe vitex) [112], *Artemisia montana* [52], *Perilla frutescens* (L.) (perilla, Korean perilla) [69,113,114] and *Sinocrassula indica* (Chinese crassula) [115]. The selenite-induced oxidative stress treated group with luteolin (isolated from *Vitex negundo*) demonstrated 80% transparency of the lenses with minor cortical vacuolization and opacity suggesting that the anticataractogenic effect was supported by the antioxidant property based on significant decrease in various antioxidant activities tested [112]. In comparison to the isolated luteolin from different botanicals, luteolin from *Platycodon grandiflorum* was identified as the highest inhibition with an IC_{50} of 0.087 µM (RLAR) [111]. The IC_{50} value increases slightly higher to 0.45 µM (*Sinocrassula indica*) in the same activity [116]. However, isolated luteolin from the same species, *Perilla frutescens* (L.) of different parts showed different values in RLAR (seeds, IC_{50} 0.6 µM [113]; IC_{50} 1.89 µM) [114] and RHAR (leaves, IC_{50} 6.34 µM) [69], almost 10 and 3.5 times higher. Luteolin (*Artemisia montana*) was found to suppress RLAR activity at an IC_{50} 0.19 µM [52]. Luteolin from *Platycodon grandiflorum* (Jacq.) exhibited comparable potent inhibitory effect of AGE (IC_{50} = 16.6 µM) with chlorogenic acid methyl ester (IC_{50} = 12.9 µM) [111].

3.4.42. Luteolin-7-*O*-β-D-glucopyranoside

Molecular formula: $C_{21}H_{20}O_{11}$ (488.38 g/mol), Melting point: >195 °C.

Luteolin-7-*O*-β-D-glucopyranoside was isolated from the leaves extract of *Stauntonia hexaphylla* (Thunb.) Decne. (stauntonia vine), traditionally used as folk medicine in China, Japan and Korea. Luteolin-7-*O*-β-D-glucopyranoside showed the highest potent inhibitory activity of RLAR at IC_{50}

value of 7.34 μM among other isolated secondary metabolites. Inhibition was reported 2.4 times higher compared to that of quercetin. This data was substantiated by inhibition of AGE at IC$_{50}$ value of 117.8 μM and it was suggested that the presence of sugar position in flavonoids enhances the activity [117].

3.4.43. Magnoflorine

Molecular formula: C$_{20}$H$_{24}$NO$_4$$^+$ (342.41 g/mol).

Magnoflorine was isolated from *Tinospora cordifolia* (heart-leaved moonseed, guduchi, giloy) [118] and *Coptidis rhizome* (coptis root, huang lian) [88] for inhibitory activities against AR. Identification of magnoflorine was conducted with spectroscopic analysis and compared with the literature for both plants. Appeared as yellow powder, this compound exhibited lowest concentration of maximum RLAR activity showing IC$_{50}$ value at 3.6 μM from isolation of *Tinospora cordifolia*. Further analysis showed that magnoflorine inhibited 72.3% of galactose-induced polyol accumulation [118]. Nevertheless, the isolated magnoflorine from *Coptidis rhizome* possessed marginal inhibition against RLAR with 18% inhibition at a concentration of 146 μM. At this point, it is not clear if the very significant differences in efficacies and potencies were due to technical differences in isolation and/or biological assay.

3.4.44. Methyl-3,5-di-*O*-caffeoylquinate

Molecular formula: C$_{26}$H$_{26}$O$_{12}$ (530.50 g/mol).

Methyl-3,5-di-*O*-caffeoylquinate or also known as 3,5-di-*O*-caffeoylquinic acid methyl ester was isolated from the flowers of *Erigeron annuus* [68] and fruits of *Xanthium strumarium* [33] with highest inhibitory activity towards RLAR among all isolated secondary metabolites. Phytochemical analysis of ethyl acetate-soluble fraction of *Erigeron annuus* methanolic flower extract afforded methyl-3,5-di-*O*-caffeoylquinate with yellow gum appearance at percentage yield of 0.0075% [68]. Isolated methyl-3,5-di-*O*-caffeoylquinate from both plants suppressed RLAR activity at IC$_{50}$ values of 0.3 to 0.81 μM. Significant results (most potent) were also observed with further assays in RHAR and galactitol accumulation in rat lenses ex vivo from *Xanthium strumarium* at IC$_{50}$ value of 0.67 μM and 117 μg/lens wet weight, respectively [33].

3.4.45. Mumeic Acid-A

Molecular formula: C$_{24}$H$_{24}$O$_{10}$ (472.44 g/mol).

In an attempt to obtain inhibitors of RLAR from *Prunus mume* (Japanese apricot), mumeic acid-A was found to be the most potent inhibitor from all isolated secondary metabolites. The IC$_{50}$ concentration of mumeic acid-A (IC$_{50}$ = 0.4 μM) was almost twice that of chlorogenic acid (IC$_{50}$ = 0.7 μM) as the positive control [119,120].

3.4.46. Puerariafuran

Molecular formula: C$_{16}$H$_{12}$O$_5$ (360.32 g/mol), Melting point: 294–296 °C.

The roots of *Pueraria lobata* has for long been used in Far Eastern Asia countries as traditional medicine [57]. The isolation of root extract affords puerariafuran, a new 2-arylbenzofuran inhibited RLAR with an IC$_{50}$ value of 22.2 μM, much lower than the positive control, 3,3-Tetramethyleneglutaric acid (IC$_{50}$ = 28.8 μM). This data was substantiated with prevention of xylose-induced lens opacity in a dose-dependent manner, with the highest dose at 15 μM [121].

3.4.47. Quercetin-3-*O*-β-D-glucoside

Molecular formula: C$_{21}$H$_{20}$O$_{12}$ (464.38 g/mol).

Quercetin-3-*O*-β-D-glucoside was isolated from *Petasites japonicus* (butterbur, fuki, sweet coltsfoot) and *Stauntonia hexaphylla* as inhibitor for RLAR and AGE activities. It inhibited RLAR activity at IC$_{50}$

values between 2.21 and 10.4 µM [117,122]. In contrast, its inhibition of AGE formation required a much higher concentration (IC$_{50}$ = 1 mM) [117].

3.4.48. Quercitrin (Quercetin 3-*O*-α-L-rhamnoside)

Molecular formula: C$_{21}$H$_{20}$O$_{11}$ (448.38 g/mol), Melting point: 177–183 °C

Quercitrin, a glycosylation of quercetin at C-3, was isolated from few botanicals including *Smilax china* L. (China root), *Agrimonia pilosa* Ledeb, *Allium victorialis* var. *platyphyllum*, and *Melastoma sanguineum* using various isolation methods. It was tested for RLAR and AGE inhibitory assays and found that quercitrin possesses significant inhibitory actions for both activities among all isolated phytoconstituents in these plants. The IC$_{50}$ values were reported at 0.17 to 0.56 µM (RLAR) and 42.0 to 58.0 µM (AGE) [99,100,123]. In contrast, quercitrin from *Melastoma sanguineum* showed IC$_{50}$ value of 0.16 µM (RLAR) and 25.11 µM (AGE), respectively [40].

3.4.49. Rhetsinine

Molecular formula: C$_{19}$H$_{17}$N$_3$O$_2$ (319.36 g/mol), Melting point: 196 °C.

Evodia rutaecarpa Bentham (Rutaceae) (wu zhu yu) is part of the Kampo-herbal medicine in Japan and has been used to relive digestion as well as painkiller. Various compounds have been isolated from this plant especially rhetsinine showed potent inhibitory activity against RLAR at an IC$_{50}$ value of 24.1 µM. Rhetsinine was also reported to inhibit sorbitol accumulation in human erythrocyte by almost 79.3% at 100 µM [124].

3.4.50. Rosmarinic Acid

Molecular formula: C$_{18}$H$_{16}$O$_8$ (360.32 g/mol), Melting point: 171–175 °C.

Rosmarinic acid, a conjugation of caffeic acid and 3,4-dihydroxyphenyllactic acid is mainly isolated from the family of Lamiaceae with good anti-cataract activity. Isolated rosmarinic acid from *Salvia grandifolia* (da ye shu wei cao) recorded the lowest IC$_{50}$ (0.30 µM) in RLAR activity [125] compared to other plants. The IC$_{50}$ values of rosmarinic acid isolated from other plants were 2.77 µM (*Prunella vulgaris* L.; woundwort, self-heal) [70], 5.38 µM (*Colocasia esculenta* (L.) Schott; taro) [126], and 11.2 µM (seeds, *Perilla frutescens* L.) [113], respectively. In RHAR assay, the activity of rosmarinic acid were shown at 2.77 µM (leaves, *Perilla frutescens* L.) [69] and 18.6 µM (*Prunella vulgaris* L.) with potent activity on galactitol accumulation in rat lenses, but low inhibitor of AGE (20.7%) for *Prunella vulgaris* L. [70].

3.4.51. Scopoletin

Molecular formula: C$_{10}$H$_8$O$_4$ (192.16 g/mol), Melting point: 204–205 °C.

The bioassay-guided fractionation of methanol extract of *Magnolia fargesii* (Shin-i) air-dried buds yielded five compounds with scopoletin showing the most potent in RLAR, AGE and xylose-induced lens opacity assays [127]. Scopoletin significantly inhibits AGE formation with an IC$_{50}$ value of 2.9 µM, approximately 327 times more potent compared to the positive control (IC$_{50}$ = 961 µM). Similarly, in the RLAR activity, scopoletin showed marked inhibitory activity with an IC$_{50}$ value of 22.5 µM. This was substantiated by the suppression of lens opacity to 72.9% (25 µM) after three days of xylose treatment [127]. A lower IC$_{50}$ concentration of RLAR activity was observed for scopoletin isolated from *Angelica gigas* (dang gui, Korean angelica) with an IC$_{50}$ value of 2.6 µM [128] showing the most potent activity among all isolated secondary metabolites. However, scopoletin from methanolic young leaves of *Artemisia montana* showed higher IC$_{50}$ value at 64.5 µM for the same activity [52,129].

3.4.52. Semilicoisoflavone B

Molecular formula: C$_{20}$H$_{16}$O$_6$ (352.34 g/mol), Melting point: 131–134 °C.

Semilicoisoflavone B is mostly found in roots and rhizomes of licorice species (*Glycyrrhiza* sp.) [130]. In searching for potential AR inhibitors, 10 secondary metabolites have been isolated from bioactivity-guided isolation of *Glycyrrhiza uralensis* with semilicoisoflavone B showed the most potent inhibition of RLAR and RHAR activities. Both inhibition rates were recorded at IC_{50} values of 1.8 and 10.6 µM, respectively. Unlike γ,γ-dimethylallyl type prenylated isoflavonoids, semilicoisoflavone B containing γ,γ-dimethylchromene ring on the aromatic ring inhibited AR more strongly. In kinetic analysis of AR inhibition, semilicoisoflavone B did not bind to any substrate and NADPH binding regions of RHAR. Ex vivo analysis showed that this compound highly inhibited sorbitol accumulation in rat lenses incubated with high glucose by 47.0% [131].

3.4.53. Sulfuretin and Butein

Molecular formula of sulfuretin: $C_{15}H_{10}O_5$ (270.24 g/mol), Melting point: 295–303 °C.
Molecular formula of butein: $C_{15}H_{12}O_5$ (272.25 g/mol), Melting point: 216 °C.
The AR and AGE guided isolation of ethanolic bark extract of *Rhus verniciflua* (lacquer tree) produced nine secondary metabolites with sulferetin and butein as the most potent phytoconstituents for AGE and RHAR, respectively. Sulferetin was isolated as white to off-white crystalline powder [132] and inhibited against AGE activity at IC_{50} value of 124 µM, 11 times lower than aminoguanidine (IC_{50} = 1450 µM). The RHAR inhibitory activity of butein was reported at IC_{50} = 0.7 µM [133]. The efficacies of both phytoconstituents have been suggested on the structure activity relationships of catechol moiety of the B ring and 4′-hydroxyl at the A ring for butein [134] and hydroxyl groups of flavones at the 3′-, 4′-, 5′-, and 7-positions for sulferetin [135].

3.4.54. Syringic Acid

Molecular formula: $C_9H_{10}O_5$ (198.17 g/mol), Melting point: 206–208 °C.
Syringic acid is a phenolic compound and a naturally occurring *O*-methylated trihydroxybenzoic acid monomer extracted from *Herba dendrobii* (shi hu). *Herba dendrobii*, found in the stem of many orchid species of the *Dendrobium* genus, has been used to improve vision centuries ago [136]. Syringic acid at medium dose (79.97%) isolated from *Herba dendrobii* improves survival of high-concentration D-galactose-injured human LEC with inhibition ratio of 20.3%. Rat lens turned clear to Grade 0 after 90 days of treatment. Syringic acid inhibits AR activity in a dose-dependent manner with an IC_{50} value of 213.17 µg/mL (1075.7 µM). Data suggest that syringic acid downregulates the expression of mRNA of AR [136]. However, the AR inhibition by syringic acid isolated from *Magnolia officinalis* was weaker with less than 10% of inhibition [137].

3.4.55. Swertisin

Molecular formula: $C_{22}H_{22}O_{10}$ (446.40 g/mol), Melting point: 243 °C.
Swertisin appears as pale yellow powdery crystals and isolated from *Enicostemma hyssopifolium* (najajihva, chota chirayita) methanol extract after repeated column chromatography over silica gel. This compound reacts with ferric chloride and turned greenish brown color as a confirmation test for flavonoids. RLAR activity was significantly inhibited by swertisin at an IC_{50} value of 0.71 µg/mL (1.6 µM; 82.3% inhibition at 10 µg/mL) indicating a higher inhibition compared to the other compound isolated, swertiamarin (IC_{50} = 7.59 µg/mL). This compound was also found to suppress polyol accumulation (41.7%) in lenses cultured in a galactitol medium [138].

3.4.56. Valoneic Acid Dilactone

Molecular formula: $C_{21}H_{10}O_{13}$ (470.29 g/mol), Melting point: 177–183 °C.
The repeated column chromatography and preparative HPLC of seed methanolic extract of *Syzygium cumini* (L.) Skeels lead to the isolation of six phytoconstituents with valoneic acid dilactone showed the highest activity against RLAR inhibitory activity at IC_{50} value of 0.075 µM [87]. Valoneic acid dilactone were the first constituents from this plant reported to possess RLAR inhibitory activity.

Table 1. Summary of relevant in vitro anti-cataract activities of phytoconstituents.

Active Ingredient	Structure	AGE	ARI	GLWW	RHAR	BLAR	HLAR	RLAR
1-O-galloyl-β-D-glucose (β-Glucogallin)		NA	NA	NA	17.00 μM [30]	NA	NA	NA
1,3-di-O-caffeoylquinic acid		24.85 μM [32]	NA	NA	0.810 μM [33]	NA	NA	0.22 μM [32]
1,5-Di-hydroxy-1,5-di-[(E)-3-(4-hydroxyphenyl)-2-propenoic]-3-pentanonyl ester (DHDP)		NA	NA	NA	194.67 μM [34]	NA	NA	NA
1,5-di-O-caffeoylquinic acid		NA	NA	NA	NA	NA	NA	2.98 μM [35]
1,3,6-trihydroxy-2-methoxymethylanthraquinone		52.72 μM [36]	NA	NA	NA	NA	NA	3.04 μM [36]

Table 1. *Cont.*

Active Ingredient	Structure	IC$_{50}$ Values						
		AGE	ARI	GLWW	RHAR	BLAR	HLAR	RLAR
1,2,3,6-tetra-*O*-galloyl-*β*-D-glucose		1.99 μM [38]	NA	NA	NA	NA	NA	0.70 μM [38]
1,3,5,8-Tetrahydroxyxanthone		NA	NA	NA	NA	NA	NA	0.0886 μM [39]
2″,4″-*O*-Diacetylquercitrin		11.46 μM [40]	NA	NA	NA	NA	NA	0.077 μM [40]
3-Isomangostin		NA	NA	NA	NA	NA	NA	3.48 μM [41]

Table 1. *Cont.*

Active Ingredient	Structure	IC_{50} Values						
		AGE	ARI	GLWW	RHAR	BLAR	HLAR	RLAR
3',5'-dimethoxy-(1,1'-biphenyl)-3,4-diol 3-O-β-D-glucopyranoside		NA	NA	NA	NA	NA	NA	3.80 μM [47]
3,5-di-O-caffeoylquinic Acid		6.06 μM [48]	NA	153 g [33]	1.34 μM [33]	NA	NA	0.19 μM [33]
4-O-butylpaeoniflorin		10.80 μM [53]	NA	NA	NA	NA	NA	36.20 μM [53]

Table 1. *Cont.*

Active Ingredient	Structure	IC$_{50}$ Values						
		AGE	ARI	GlWW	RHAR	BLAR	HLAR	RLAR
4,5-Di-O-trans-caffeoyl-D-quinic acid		NA	NA	NA	NA	NA	NA	0.29 μM [55]
5-O-Feruloly quinic acid		NA	NA	NA	NA	NA	NA	14.19 μM [56]
5,7,4'-trihydroxyisoflavone (Genistein)		NA	NA	NA	NA	NA	NA	9.48 μM [60]
20(S)-Ginsenoside Rh2		NA	NA	NA	147.40 μM [61]	NA	NA	NA
Acteoside		5.11 μM [63]	NA	NA	NA	NA	NA	0.83 μM [63]

Table 1. *Cont.*

Active Ingredient	Structure	IC$_{50}$ Values								
		AGE	ARI	GLWW	RHAR	BLAR	HLAR	RLAR		
Basilicumin [7-(3-hydroxypropyl)-3-methyl-8-β-O-D-glucoside-2H-chromen-2-one]		NA	NA	NA	NA	2.09 μM	NA	NA		
Caffeic acid		7.56 μM [68]	NA	NA	210.28μM [66]	NA	NA	16.71 μM [65]		
Canangafruiticoside E	 Glc=β-D-glucopyranoside	NA	NA	NA	NA	NA	NA	0.80 μM [71]		
Capsofulvesin A [(2S)-1-O-(6Z,9Z,12Z,15Zoctadecatetraenoyl)-2-O-(4Z,10Z,13Zhexadecatetraenoyl)-3-O-β-D-galactopyranosyl glycerol)]		NA	NA	NA	NA	NA	NA	52.53 μM [72]		
Davallialactone		NA	NA	NA	0.56 μM [67]	NA	NA	0.33 μM [67]		

Table 1. *Cont.*

Active Ingredient	Structure	IC$_{50}$ Values								
		AGE	ARI	GLWW	RHAR	BLAR	HLAR	RLAR		
Delphinidin 3-*O*-β-galactopyranoside -3'-*O*-β-glucopyranoside	 Glc= β-glucopyranoside, Gal= β-galactopyranoside	NA	NA	NA	NA	NA	NA	0.37 μM [83]		
Desmethylanhydroicaritin		294.60 μM [84]	NA	NA	0.45 μM [84]	NA	NA	0.95 μM [84]		
Ellagic acid		26.0 μM [86]	NA	42.47% [85]	NA	NA	1.37 μM [67]	0.12 μM [87]		

Table 1. *Cont.*

Active Ingredient	Structure	IC$_{50}$ Values								
		AGE	ARI	GLWW	RHAR	BLAR	HLAR	RLAR		
Epiberberine		NA	NA	NA	168.10 μM [88]	NA	NA	100.07 μM [88]		
Geraniin		21.00 μM 96% * [89]	0.15 μM [89]	39.87% [85]	NA	NA	NA	NA		
Hipolon		NA	NA	NA	NA	NA	NA	9.47 μM [90]		

Table 1. *Cont.*

Active Ingredient	Structure	IC$_{50}$ Values						
		AGE	ARI	GLWW	RHAR	BLAR	HLAR	RLAR
Hirsutrin		NA	NA	33.78% [91]	NA	NA	NA	4.78 µM [91]
Hopeafuran		NA	NA	NA	NA	NA	NA	14.80 µM [92]
Hypolaetin 7-O-[6‴-O-acetyl-β-D-allopyranosyl-(1→2)]-6″-O-acetyl-β-D-glucopyranoside		NA	0.66 µM [93]	NA	NA	NA	NA	NA
Isocampneoside II		NA	NA	NA	9.72 µM [66]	NA	NA	NA

Table 1. *Cont.*

Active Ingredient	Structure	IC$_{50}$ Values						
		AGE	ARI	GLWW	RHAR	BLAR	HLAR	RLAR
Kaempferol		36.01 µM [100]	NA	NA	45.58 µM [66]	NA	NA	1.10 µM [98,100]
Kakkalide		NA	NA	NA	NA	NA	NA	0.56 µM [101]
Lucidumol A [(24S)-24,25-Dihydroxylanost-8-ene-3,7-dione]		NA	NA	NA	NA	19.10 µM [102]	NA	NA
Lupeol		NA	NA	NA	3.60 µM [107]	NA	NA	NA

Table 1. *Cont.*

Active Ingredient	Structure	IC$_{50}$ Values						
		AGE	ARI	GLWW	RHAR	BLAR	HLAR	RLAR
Luteolin (2-(3,4-dihydroxyphenyl)-5,7-dihydroxy-4-chromenone)		16.60 µM [111]	NA	NA	6.34 µM [52]	NA	NA	0.087 µM [111]
Luteolin-7-O-β-D-glucopyranoside		117.80 µM [117]	NA	NA	NA	NA	NA	7.34 µM [117]
Magnoflorine		NA	NA	NA	NA	NA	NA	3.60 µM [118]
Methyl-3,5-di-O-caffeoylquinate		NA	NA	117 g [33]	0.67 µM [33]	NA	NA	0.30 µM [33]

Table 1. *Cont.*

Active Ingredient	Structure	IC$_{50}$ Values						
		AGE	ARI	GLWW	RHAR	BLAR	HLAR	RLAR
Mumeic acid-A		NA	NA	NA	NA	NA	NA	0.40 µM [119]
Palbinone		>500 µM [53]	NA	NA	NA	NA	NA	11.40 µM [53]
Puerariafuran		NA	NA	NA	NA	NA	NA	22.20 µM [57,121]
Quercetin-3-*O*-β-D-glucoside		>1000 µM [117]	NA	NA	NA	NA	NA	2.21 µM [122]

Table 1. *Cont.*

Active Ingredient	Structure	IC$_{50}$ Values						
		AGE	ARI	GLWW	RHAR	BLAR	HLAR	RLAR
Quercitrin (Quercetin 3-O-α-L-rhamnoside)		4.20 µM [100]	NA	NA	NA	NA	NA	0.17 µM [40]
Rhetsinine		NA	NA	NA	NA	NA	NA	24.10 µM [124]
Rosmarinic acid		NA	NA	532.38g [70]	2.77 µM [69]	NA	NA	0.30 µM [125]
Scopoletin		2.93 µM [127]	NA	NA	NA	NA	NA	2.59 µM [128]
Semilicoisoflavone B		NA	NA	NA	10.60 µM [131]	NA	NA	1.80 µM [131]

Table 1. *Cont.*

Active Ingredient	Structure	IC$_{50}$ Values						
		AGE	ARI	GLWW	RHAR	BLAR	HLAR	RLAR
Sulfuretin		124.00 µM [133]	NA	NA	1.30 µM [133]	NA	NA	NA
Syringic Acid		NA	NA	NA	NA	NA	NA	1081.1 µm [136]
Swertisin (C-glycosidic flavonoid)		NA	NA	NA	NA	NA	NA	1.60 µm [138]
Valoneic acid dilactone		NA	NA	NA	NA	NA	NA	0.075 µM [87]

Table 2. Summary of relevant ex vivo and in vivo activities of phytoconstituents.

Constituent Name (Class of Constituent)	Structure	Doses (IC$_{50}$/EC$_{50}$)					Ref
		AR Transgenic Mice	Selenite-Induced	AR Rat Lens	Galactose-Induced Lens Opacity	Xylose-Induced Lens Opacity	
1-O-galloyl-β-D-glucose (β-Glucogallin)		Ex vivo: 30.00 μM	NA	NA	NA	NA	[29]
1,2,3,6-Tetra-O-galloyl-β-D-glucose		NA	NA	NA	NA	Ex vivo: 80.00 μM	[38]
3,5-di-O-caffeoyl-epi-quinic Acid		NA	NA	NA	NA	Ex vivo: 10.00 μM	[48]
5,7,4'-trihydroxyisoflavone (Genistein)		NA	Ex vivo: 52.25 μM	NA	NA	Ex vivo: 18.50 μM	[60]
Isorhamnetin-3-glucoside		NA	NA	NA	NA	NA	[97]

Table 2. *Cont.*

Constituent Name (Class of Constituent)	Structure	Doses (IC$_{50}$/EC$_{50}$)					Ref
		AR Transgenic Mice	Selenite-Induced	AR Rat Lens	Galactose-Induced Lens Opacity	Xylose-Induced Lens Opacity	
Lupeol		NA	In vivo: 126.15 µM	NA	NA	NA	[108]
Luteolin (2-(3,4-dihydroxyphenyl)-5,7-dihydroxy-4-chromomenone)		NA	Ex vivo: 6.98 µM	NA	NA	NA	[112]
Puerariafuran		NA	NA	NA	NA	Ex vivo: 15.00 µM	[121]
Scopoletin		NA	NA	NA	NA	Ex vivo: 25.00 µM	[127]
Syringic acid		NA	NA	NA	Ex vivo: 1075.70 µM In vivo: 2% syringic acid eye drop (131,197.80 µM)	NA	[136]

4. Discussion and Outlook

Despite the success in surgical replacement of the cataractous lens with an artificial intraocular lens, discovery of pharmacological prevention and treatment of this blinding disorder has been an earnest, continuous effort in ophthalmology research. In this review manuscript, we summarize findings of phytoconstitutents and their pharmacological effects as potential anti-cataract agents. The large number of interesting compounds is exciting. It raises hope that clinically useful medication may have a good chance to be derived from this sizable collection of chemicals with diverse structural scaffolds.

Many of the compounds have potent and efficacious in vitro pharmacological activities that are consistent with potential anti-cataract effects. For example, 1,2,3,6-tetra-*O*-galloyl-β-D-glucose inhibits AGE formation with an IC_{50} of 2 M. Both 1,3,5,8-tetrahydroxyxanthone and 2″,4″-*O*-diacetylquercitrin inhibit AR with IC_{50} values below 0.1 M. However, a major limitation of the listed compounds is that although they have been shown to have the appropriate biological actions in a variety of in vitro or ex vivo assays, many of them were not tested in animal cataract models. Additionally, a few have been evaluated in only one animal model. Without relevant in vivo data, it is obviously very difficult to develop the compounds into meaningful treatments for cataract patients.

In addition to the lack of in vivo data, there are other challenges facing development of anti-cataract pharmaceuticals. For example, cataract medication has to compete with the very successfully and generally affordable (as least in developed countries) surgical procedure. Moreover, pharmacological prevention of cataract formation is expected to require a long-term, likely multi-year, administration of medicine, which, to some, is undesirable. Overcoming these challenges necessitates careful considerations of drug safety, convenience of administration, and cost. These concerns may have previously prohibited the development of certain agents. Nonetheless, we feel that phytoconstituents are advantageous compared to conventional synthetic drugs. Many societies have been using plant products from where some of the ingredients are derived for centuries, indicating long-term safety and acceptance. The development path and clinical use will be similar to vitamins and phytochemicals such as lutein and zeaxanthine. If proven safe, cost-effective, and most importantly, efficacious in preventing or reversing cataract formation, phytoconstituents can be a revolutionary approach in the treatment of cataract.

5. Conclusions

Despite the success in surgical replacement of the cataractous lens with artificial intraocular lens, pharmacological prevention and treatment of this blinding disorder have been an earnest, continuous effort in ophthalmology research. In this review manuscript, we summarize findings of 56 entries of phytoconstitutents and their pharmacological effects as potential anti-cataract agents. The large number of interesting compounds is exciting. It raises hope that clinically useful medication may have a good chance to be derived from this sizable collection of chemicals with diverse structural scaffolds.

Many of the compounds have potent and efficacious in vitro pharmacological activities that are consistent with potential anti-cataract effects. For example, 1,2,3,6-tetra-*O*-galloyl-β-D-glucose inhibits AGE formation with an IC_{50} of 2 μM. Additionally, both 1,3,5,8-tetrahydroxyxanthone and 2″,4″-*O*-diacetylquercitrin inhibit AR with IC_{50} values below 0.1 μM. However, a major limitation of the listed compounds is that although they have been shown to have the appropriate biological actions in a variety of in vitro or ex vivo assays, many of them were not tested in animal cataract models. And a few have been evaluated in only one animal model. Without relevant in vivo data, it is obviously very difficult to develop the compounds into meaningful treatments for cataract patients. We feel that, by listing the comprehensive collection of phytocontituents in one place, this manuscript serves as an overview and perhaps an inspiration to prompt additional studies in this important research area. Collaborative efforts between phytochemists and cataract researchers are promisingly fruitful.

In addition to the lack of in vivo data, there are other challenges facing development of anti-cataract pharmaceuticals. For example, cataract medication has to compete with the very successfully and generally affordable (as least in developed countries) surgical procedure. Moreover, pharmacological prevention of cataract formation is expected to require a long-term, likely multi-year, administration of medicine, which, to some, is undesirable. Overcoming these challenges necessitates careful considerations of drug safety, convenience of administration, and cost. These concerns may have previously prohibited the development of certain agents. Nonetheless, we feel that phytoconstituents are advantageous compared to conventional synthetic drugs. Many societies have been using plant products where some of the ingredients are derived from for centuries, indicating long-term safety and acceptance. The development path and clinical use will be similar to vitamins and phytochemicals such as lutein and zeaxanthine. If proven safe, cost-effective, and most importantly, efficacious in preventing or reversing cataract formation, phytoconstituents can be a revolutionary approach in the treatment of cataract.

Author Contributions: V.L. and H.W. drafted the review. E.S. provided key reviewing. I.-H.P. critically revised the drafted manuscript. All authors were involved in writing the paper and had final approval of the submitted and published versions.

Funding: Funding in part supported by the Frick Foundation, BrightFocus Foundation for Macular Degeneration (Grant No. M2015180), California Table Grape Commission and Bridging Grant (304/CIIPT/6316033).

Acknowledgments: This review was supported in part by a grant from the Community Solutions Program, a program of the Bureau of Educational and Cultural Affairs (ECA) of the United States Department of State, implemented by IREX (International Research and Exchanges Board) and from the Department of Pharmaceutical Science, University of North Texas, System College of Pharmacy. The views expressed are the author's own and do not represent the Community Solutions Program, the U.S. Department of State, or IREX. VL would like to thank Universiti Sains Malaysia for providing research leave to complete the Community Solutions Program. We also sincerely appreciate the effort of the staffs in BRIT and UNTHSC for the support provided to us.

Conflicts of Interest: The authors declare no conflict of interest.

Abbreviations

AGE	Advanced glycation end-product
AR	Aldose reductase
ARI	Aldose reductase inhibition
BLAR	Bovine lens aldose reductase
GC	Glucocorticoid
GSH	Glutathione
HLAR	Human lens aldose reductase
LEC	Lens epithelial cell
NADPH	Nicotinamide adenine dinucleotide phosphate
RHAR	Recombinant human aldose reductase
RLAR	Rat lens aldose reductase
ROS	Reactive oxygen species
SDH	Sorbitol dehydrogenase
STZ	Streptozotocin
UV	Ultraviolet

References

1. Lou, M.F. Redox regulation in the lens. *Prog. Retin. Eye Res.* **2003**, *22*, 657–682. [CrossRef]
2. Liu, Y.C.; Wilkins, M.; Kim, T.; Malyugin, B.; Mehta, J.S. Cataracts. *Lancet* **2017**, *390*, 600–612. [CrossRef]
3. Fischel, J.D.; Lipton, J.R. Cataract surgery and recent advances: A review. *Nurs. Stand.* **1996**, *10*, 39–43. [CrossRef] [PubMed]
4. Spector, A. Oxidative stress-induced cataract: Mechanism of action. *FASEB J.* **1995**, *9*, 1173–1182. [CrossRef] [PubMed]
5. Spector, A.; Garner, W.H. Hydrogen peroxide and human cataract. *Exp. Eye Res.* **1981**, *33*, 673–681. [CrossRef]
6. Cui, X.L.; Lou, M.F. The effect and recovery of long-term H_2O_2 exposure on lens morphology and biochemistry. *Exp. Eye Res.* **1993**, *57*, 157–167. [CrossRef] [PubMed]

7. Zigler, J.S., Jr.; Jernigan, H.M., Jr.; Garland, D.; Reddy, V.N. The effects of "oxygen radicals" generated in the medium on lenses in organ culture: Inhibition of damage by chelated iron. *Arch. Biochem. Biophys.* **1985**, *241*, 163–172. [CrossRef]

8. Caird, F.I.; Hutchinson, M.; Pirie, A. Cataract and Diabetes. *Br. Med. J.* **1964**, *2*, 665–668. [CrossRef] [PubMed]

9. Chodos, J.B.; Habegger-Chodos, H.E. Cataract formation in human and experimental diabetes. Part I. *Surv. Ophthalmol.* **1960**, *5*, 129–159. [PubMed]

10. Snow, A.; Shieh, B.; Chang, K.C.; Pal, A.; Lenhart, P.; Ammar, D.; Ruzycki, P.; Palla, S.; Reddy, G.B.; Petrash, J.M. Aldose reductase expression as a risk factor for cataract. *Chem. Biol. Interact.* **2015**, *234*, 247–253. [CrossRef] [PubMed]

11. Hayman, S.; Kinoshita, J.H. Isolation and Properties of Lens Aldose Reductase. *J. Biol. Chem.* **1965**, *240*, 877–882. [PubMed]

12. Obazawa, H.; Merola, L.O.; Kinoshita, J.H. The effects of xylose on the isolated lens. *Investig. Ophthalmol.* **1974**, *13*, 204–209.

13. Ai, Y.; Zheng, Z.; O'Brien-Jenkins, A.; Bernard, D.J.; Wynshaw-Boris, T.; Ning, C.; Reynolds, R.; Segal, S.; Huang, K.; Stambolian, D. A mouse model of galactose-induced cataracts. *Hum. Mol. Genet.* **2000**, *9*, 1821–1827. [CrossRef] [PubMed]

14. Franke, S.; Dawczynski, J.; Strobel, J.; Niwa, T.; Stahl, P.; Stein, G. Increased levels of advanced glycation end products in human cataractous lenses. *J. Cataract Refract. Surg.* **2003**, *29*, 998–1004. [CrossRef]

15. Ramalho, J.S.; Marques, C.; Pereira, P.C.; Mota, M.C. Role of glycation in human lens protein structure change. *Eur. J. Ophthalmol.* **1996**, *6*, 155–161. [CrossRef] [PubMed]

16. Abiko, T.; Abiko, A.; Ishiko, S.; Takeda, M.; Horiuchi, S.; Yoshida, A. Relationship between autofluorescence and advanced glycation end products in diabetic lenses. *Exp. Eye Res.* **1999**, *68*, 361–366. [CrossRef] [PubMed]

17. Joon, T.L.; Foo, K.; Panagiotopoulos, S.; Jerums, G.; Taylor, H.R. In vivo assessment of an animal model of diabetic cataract: Medical intervention studies. *Dev. Ophthalmol.* **1994**, *26*, 57–62. [PubMed]

18. Patil, M.A.; Suryanarayana, P.; Putcha, U.K.; Srinivas, M.; Reddy, G.B. Evaluation of neonatal streptozotocin induced diabetic rat model for the development of cataract. *Oxid. Med. Cell Longev.* **2014**, *2014*, 463264. [CrossRef] [PubMed]

19. Lee, A.Y.; Chung, S.S. Contributions of polyol pathway to oxidative stress in diabetic cataract. *FASEB J.* **1999**, *13*, 23–30. [CrossRef] [PubMed]

20. Shearer, T.R.; David, L.L.; Anderson, R.S. Selenite cataract: A review. *Curr. Eye Res.* **1987**, *6*, 289–300. [CrossRef] [PubMed]

21. Shearer, T.R.; David, L.L.; Anderson, R.S.; Azuma, M. Review of selenite cataract. *Curr. Eye Res.* **1992**, *11*, 357–369. [CrossRef] [PubMed]

22. Shearer, T.R.; Ma, H.; Fukiage, C.; Azuma, M. Selenite nuclear cataract: Review of the model. *Mol. Vis.* **1997**, *3*, 8. [PubMed]

23. Dillon, J. UV-B as a pro-aging and pro-cataract factor. *Doc. Ophthalmol.* **1994**, *88*, 339–344. [CrossRef] [PubMed]

24. Wolff, S.P. Cataract and UV radiation. *Doc. Ophthalmol.* **1994**, *88*, 201–204. [CrossRef] [PubMed]

25. Kronschlager, M.; Lofgren, S.; Yu, Z.; Talebizadeh, N.; Varma, S.D.; Soderberg, P. Caffeine eye drops protect against UV-B cataract. *Exp. Eye Res.* **2013**, *113*, 26–31. [CrossRef] [PubMed]

26. James, E.R. The etiology of steroid cataract. *J. Ocul. Pharmacol. Ther.* **2007**, *23*, 403–420. [CrossRef] [PubMed]

27. Shui, Y.B.; Kojima, M.; Sasaki, K. A new steroid-induced cataract model in the rat: Long-term prednisolone applications with a minimum of X-irradiation. *Ophthalmic Res.* **1996**, *28*, 92–101. [CrossRef] [PubMed]

28. Shui, Y.B.; Vrensen, G.F.; Kojima, M. Experimentally induced steroid cataract in the rat: A scanning electron microscopic study. *Surv. Ophthalmol.* **1997**, *42*, S127–S132. [CrossRef]

29. Variya, B.C.; Bakrania, A.K.; Patel, S.S. *Emblica officinalis* (Amla): A review for its phytochemistry, ethnomedicinal uses and medicinal potentials with respect to molecular mechanisms. *Pharmacol. Res.* **2016**, *111*, 180–200. [CrossRef] [PubMed]

30. Puppala, M.; Ponder, J.; Suryanarayana, P.; Reddy, G.B.; Petrash, M.; LaBarbera, D.V. The isolation and characterization of β-glucogallin as a novel aldose reductase inhibitor from *Emblica officinalis*. *PLoS ONE* **2012**, *7*, e31399. [CrossRef] [PubMed]

31. Chang, K.C.; Laffin, B.; Ponder, J.; Énzsöly, A.; Németh, J.; Labarbera, D.V.; Petrash, J.M. Beta-glucogallin reduces the expression of lipopolysaccharide-induced inflammatory markers by inhibition of aldose reductase in murine macrophages and ocular tissues. *Chem. Biol. Interact.* **2013**, *202*, 283–287. [CrossRef] [PubMed]

32. Lee, Y.K.; Hong, E.Y.; Whang, W.K. Inhibitory Effect of Chemical Constituents Isolated from *Artemisia iwayomogi* on Polyol Pathway and Simultaneous Quantification of Major Bioactive Compounds. *BioMed Res. Int.* **2017**, *2017*. [CrossRef] [PubMed]

33. Yoon, H.N.; Lee, M.Y.; Kim, J.K.; Suh, H.W.; Lim, S.S. Aldose reductase inhibitory compounds from *Xanthium strumarium*. *Arch. Pharm. Res.* **2013**, *36*, 1090–1095. [CrossRef] [PubMed]

34. Wang, Z.; Hwang, S.H.; Lim, S.S. Characterization of DHDP, a novel aldose reductase inhibitor isolated from *Lysimachia christinae*. *J. Funct. Foods* **2017**, *37*, 241–248. [CrossRef]

35. Paek, J.H.; Lim, S.S. Preparative isolation of aldose reductase inhibitory compounds from *Nardostachys chinensis* by elution–extrusion counter-current chromatography. *Arch. Pharm. Res.* **2014**, *37*, 1271–1279. [CrossRef] [PubMed]

36. Yoo, N.H.; Jang, D.S.; Lee, Y.M.; Jeong, I.H.; Cho, J.H.; Kim, J.H.; Kim, J.S. Anthraquinones from the Roots of *Knoxia valerianoides* inhibit the formation of advanced glycation end products and rat lens aldose reductase in vitro. *Arch. Pharm. Res.* **2010**, *33*, 209–214. [CrossRef] [PubMed]

37. Duan, D.; Li, Z.; Luo, H.; Zhang, W.; Chen, L.; Xu, X. Antiviral compounds from traditional Chinese medicines *Galla Chinese* as inhibitors of HCV NS3 protease. *Bioorg. Med. Chem. Lett.* **2004**, *14*, 6041–6044. [CrossRef] [PubMed]

38. Lee, J.; Jang, D.S.; Kim, N.H.; Lee, Y.M.; Kim, J.; Kim, J.S. Galloyl glucoses from the seeds of *Cornus officinalis* with inhibitory activity against protein glycation, aldose reductase, and cataractogenesis ex vivo. *Biol. Pharm. Bull.* **2011**, *34*, 443–446. [CrossRef] [PubMed]

39. Zheng, H.H.; Luo, C.T.; Chen, H.; Lin, J.N.; Ye, C.L.; Mao, S.S.; Li, Y.L. Xanthones from *Swertia mussotii* as multitarget-directed antidiabetic agents. *Chem. Med. Chem.* **2014**, *9*, 1374–1377. [CrossRef] [PubMed]

40. Lee, I.S.; Kim, I.S.; Lee, Y.M.; Lee, Y.; Kim, J.H.; Kim, J.S. 2″,4″-*O*-diacetylquercitrin, a novel advanced glycation end-product formation and aldose reductase inhibitor from *Melastoma sanguineum*. *Chem. Pharm. Bull. (Tokyo)* **2013**, *61*, 662–665. [CrossRef] [PubMed]

41. Fatmawati, S.; Ersam, T.; Shimizu, K. The inhibitory activity of aldose reductase in vitro by constituents of *Garcinia mangostana* Linn. *Phytomedicine* **2015**, *22*, 49–51. [CrossRef] [PubMed]

42. Fan, Y.; Han, H.; He, C.; Yang, L.; Wang, Z. Identification of the metabolites of gigantol in rat urine by ultra-performance liquid chromatography combined with electrospray ionization quadrupole time-of-flight tandem mass spectrometry. *Biomed. Chromatogr.* **2014**, *28*, 1808–1815. [CrossRef] [PubMed]

43. Fang, H.; Hu, X.; Wang, M.; Wan, W.; Yang, Q.; Sun, X.; Gu, Q.; Gao, X.X.; Wang, Z.T.; Gu, L.Q.; et al. Anti-osmotic and antioxidant activities of gigantol from *Dendrobium aurantiacum var. denneanum* against cataractogenesis in galactosemic rats. *J. Ethnopharmacol.* **2015**, *172*, 38–46. [CrossRef] [PubMed]

44. Yang, L.; Wang, Z.; Xu, L. Phenols and a triterpene from *Dendrobium aurantiacum var. denneanum* (Orchidaceae). *Biochem. Syst. Ecol.* **2006**, *34*, 658–660. [CrossRef]

45. Hu, J.; Fan, W.; Dong, F.; Miao, Z.; Zhou, J. Chemical components of *Dendrobium chrysotoxum*. *Chin. J. Chem.* **2012**, *30*, 1327–1330. [CrossRef]

46. Wu, J.; Li, X.; Wan, W.; Yang, Q.; Ma, W.; Chen, D.; Hu, J.M.; Chen, C.Y.O.; Wei, X.Y. Gigantol from *Dendrobium chrysotoxum* Lindl. binds and inhibits aldose reductase gene to exert its anti-cataract activity: An in vitro mechanistic study. *J. Ethnopharmacol.* **2017**, *198*, 255–261. [CrossRef] [PubMed]

47. Lee, I.S.; Jung, S.H.; Lee, Y.M.; Choi, S.J.; Sun, H.; Kim, J.S. Phenolic Compounds from the Leaves and Twigs of *Osteomeles schwerinae* That Inhibit Rat Lens Aldose Reductase and Vessel Dilation in Zebrafish Larvae. *J. Nat. Prod.* **2015**, *78*, 2249–2254. [CrossRef] [PubMed]

48. Jang, D.S.; Yoo, N.H.; Kim, N.H.; Lee, Y.M.; Kim, C.S.; Kim, J.; Kim, J.H.; Kim, J.S. 3,5-Di-*O*-caffeoyl-epi-quinic acid from the leaves and stems of *Erigeron annuus* inhibits protein glycation, aldose reductase, and cataractogenesis. *Biol. Pharm. Bull.* **2010**, *33*, 329–333. [CrossRef] [PubMed]

49. Pauli, G.F.; Poetsch, F.; Nahrstedt, A. Structure assignment of natural quinic acid derivatives using proton nuclear magnetic resonance techniques. *Phytochem. Anal.* **1998**, *9*, 177–185. [CrossRef]

50. Kim, H.; Lee, Y.S. Identification of new dicaffeoylquinic acids from *Chrysanthemum morifolium* and their antioxidant activities. *Planta Med.* **2005**, *71*, 871–876. [CrossRef] [PubMed]

51. Lee, J.; Lee, Y.M.; Lee, B.W.; Kim, J.H.; Kim, J.S. Chemical constituents from the aerial parts of *Aster koraiensis* with protein glycation and aldose reductase inhibitory activities. *J. Nat. Prod.* **2012**, *75*, 267–270. [CrossRef] [PubMed]

52. Jung, H.A.; Islam, M.D.N.; Kwon, Y.S.; Jin, S.E.; Son, Y.K.; Park, J.J.; Sohn, H.S.; Choi, J.S. Extraction and identification of three major aldose reductase inhibitors from *Artemisia montana*. *Food Chem. Toxicol.* **2011**, *49*, 376–384. [CrossRef] [PubMed]

53. Do, T.H.; Tran, M.N.; Lee, I.S.; Yun, M.L.; Jin, S.K.; Jung, H.J.; Lee, S.M.; Na, M.K.; Bae, K. Inhibitors of aldose reductase and formation of advanced glycation end-products in Moutan cortex (*Paeonia suffruticosa*). *J. Nat. Prod.* **2009**, *72*, 1465–1470.

54. Kadota, S.; Terashima, S.; Basnet, P.; Kikuchi, T.; Namba, T.; Palbinone, A. Novel Terpenoid from *Peaonia albiflora*; Potent Inhibitory Activity on 3α-Hydroxysteroid Dehydrogenase. *Chem. Pharm. Bull.* **1993**, *41*, 487–489. [CrossRef] [PubMed]

55. Liu, J.; Nakamura, S.; Zhuang, Y.; Yoshikawa, M.; Mohamed, G.; Hussein, E.; Matsuo, K.; Matsuda, H. Medicinal Flowers. XXXX. 1) Structures of Dihydroisocoumarin Glycosides and Inhibitory Effects on Aldose Reducatase from the Flowers of *Hydrangea macrophylla var. thunbergii*. *Chem. Pharm. Bull.* **2013**, *61*, 655–661. [CrossRef] [PubMed]

56. Jung, H.J.; Jung, H.A.; Kang, S.S.; Lee, J.H.; Cho, Y.S.; Moon, K.H.; Choi, J.S. Inhibitory activity of *aralia continentalis* roots on protein tyrosine phosphatase 1B and rat lens aldose reductase. *Arch. Pharm. Res.* **2012**, *35*, 1771–1777. [CrossRef] [PubMed]

57. Jang, D.S.; Kim, J.M.; Lee, Y.M.; Kim, Y.S.; Kim, J.H.; Kim, J.S. Puerariafuran, a new inhibitor of advanced glycation end products (AGEs) isolated from the roots of *Pueraria lobata*. *Chem. Pharm. Bull.* **2006**, *54*, 1315–1317. [CrossRef] [PubMed]

58. Kinjo, J.E.; Furusawa, J.I.; Baba, J.; Takeshita, T.; Yamasaki, M.; Nohara, T. Studies on the constituents of *Pueraria lobata*. III. Isoflavonoids and related compounds in the roots and the voluble stems. *Chem. Pharm. Bull. (Tokyo)* **1987**, *35*, 4846–4850. [CrossRef]

59. Kim, S.B.; Hwang, S.H.; Wang, Z.; Yu, J.M.; Lim, S.S. Rapid identification and isolation of inhibitors of rat lens aldose reductase and antioxidant in *Maackia amurensis*. *BioMed Res. Int.* **2017**, *2017*. [CrossRef] [PubMed]

60. Kim, Y.S.; Kim, N.H.; Jung, D.H.; Jang, D.S.; Lee, Y.M.; Kim, J.M.; Kim, J.S. Genistein inhibits aldose reductase activity and high glucose-induced TGF-beta2 expression in human lens epithelial cells. *Eur. J. Pharmacol.* **2008**, *594*, 18–25. [CrossRef] [PubMed]

61. Fatmawati, S.; Ersam, T.; Yu, H.; Zhang, C.; Jin, F.; Shimizu, K. 20(S)-Ginsenoside Rh2 as aldose reductase inhibitor from *Panax ginseng*. *Bioorg. Med. Chem. Lett.* **2014**, *24*, 4407–4409. [CrossRef] [PubMed]

62. Li, H.M.; Kim, J.K.; Jang, J.M.; Cui, C.B.; Lim, S.S. Analysis of the inhibitory activity of *Abeliophyllum distichum* leaf constituents against aldose reductase by using high-speed counter current chromatography. *Arch. Pharm. Res.* **2013**, *36*, 1104–1112. [CrossRef] [PubMed]

63. Yu, S.Y.; Lee, I.S.; Jung, S.H.; Lee, Y.M.; Lee, Y.R.; Kim, J.H.; Sun, H.; Kim, J.S. Caffeoylated phenylpropanoid glycosides from *Brandisia hancei* inhibit advanced glycation end product formation and aldose reductase in vitro and vessel dilation in larval zebrafish in vivo. *Planta Med.* **2013**, *79*, 1705–1709. [CrossRef] [PubMed]

64. Aslam, H.; Tehseen, Y.; Maryam, K.; Uroos, M.; Siddiqui, B.S.; Hameed, A.; Iqbal, J. Bioorganic Chemistry Identification of new potent inhibitor of aldose reductase from *Ocimum basilicum*. *Bioorg. Chem.* **2017**, *75*, 62–70.

65. Koo, D.C.; Baek, S.Y.; Jung, S.H.; Shim, S.H. Aldose reductase inhibitory compounds from extracts of *Dipsacus asper*. *Biotechnol. Bioprocess Eng.* **2013**, *18*, 926–931. [CrossRef]

66. Kim, J.K.; Lee, Y.S.; Kim, S.H.; Bae, Y.S.; Lim, S.S. Inhibition of aldose reductase by phenylethanoid glycoside isolated from the seeds of *Paulownia coreana*. *Biol. Pharm. Bull.* **2011**, *34*, 160–163. [CrossRef] [PubMed]

67. Lee, Y.S.; Kang, Y.H.; Jung, J.Y.; Kang, I.J.; Han, S.N.; Chung, J.S.; Shin, H.K.; Lim, S.S. Inhibitory constituents of aldose reductase in the fruiting body of *Phellinus linteus*. *Biol. Pharm. Bull.* **2008**, *31*, 765–768. [CrossRef] [PubMed]

68. Jang, D.S.; Yoo, N.H.; Lee, Y.M.; Yoo, J.L.; Kim, Y.S.; Kim, J.S. Constituents of the flowers of *Erigeron annuus* with inhibitory activity on the formation of advanced glycation end products (AGEs) and aldose reductase. *Arch. Pharm. Res.* **2008**, *31*, 900–904. [CrossRef] [PubMed]

69. Paek, J.H.; Shin, K.H.; Kang, Y.; Lee, J.; Lim, S.S. Rapid Identification of Aldose Reductase Inhibitory Compounds from *Perilla frutescens*. *BioMed Res. Int.* **2013**, *2013*. [CrossRef] [PubMed]

70. Li, H.M.; Kim, J.K.; Jang, J.M.; Kwon, S.O.; Cui, C.B.; Lim, S.S. The inhibitory effect of *Prunella vulgaris* L. on aldose reductase and protein glycation. *J. Biomed. Biotechnol.* **2012**, *2012*. [CrossRef] [PubMed]

71. Matsumoto, T.; Nakamura, S.; Fujimoto, K.; Ohta, T.; Ogawa, K.; Yoshikawa, M.; Matsuda, H. Structure of constituents isolated from the flower buds of *Cananga odorata* and their inhibitory effects on aldose reductase. *J. Nat. Med.* **2014**, *68*, 709–716. [CrossRef] [PubMed]

72. Islam, M.N.; Choi, S.H.; Moon, H.E.; Park, J.J.; Jung, H.A.; Woo, M.H.; Woo, H.C.; Choi, J.S. The inhibitory activities of the edible green alga *Capsosiphon fulvescens* on rat lens aldose reductase and advanced glycation end products formation. *Eur. J. Nutr.* **2014**, *53*, 233–242. [CrossRef] [PubMed]

73. Abdallah, H.M.; Salama, M.M.; Abd-Elrahman, E.H.; El-Maraghy, S.A. Antidiabetic activity of phenolic compounds from Pecan bark in streptozotocin-induced diabetic rats. *Phytochem. Lett.* **2011**, *4*, 337–341. [CrossRef]

74. Chethan, S.; Dharmesh, S.M.; Malleshi, N.G. Inhibition of aldose reductase from cataracted eye lenses by finger millet (*Eleusine coracana*) polyphenols. *Bioorg. Med. Chem.* **2008**, *16*, 10085–10090. [CrossRef] [PubMed]

75. Matsuda, H.; Morikawa, T.; Toguchida, I.; Yoshikawa, M. Structural requirements of flavonoids and related compounds for aldose reductase inhibitory activity. *Chem. Pharm. Bull. (Tokyo)* **2002**, *50*, 788–795. [CrossRef] [PubMed]

76. Abdo, S.M.; Hetta, M.H.; Samhan, F.A.; El Din, R.A.S.; Ali, G.H. Phytochemical and antibacterial study of five freshwater algal species. *Asian J. Plant Sci. Res.* **2012**, *11*, 109–116.

77. Kumari, R.P.; Anbarasu, K. Protective role of C-phycocyanin against secondary changes during sodium selenite mediated cataractogenesis. *Nat. Products Bioprospect.* **2014**, *4*, 81–89. [CrossRef] [PubMed]

78. Kumari, R.P.; Sivakumar, J.; Thankappan, B.; Anbarasu, K. C-phycocyanin modulates selenite-induced cataractogenesis in rats. *Biol. Trace Elem. Res.* **2013**, *151*, 59–67. [CrossRef] [PubMed]

79. Liu, Q.; Huang, Y.; Zhang, R.; Cai, T.; Cai, Y. Medical Application of *Spirulina platensis* Derived C-Phycocyanin. *Evid.-Based Complement Altern. Med.* **2016**, *2016*. [CrossRef]

80. Kumari, R.P.; Ramkumar, S.; Thankappan, B.; Natarajaseenivasan, K.; Balaji, S.; Anbarasu, K. Transcriptional regulation of crystallin, redox, and apoptotic genes by C-Phycocyanin in the selenite-induced cataractogenic rat model. *Mol. Vis.* **2015**, *21*, 26. [PubMed]

81. Kothadia, A.D.; Shenoy, A.M.; Shabaraya, A.R.; Rajan, M.S.; Viradia, U.M.; Patel, N.H. Evaluation of Cataract Preventive Action of Phycocyanin. *Int. J. Pharm. Sci. Drug Res.* **2011**, *3*, 42–44.

82. Ou, Y.; Yuan, Z.; Li, K.; Yang, X. Phycocyanin may suppress d-galactose-induced human lens epithelial cell apoptosis through mitochondrial and unfolded protein response pathways. *Toxicol. Lett.* **2012**, *215*, 25–30. [CrossRef] [PubMed]

83. Lee, S.J.; Park, W.H.; Park, S.D.; Moon, H.I. Aldose reductase inhibitors from *Litchi chinensis* Sonn. *J. Enzyme Inhib. Med. Chem.* **2009**, *24*, 957–959. [CrossRef] [PubMed]

84. Jung, H.A.; Yoon, N.Y.; Kang, S.S.; Kim, Y.S.; Choi, J.S. Inhibitory activities of prenylated flavonoids from *Sophora flavescens* against aldose reductase and generation of advanced glycation endproducts. *J. Pharm. Pharmacol.* **2008**, *60*, 227–236. [CrossRef] [PubMed]

85. Choi, S.J.; Kim, J.K.; Jang, J.M.; Lim, S.S. Inhibitory Effect of the Phenolic Compounds from *Geranium thunbergii* on Rat Lens Aldose Reductase and Galactitol Formation. *Korean J. Med. Crop. Sci.* **2012**, *20*, 222–230. [CrossRef]

86. Kim, J.M.; Jang, D.S.; Lee, Y.M.; Yoo, J.L.; Kim, Y.S.; Kim, J.H.; Kim, J.S. Aldose-reductase- and protein-glycation-inhibitory principles from the whole plant of *Duchesnea chrysantha*. *Chem. Biodivers.* **2008**, *5*, 352–356. [CrossRef] [PubMed]

87. Sawant, L.; Singh, V.K.; Dethe, S.; Bhaskar, A.; Balachandran, J.; Mundkinajeddu, D.; Agarwal, A. Aldose reductase and protein tyrosine phosphatase 1B inhibitory active compounds from *Syzygium cumini* seeds. *Pharm. Biol.* **2015**, *53*, 1176–1182. [CrossRef] [PubMed]

88. Jung, H.A.; Yoon, N.Y.; Bae, H.J.; Min, B.S.; Choi, J.S. Inhibitory activities of the alkaloids from *Coptidis Rhizoma* against aldose reductase. *Arch. Pharm. Res.* **2008**, *31*, 1405–1412. [CrossRef] [PubMed]

89. Palanisamy, U.D.; Ling, L.T.; Manaharan, T.; Appleton, D. Rapid isolation of geraniin from *Nephelium lappaceum* rind waste and its anti-hyperglycemic activity. *Food Chem.* **2011**, *127*, 21–27. [CrossRef]

90. Huang, G.J.; Hsieh, W.T.; Chang, H.Y.; Huang, S.S.; Lin, Y.C.; Kuo, Y.H. α-glucosidase and aldose reductase inhibitory activities from the fruiting body of *Phellinus merrillii*. *J. Agric. Food Chem.* **2011**, *59*, 5702–5706. [CrossRef] [PubMed]

91. Kim, T.H.; Kim, J.K.; Kang, Y.H.; Lee, J.Y.; Kang, I.J.; Lim, S.S. Aldose Reductase Inhibitory Activity of Compounds from *Zea mays* L. *BioMed Res. Int.* **2013**, *2013*, 1–8.

92. Morikawa, T.; Chaipech, S.; Matsuda, H.; Hamao, M.; Umeda, Y.; Sato, H.; Tamura, H.; Kon'i, H.; Ninomiya, K.; Yoshikawa, M.; et al. Antidiabetogenic oligostilbenoids and 3-ethyl-4-phenyl-3,4-dihydroisocoumarins from the bark of *Shorea roxburghii*. *Bioorg. Med. Chem.* **2012**, *20*, 832–840. [CrossRef] [PubMed]

93. Güvenç, A.; Okada, Y.; Akkol, E.K.; Duman, H.; Okuyama, T.; Çalis, I. Investigations of anti-inflammatory, antinociceptive, antioxidant and aldose reductase inhibitory activities of phenolic compounds from *Sideritis brevibracteata*. *Food Chem.* **2010**, *118*, 686–692. [CrossRef]

94. Rodriguez-Lyon, M.L.; Diaz-Lanza, A.M.; Bernabé, M.; Villaescusa-castillo, L. Flavone glycosides containing acetylated sugars from *Sideritis hyssopifolia*. *Magn. Reson. Chem.* **2000**, *38*, 684–687. [CrossRef]

95. Lenherr, A.; Mabry, T.J. Acetylated allose-containing flavonoid glucosides from *Stachys anisochila*. *Phytochemistry* **1987**, *26*, 1185–1188. [CrossRef]

96. Patel, D.K.; Prasand, S.K.; Kumar, R.; Hemalatha, S. Cataract: A major secondary complication of diabetes, its epidemiology and an overview on major medicinal plants screened for anticataract activity. *Asian Pac. J. Trop. Dis.* **2011**, *1*, 323–329. [CrossRef]

97. Devi, V.G.; Rooban, B.N.; Sasikala, V.; Sahasranamam, V.; Abraham, A. Isorhamnetin-3-glucoside alleviates oxidative stress and opacification in selenite cataract in vitro. *Toxicol. In Vitro* **2010**, *24*, 1662–1669. [CrossRef] [PubMed]

98. Lee, I.S.; Kim, Y.J.; Jung, S.H.; Kim, J.H.; Kim, J.S. Flavonoids from *Litsea japonica* Inhibit AGEs Formation and Rat Lense Aldose Reductase in Vitro and Vessel Dilation in Zebrafish. *Planta Med.* **2017**, *83*, 318–325. [CrossRef] [PubMed]

99. Kim, S.B.; Hwang, S.H.; Suh, H.W.; Lim, S.S. Phytochemical analysis of *Agrimonia pilosa* Ledeb, its antioxidant activity and aldose reductase inhibitory potential. *Int. J. Mol. Sci.* **2017**, *18*, 379. [CrossRef] [PubMed]

100. Kim, Y.S.; Jung, D.H.; Lee, I.S.; Choi, S.J.; Yu, S.Y.; Ku, S.K.; Kim, M.H.; Kim, J.S. Effects of *Allium victorialis* leaf extracts and its single compounds on aldose reductase, advanced glycation end products and TGF-β1 expression in mesangial cells. *BMC Complement Altern. Med.* **2013**, *13*, 251. [CrossRef] [PubMed]

101. Chung, I.M.; Kim, M.Y.; Park, W.H.; Moon, H.I. Aldose reductase inhibitors from *Viola hondoensis* W. Becker et H Boss. *Am. J. Chin. Med.* **2008**, *36*, 799–803. [CrossRef] [PubMed]

102. Chen, B.; Tian, J.; Zhang, J.; Wang, K.; Liu, L.; Yang, B.; Bao, L.; Liu, H. Triterpenes and meroterpenes from *Ganoderma lucidum* with inhibitory activity against HMGs reductase, aldose reductase and α-glucosidase. *Fitoterapia* **2017**, *120*, 6–16. [CrossRef] [PubMed]

103. Yang, M.; Wang, X.; Guan, S.; Xia, J.; Sun, J.; Guo, H.; Guo, D. Analysis of Triterpenoids in *Ganoderma lucidum* Using Liquid Chromatography Coupled with Electrospray Ionization Mass Spectrometry. *J. Am. Soc. Mass Spectrom.* **2007**, *18*, 927–939. [CrossRef] [PubMed]

104. Fatmawati, S.; Shimizu, K.; Kondo, R. Ganoderic acid Df, a new triterpenoid with aldose reductase inhibitory activity from the fruiting body of *Ganoderma lucidum*. *Fitoterapia* **2010**, *81*, 1033–1036. [CrossRef] [PubMed]

105. Fatmawati, S.; Shimizu, K.; Kondo, R. Inhibition of aldose reductase in vitro by constituents of *Ganoderma lucidum*. *Planta Med.* **2010**, *76*, 1691–1693. [CrossRef] [PubMed]

106. Fatmawati, S.; Kondo, R.; Shimizu, K. Structure-activity relationships of lanostane-type triterpenoids from *Ganoderma lingzhi* as α-glucosidase inhibitors. *Bioorg. Med. Chem. Lett.* **2013**, *23*, 5900–5903. [CrossRef] [PubMed]

107. Ramu, R.; Shirahatti, P.S.; Zameer, F.; Ranganatha, L.V.; Prasad, M.N.N. Inhibitory effect of banana (*Musa* sp. var. Nanjangud rasa bale) flower extract and its constituents Umbelliferone and Lupeol on α-glucosidase, aldose reductase and glycation at multiple stages. *S. Afr. J. Bot.* **2014**, *95*, 54–63. [CrossRef]

108. Asha, R.; Devi, V.G.; Abraham, A. Chemico-Biological Interactions Lupeol, a pentacyclic triterpenoid isolated from *Vernonia cinerea* attenuate selenite induced cataract formation in Sprague Dawley rat pups. *Chem. Biol. Interact.* **2016**, *245*, 20–29. [CrossRef] [PubMed]

109. Chirumbolo, S. Anticataractogenic Activity of Luteolin. *Chem. Biodivers.* **2016**, *13*, 343–344. [CrossRef] [PubMed]

110. Vit, P.; Jacob, T.J. Putative Anticataract Properties of Honey Studied by the Action of Flavonoids on a Lens Culture Model. *J. Heal. Sci.* **2008**, *54*, 196–202. [CrossRef]

111. Jang, D.S.; Lee, Y.M.; Jeong, I.H.; Kim, J.S. Constituents of the flowers of *Platycodon grandiflorum* with inhibitory activity on advanced glycation end products and rat lens aldose reductase in vitro. *Arch. Pharm. Res.* **2010**, *33*, 875–880. [CrossRef] [PubMed]

112. Rooban, B.N.; Sasikala, V.; Devi, V.G.; Sahasranamam, V.; Abraham, A. Prevention of selenite induced oxidative stress and cataractogenesis by luteolin isolated from *Vitex negundo*. *Chem. Biol. Interact.* **2012**, *196*, 30–38. [CrossRef] [PubMed]

113. Ha, T.J.; Lee, J.H.; Lee, M.H.; Lee, B.W.; Kwon, H.S.; Park, C.H.; Shim, K.B.; Kim, H.T.; Baek, I.Y.; Jang, D.S. Isolation and identification of phenolic compounds from the seeds of *Perilla frutescens* (L.) and their inhibitory activities against α-glucosidase and aldose reductase. *Food Chem.* **2012**, *135*, 1397–1403. [CrossRef] [PubMed]

114. Lee, J.; Rodriguez, J.P.; Quilantang, N.G.; Lee, M.H.; Cho, E.J.; Jacinto, S.D.; Lee, S. Determination of flavonoids from *Perilla frutescens* var. *japonica* seeds and their inhibitory effect on aldose reductase. *Appl. Biol. Chem.* **2017**, *60*, 155–162. [CrossRef]

115. Morikawa, T.; Xie, H.; Wang, T.; Matsuda, H.; Yoshikawa, M. Bioactive constituents from Chinese natural medicines. XXXII. Aminopeptodase N and aldose reductase inhibitors from *Sinocrassula indica*: Structures of Sinocrassosides B_4, B_5, C_1 and D_1-D_3. *Chem. Pharm. Bull. (Tokyo)* **2008**, *56*, 1438–1444. [CrossRef] [PubMed]

116. Matsuda, H.; Morikawa, T.; Tao, J.; Ueda, K.; Yoshikawa, M. Bioactive constituents of Chinese natural medicines. VII. Inhibitors of degranulation in RBL-2H3 cells and absolute stereostructures of three new diarylheptanoid glycosides from the bark of *Myrica rubra*. *Chem. Pharm. Bull. (Tokyo)* **2002**, *50*, 208–215. [CrossRef] [PubMed]

117. Hwang, S.H.; Kwon, S.H.; Kim, S.B.; Lim, S.S. Inhibitory Activities of *Stauntonia hexaphylla* Leaf Constituents on Rat Lens Aldose Reductase and Formation of Advanced Glycation End Products and Antioxidant. *BioMed Res. Int.* **2017**, *2017*, 4273257. [CrossRef] [PubMed]

118. Patel, M.B.; Mishra, S. Isoquinoline alkaloids from *Tinospora cordifolia* inhibit rat lens aldose reductase. *Phyther. Res.* **2012**, *26*, 1342–1347. [CrossRef] [PubMed]

119. Nakamura, S.; Fujimoto, K.; Matsumoto, T.; Nakashima, S.; Ohta, T.; Ogawa, K.; Tamura, H.; Matsuda, H.; Yoshikawa, M. Structures of Acylated Sucroses and Inhibitory Effects of Constituents on Aldose Reducatase from the Flower Buds of *Prunus mume*. *Chem. Pharm. Bull.* **2013**, *61*, 445–451.

120. Nakamura, S.; Fujimoto, K.; Matsumoto, T.; Ohta, T.; Ogawa, K.; Tamura, H.; Matsuda, H.; Yoshikawa, M. Structures of acylated sucroses and an acylated flavonol glycoside and inhibitory effects of constituents on aldose reductase from the flower buds of *Prunus mume*. *J. Nat. Med.* **2013**, *67*, 799–806. [CrossRef] [PubMed]

121. Kim, N.H.; Kim, Y.S.; Lee, Y.M.; Jang, D.S.; Kim, J.S. Inhibition of aldose reductase and xylose-induced lens opacity by puerariafuran from the roots of *Pueraria lobata*. *Biol. Pharm. Bull.* **2010**, *33*, 1605–1609. [CrossRef] [PubMed]

122. Stobiecki, M.; Kachlicki, P. Isolation and identification of flavonoids. *Sci. Flavonoids* **2006**, *27*, 47–69.

123. Lee, H.E.; Kim, J.A.; Whang, W.K. Chemical constituents of smilax China l. stems and their inhibitory activities against glycation, aldose reductase, α-glucosidase, and lipase. *Molecules* **2017**, *22*, 451. [CrossRef] [PubMed]

124. Kato, A.; Yasuko, H.; Goto, H.; Hollinshead, J.; Nash, R.J.; Adachi, I. Inhibitory effect of rhetsinine isolated from *Evodia rutaecarpa* on aldose reductase activity. *Phytomedicine* **2009**, *16*, 258–261. [CrossRef] [PubMed]

125. Kang, J.; Tang, Y.; Liu, Q.; Guo, N.; Zhang, J.; Xiao, Z.; Chen, R.; Shen, Z. Isolation, modification, and aldose reductase inhibitory activity of rosmarinic acid derivatives from the roots of *Salvia grandifolia*. *Fitoterapia* **2016**, *112*, 197–204. [CrossRef] [PubMed]

126. Li, H.M.; Hwang, S.H.; Kang, B.G.; Hong, J.S.; Lim, S.S. Inhibitory effects of *Colocasia esculenta* (L.) Schott constituents on aldose reductase. *Molecules* **2014**, *19*, 13212–13224. [CrossRef] [PubMed]

127. Lee, J.; Kim, N.H.; Nam, J.W.; Lee, Y.M.; Jang, D.S.; Kim, Y.S.; Nam, S.H.; Seo, E.K.; Yang, M.S.; Kim, J.S. Scopoletin from the flower buds of *Magnolia fargesii* inhibits protein glycation, aldose reductase, and cataractogenesis Ex Vivo. *Arch. Pharm. Res.* **2010**, *33*, 1317–1323. [CrossRef] [PubMed]

128. Park, H.Y.; Kwon, S.B.; Heo, N.K.; Chun, W.J.; Kim, M.J.; Kwon, Y.S. Constituents of the stem of *Angelica gigas* with rat lens aldose reductase inhibitory activity. *J. Appl. Biol. Chem.* **2011**, *54*, 194–199. [CrossRef]

129. Abou Assi, R.; Darwis, Y.; Abdulbaqi, I.; Khan, A.A.; Lim, V.; Laghari, M.H. *Morinda citrifolia* (Noni): A comprehensive review on its industrial uses, pharmacological activities, and clinical trials. *Arab. J. Chem.* **2017**, *10*, 691–707. [CrossRef]

130. Wang, Q.; Qian, Y.; Wang, Q.; Yang, Y.F.; Ji, S.; Song, W.; Qiao, X.; Guo, D.; Liang, H.; Ye, M. Metabolites identification of bioactive licorice compounds in rats. *J. Pharm. Biomed. Anal.* **2015**, *115*, 515–522. [CrossRef] [PubMed]

131. Lee, Y.S.; Kim, S.H.; Jung, S.H.; Kim, J.K.; Pan, C.H.; Lim, S.S. Aldose Reductase Inhibitory Compounds from *Glycyrrhiza uralensis*. *Biol. Pharm. Bull.* **2010**, *33*, 917–921. [CrossRef] [PubMed]

132. Lee, Y.R.; Hwang, J.K.; Koh, H.W.; Jang, K.Y.; Lee, J.H.; Park, J.W.; Park, B.H. Sulfuretin, a major flavonoid isolated from *Rhus verniciflua*, ameliorates experimental arthritis in mice. *Life Sci.* **2012**, *90*, 799–807. [CrossRef] [PubMed]

133. Lee, E.H.; Song, D.G.; Lee, J.Y.; Pan, C.H.; Um, B.H.; Jung, S.H. Inhibitory Effect of the Compounds Isolated from *Rhus verniciflua* on Aldose Reductase and Advanced Glycation Endproducts. *Biol. Pharm. Bull.* **2008**, *31*, 1626–1630. [CrossRef] [PubMed]

134. Lim, S.S.; Jung, S.H.; Ji, J.; Shin, K.H.; Keum, S.R. Inhibitory effects of 2′-hydroxychalcones on rat lens aldose reductase and rat platelet aggregation. *Chem. Pharm. Bull.* **2000**, *48*, 1786–1789. [CrossRef] [PubMed]

135. Lou, H.; Yuan, H.; Yamazaki, Y.; Sasaki, T.; Oka, S. Alkaloids and Flavonoids from Peanut Skins. *Planta Med.* **2001**, *67*, 345–349. [CrossRef] [PubMed]

136. Wei, X.; Chen, D.; Yi, Y.; Qi, H.; Gao, X.; Fang, H.; Gu, Q.; Wang, L.; Gu, L. Syringic acid extracted from *Herba dendrobii* prevents diabetic cataract pathogenesis by inhibiting aldose reductase activity. *Evid.-Based Complement Altern. Med.* **2012**, *2012*. [CrossRef] [PubMed]

137. Yan, R.Y.; Liu, H.L.; Zhang, J.Y.; Yang, B. Phenolic glycosides and other constituents from the bark of *Magnolia officinalis*. *J. Asian Nat. Prod. Res.* **2014**, *16*, 400–405. [CrossRef] [PubMed]

138. Patel, M.B.; Mishra, S.M. Aldose Reductase Inhibitory Activity of a Cglycosidic Flavonoid Derived from *Enicostemma hyssopifolium*. *Inf. J. Complement. Integr. Med.* **2009**, *6*, 1553–3840.

MDPI

St. Alban-Anlage 66

4052 Basel

Switzerland

Tel. +41 61 683 77 34

Fax +41 61 302 89 18

www.mdpi.com

Nutrients Editorial Office

E-mail: nutrients@mdpi.com

www.mdpi.com/journal/nutrients

www.ingramcontent.com/pod-product-compliance
Lightning Source LLC
Chambersburg PA
CBHW051730210326
41597CB00032B/5668